OXFORD STATISTICAL SCIENCE SERIES

SERIES EDITORS

A. C. ATKINSON   J. B. COPAS

D. A. PIERCE   M. J. SCHERVISH

D. M. TITTERINGTON

OXFORD STATISTICAL SCIENCE SERIES

Models for Repeated Measurements

Department of Biostatistics,
Limburgs Universitair Centrum,
Diepenbeek, Belgium

CLARENDON PRESS · OXFORD

Oxford University Press, Great Clarendon Street, Oxford OX2 6DP
Oxford New York
Athens Auckland Bangkok Bogota Bombay Buenos Aires
Calcutta Cape Town Dar es Salaam Delhi Florence Hong Kong
Istanbul Karachi Kuala Lumpur Madras Madrid Melbourne
Mexico City Nairobi Paris Singapore Taipei Tokyo Toronto
and associated companies in
Berlin Ibadan

Oxford is a trade mark of Oxford University Press

Published in the United States by
Oxford University Press Inc., New York

© J. K. Lindsey, 1993

First published 1993
Reprinted 1994 (with corrections), 1997

A catalogue record for this book is available from the British Library

Library of Congress Cataloging in Publication Data
Lindsey, James K.
Models for repeated measurements / J. K. Lindsey.—1st ed.
p.   cm.—(Oxford statistical science series ; 10)
Includes bibliographical references and index.
1. Multivariate analysis.   I. Title.   II. Series.
QA278.L554   1994   519.5'35—dc20   93–8637
ISBN 0 19 852299 1

Printed in Great Britain by
Bookcraft (Bath) Ltd, Midsomer Norton, Avon

Schnabel nonlinear optimization solver to be invaluable for finding maximum likelihood estimates.

Many of the examples have been analysed with GLIM or MatLab. A number of the algebraic and calculus problems were checked using the symbolic computation program, Maple, which was also used to produce the three-dimensional plots. The two-dimensional graphics were done with MultiPlot and the the contour plots with XLisp-Stat, for which I thank respectively Alan Baxter and Luke Tierney. Thanks also go to Richard Jones for CARMA and Daniel Stram for REML, used in some of the more complex normal theory models, to Dan Heitjan for the nonlinear growth curve program, GROWTH, to Brian Francis for SABRE and MIXTURE which handle normal compound distributions, and to A.C. Harvey, Cristiano Fernandes (DISCOUNT), and Genshiro Kitagawa for dynamic generalized linear model programs.

I also thank all of the contributors of data sets; they are individually cited when each table is first presented.

I would like to acknowledge the generous help of the following people in providing comments: Adelin Albert, Emmanuel Lesaffre, and Jean-Pierre Urbain for reading an early draft of the first four chapters, Dan Heitjan for Chapter 4, Alan Agresti and Bent Jørgensen for reading Chapters 5 and 6, Tom Fleming for illuminating discussions about counting processes, and Philippe Lambert and Franz Palm for a detailed reading of the whole manuscript. Finally, I would like to thank my biostatistical colleagues at Limburgs University, Diepenbeek, for the cordial atmosphere they provided throughout this work, without which it would never have been possible, as well as those at the Seminar für Statistik, the Eidgenössische Technische Hochschule, Zürich, for their hospitality during a month's visit in the hectic final phase of the work.

Diepenbeek   J.K.L.

January, 1993

Contents

II   NORMAL DISTRIBUTION MODELS

III   MODELS FOR CATEGORICAL DATA

APPENDICES

Notation and Symbols

Vectors are bold lower case and matrices bold upper case Greek or Roman letters, although vectors as special cases of matrices are upper case. An exception is the response variable, y, where bold upper case may mean either the vector random variable or the matrix of observed values, depending on the context. Primes denote derivatives of a function and T the transpose of a vector or matrix.

Some of the notation is only applicable to members of the exponential family.

Y	–	random response variable
$y, \mathbf{Y}$	–	observed response variable
$x, \mathbf{X}$	–	between unit explanatory variables
$z, \mathbf{Z}$	–	within unit explanatory variables
$u, \mathbf{U}$	–	between unit random parameter variables
$v, \mathbf{V}$	–	within unit random parameter variables
N_t	–	cumulated event counts
M_t	–	martingale
$\mathcal{F}_t$	–	complete history up to time t
n	–	binomial denominator
i	–	index for units (of N)
j	–	index for conditions across units (of C)
k	–	index for repetitions within units (of R)
l	–	index for parameters in the location model (of P)
t	–	index for time (of R)
M	–	order of a Markov process
h, m	–	other arbitrary indices
K	–	arbitrary constant
$f(y)$	–	probability density function
$F(y)$	–	cumulative distribution function
$S(y)$	–	survival function
$\omega(\cdot)$	–	event rate, hazard, or intensity (function)
$\mathbf{T}$	–	transition matrix

$\Pr(y)$	–	probability of response
$p(\cdot)$	–	probability of a random parameter
U	–	score
$\theta, \boldsymbol{\Theta}$	–	canonical parameter
$b(\theta)$	–	cumulant function
$\mu = \mathrm{E}[Y] = b'(\theta), \mathbf{M}$	–	mean or expected value
$\eta, \mathbf{H}$	–	linear predictor
$g(\cdot)$	–	link function
$\alpha, \beta, \mathbf{B}$	–	location model parameters
π	–	probability parameter
ϕ	–	scale parameter
$\tau^2 = b''(\theta)$	–	variance function
w	–	prior weight
$\sigma^2 = \mathrm{var}[Y] = a(\phi)\tau^2, \boldsymbol{\Sigma}$	–	total variance
$\psi^2, \boldsymbol{\Psi}$	–	intra-unit response variance component
$\delta, \boldsymbol{\Delta}$	–	heterogeneity variance component
$\xi, \boldsymbol{\Xi}$	–	longitudinal variance component
$\rho, \mathbf{R}$	–	correlation, autocorrelation, autoregression coefficient, intra-unit correlation
$\gamma(\cdot)$	–	autocovariance function, local dependence function
$s(\cdot)$	–	spectral density function
ω	–	spectral frequency
$I(\cdot)$	–	periodogram
$\lambda, \boldsymbol{\Lambda}$	–	random parameter
ε	–	residual or 'error'
$\kappa, \upsilon, \nu, \zeta$	–	other arbitrary parameters
$\mathbf{I}$	–	identity matrix
$\mathbf{J}$	–	row vector of ones
$\mathbf{A}, \mathbf{C}, \mathbf{D}, \mathbf{F}$	–	other arbitrary matrices
$\mathbf{a}, \mathbf{c}, \mathbf{d}, \mathbf{f}$	–	other arbitrary vectors
$B(\cdot, \cdot)$	–	Beta function
$\Gamma(\cdot)$	–	gamma function
$\Psi^{n-1}(\cdot)$	–	polygamma function
$\mathrm{I}(\cdot)$	–	indicator function
$\mathrm{J}(\cdot)$	–	Jacobian
$\mathrm{L}(\cdot)$	–	likelihood function
$c(\cdot), d(\cdot)$	–	other arbitrary functions

Part I

Introduction

1
Basic concepts

1.1 Introduction

Repeated measurements, as the name suggests, are observations of the same characteristic which are made several times. What distinguishes such observations from those in more traditional statistical data modelling is that

- the same variable is measured on the same observational unit more than once: the responses are not independent as in the usual regression analysis and
- more than one observational unit is involved: the responses do not form a simple time series.

To many statisticians, a mention of the term, 'repeated measurements', evokes the idea either of the biological study of growth curves or of split plot designs. However, once one begins to delve into the subject, one realizes that these two subjects, in no way, completely cover the field of repeated observations. In fact, repeated measurements are very frequent in almost all scientific fields where statistical models are used. A few common examples include

Agriculture Crop yields in different fields over different years.

Biology Growth curves.

Business Survival patterns of small businessmen.

Commerce Consumer purchasing behaviour with regard to competing products.

Criminology Recidivism.

Demography Intervals between successive births.

Economics Patterns of employment and unemployment.

Education Student progress under various learning conditions.

Engineering The breakdown and repair sequences of machines.

Geography Migration among urban centres.

Industry Quality control in batch production.

Insurance Evolving relationships between premiums and claims for different firms.

Labour Relations Frequency or length of strikes in different firms.

Medicine Successive periods of illness and recovery under different treatment regimes.

Meteorology Patterns of rainfall in different regions.

Politics Comparative histories of political regimes.

Sociology Social mobility.

Transport Series of accidents.

Zoology Sequences of behavioural patterns, such as bird songs.

Let us consider a few examples in a little more detail.

- Very often, an organization using sample surveys, in any scientific or business field, must be concerned about the quality of the responses obtained by the interviewers. Such responses will often be indications of choices from sets of categories. Different interviewers administering the same questionnaire to the same persons under similar circumstances may obtain substantially different responses. The dress, the accent, and many other personal characteristics, as well as the time of day, recent current or personal events may influence the response. None of these factors may be considered directly pertinent for explaining the responses in the study at hand, but they do influence the results. Some can be taken as completely random, across interviewers, and might be called measurement error. However, others, no more pertinent to the study, will vary systematically with the interviewer. For a given question, some interviewers will tend to elicit one response more often across all respondents, another interviewer some other response. In other words, a set of questionnaires administered by one person will tend to be more homogeneous in the responses obtained than a random set taken from all interviewers. In order to obtain reliable results from a sample survey, it may be necessary to take this problem directly into account in modelling the results. Here, the repeated response arises from the same interviewer administering more than one, but not all, questionnaires.

  One can go a step further, and study directly the heterogeneity, or lack of agreement, among interviewers. Although there is no perfect way of doing this, the usual approach is to overlap the people interviewed. Each respondent is interviewed two or more times by different interviewers. The situation immediately becomes much more complex, because we are now faced with two types of repetition: that above, but also several interview responses of the same people to the same questionnaire. Problems such as the order of the interviews complicate further the situation.

- In medical survival studies, patients are followed during some critical type of illness, such as cancer or AIDS, until death. Competing treatments are available, and knowledge is sought on their relative efficacy in retarding or preventing death. Such a study could already be thought of as a repeated measurement, because the patient must be continuously observed until the terminal event occurs. However, the recorded response is unique for a patient: the time to death. In spite of the enormous usefulness of such studies, and the accompanying models, they cannot really be imagined to be completely satisfactory. The patient *is* being observed through time, and this must be continuously providing further relevant information about survival. A first step is to include time-varying variables in a model to describe survival time (Andersen, 1992).

However, this period to death is not a homogenous time interval. The patient may apparently recover, then relapse, have some unrelated minor sickness for some time, and so on. The total survival time will, in fact, be made up of a sequence of distinct subperiods of time between successive events. These should be modelled. In many studies, the final event may not even be death, but, say, recovery. In such an event or life history study, the repeated measurements are the successive durations or events.

- An agricultural research worker wishes to compare several different feed preparations to see which is most conducive to the growth of pigs. A simple procedure is to assign the available pigs at random to the different mixtures and to record, at the end, which produces the heaviest animals. However, interest may not only centre on this final response. It may be important to follow the pigs through their life cycle and record how they grow at each stage. This may be important in order to determine when is the best point to slaughter them. Then, interest lies in the profiles of growth curves under the different treatments, where one treatment may be better as preparation for slaughter at one age, and another better for a different age. The repeated measurements are the weights at successive ages.

A further complication to the design may be useful in certain circumstances. Interest may centre, not on the profiles of growth, but on obtaining very precise estimates of the differences in average growth among the treatments. There may be considerable variability in the reaction of the animals to them. To control for this, the pigs are randomly assigned different feed preparations at different stages of their life, not all in the same order. Now, all of the feeds can be compared on the same animal; each pig acts as its own control. However, the reaction of the pig to the feed may depend on the point in its life when it receives it, and on what other feed mixtures it received before. These must be taken into account in the model. The repeated measurement is the weight gain under each treatment for each pig, but the overall growth curve is no longer estimable in a meaningful sense.

Throughout the later parts of this book, examples, similar to those just outlined, will be analysed in detail as illustrations of some of the possible ways in which models for repeated measurements may be constructed.

As is usually the case in statistics, similar models can often be applied in widely varying fields. As we proceed, we shall see that two factors provide unifying themes:

(1) the two types of *stochastic dependence* among measurements on the same observational unit,
   - homogeneity of responses on a unit/heterogeneity across units,
   - distance, in time or space, among responses on a unit;
(2) the three basic types of responses which may be measured,
   - general continuous data,
   - categorical and count data,
   - duration and survival data.

In the present chapter, I discuss the general conditions under which repeated measurements arise. In this way, I begin to elaborate on the above classification which will, then, structure the book. The second chapter outlines some general basic principles of model construction, those which are pertinent to the study of repeated measurements. In some ways, the latter may be the most difficult chapter, especially for the reader impatient to get on to the nitty-gritty of repeated measurements. However, it provides the 'glue' which binds what follows together. Thus, on a first reading, Chapter 2 might be skipped, and referred back to as a reference manual. This will generally be its function in the everyday use of the book in statistical practice.

The remainder of the book is structured into six chapters, according to the above cross-classification. Each of the three types of responses is treated in two chapters, homogeneity/heterogeneity and longitudinal models. The first chapter of each pair first looks at any general multivariate models which may be available and appropriate for the type of responses concerned, before going on to cover heterogeneity among observational units. The second chapter of the pair covers models for longitudinal data and, then, goes on to combine these with results of the first to yield more globally applicable models. Typical data sets are analysed for each circumstance. Given the wide variety of conditions under which repeated measurements may be produced, I necessarily must look at many small data sets, illustrating typical models. This has the added advantages that other, alternative approaches and models for a data set are usually available in the literature and that the reader, him or herself, can try fitting the models to the data sets. As already stated, the unity of these last six chapters is to be found in the second chapter, which precedes them. However, for the scientist or statistician simply looking for a method of treating some particular problem, these last six chapters might possibly be used as a sort of 'cookbook'.

At the end of each of the four parts, a few selected references are given as suggested further reading.

1.2   The unit of observation

The one special characteristic of repeated measurements which distinguishes them most is that more than one observation on the *same* response variable is available on each observational unit. For a set of responses on each of several units, those on the same one may often, but not always, be expected to be more closely related than those among different ones. Thus, we are in a situation of stochastically dependent data which must be modelled by some form of multivariate methods. This may be distinguished from more general multivariate methods which treat interdependence among *different* types of response variables.

Repeated measurement of the same variable on the same unit may be necessary for a number of reasons.

- Repeated observation may be the only way of obtaining the required measurement, as in counting the occurrences of some phenomenon.
- Interest may centre on the evolution of some response, given initial conditions which may or may not be fixed experimentally. Simple growth curves are the most common example.
- The investigator may wish to compare the effects of continued administration of some treatment over time.
- Different treatments may need to be compared in a situation where variability among units is an important uncontrollable factor. To increase precision, intra-unit comparisons of the different treatments are necessary.
- One may want to study the total effects of different sequences of treatments, as in the study of crop rotations in agriculture.

An important preliminary question is what we mean by the observational unit. The biologist may call it the subject or animal, the medical research worker, the case, the social scientist, the person or individual, the agricultural worker, the plot, and so on. In the clearest cases, the same form of observation is made more than once on the same object. Thus, in a growth study, measures of weight may be made on each animal on successive days. Here, the measurements follow each other in time, and *change* with time is of primary interest. It is clear what is the object we are observing, i.e. what is the *unit of observation*. This may not always be so.

Consider a very simple case: count data. As an example, suppose that we record the number of eggs laid by each fish in an experiment. Is the individual unit of observation the fish or an egg? Because we do not distinguish among the eggs from the same fish, it is useful to consider our unit to be the fish. But our egg count is, in fact, a repeated measurement, because we must observe the fish once for each egg laid, even although the final result is an aggregate count.

Often, it may be necessary to go even further in this direction. Suppose that a global unit of observation has several similar *subunits*, like the eggs, but that we now distinguish the observations made on them. We might be measuring, for example, the size of each egg. Thus, we distinguish units and subunits. In agriculture, these are often known as plots and subplots. For our egg example, with counts, the unit was the fish whereas the subunit did not explicitly enter into the model, because the count results in a single number. On the other hand, with size measurements, the unit is the fish and the subunit, the egg, must be modelled directly.

One important reason for making the distinction as we have done is to make clear how observations on a single unit, as we have defined it, will often be more closely related, in various ways still to be elaborated, than those across different units. As we shall see, we are faced with three types of relationships among our observations, one across units and two within.

(1) Different units will be assumed to be sampled in such a way that responses are independent.

(2) All responses on a given unit will usually be generally more closely related than those across different units. Some, but not all, of this variability may be accounted for by the use of covariates describing the different units. That which remains must be accounted for as stochastic (positive or negative) dependence.

(3) When some continuum, such as time or space, is involved in distinguishing among observations, those made more closely together on the same unit will usually be more closely related, a second form of stochastic (positive or negative) dependence.

The reader may, by now, be wondering why the study of stochastic interdependence among responses on a unit should be so important. Consider repeated measurement of some binary response (Chapters 5 and 6) such as whether an animal is awake or asleep. Suppose a subject is asleep 30% of the time. If responses on the unit were independent, the probability of observing an animal asleep in four consecutive periods would be $0.3 \times 0.3 \times 0.3 \times 0.3 = 0.0081$. Now suppose that the probability of the animal being asleep depends on its state in the previous period, so that, if it is asleep in one period, the (transition) probability of it still being asleep in the next is 0.9 . The probability of four consecutive periods of sleep is now $0.3 \times 0.9 \times 0.9 \times 0.9 = 0.2187$. The corresponding probabilities, under the two models, of four consecutive periods awake, if the transition probability is the same, are 0.2401 and 0.5103. Thus, the model of dependence among responses can have a great influence on the ability of the complete model to describe the observations.

In conclusion, we shall find it useful to take the unit of observation to be any object upon which more than one interdependent response is made, whether these occur successively for the same object or more or less simultaneously on subunits. Throughout the text, we shall call this the *unit*, avoiding the less neutral term, individual, which implies indivisibility.

1.3 Types of designs

The design of a study defines how the data are to be collected. This is especially important in a context such as ours, where units are to be observed more than once.

In most statistical studies, one must try to conciliate two conflicting goals in choosing the design specifying the units to observe. To the extent that the results are meant to be representative of a population, the units should show as much variability as that population. To the extent that the results are not to be influenced by unobservable explanatory variables, the units should be as homogeneous as possible.

A first distinction which should be made is between experimental and sample survey designs. In fact, it is more useful to distinguish between types of explanatory variables, because the two kinds of resulting observations may be mixed within the same study. Thus, an *experimental* variable is a characteristic which has been assigned to the unit, or subunit, *at random* by the investigator. I shall call this a *treatment*. In contrast, a *sample* variable is one which is simply observed and taken as is. In an experimental situation, it is often called a *covariate*. It is very important to note that time is always a sample variable, because it cannot be randomly assigned.

An experimental variable is one which has been actively manipulated by the researcher. Change is the basis of such variables, change controlled by the observer.

Randomization has a number of important functions.

- It allows statements of *causality*, because we know the origin of the value of an experimental variable on a unit. Thus, the responses measured cannot have influenced a treatment; only the random choice of the research worker has. Such is almost always not the case with sample variables.
- It can minimize the effects of inter-response variability, by distributing it randomly over treatments. In other words, it can help to ensure homogeneity of variability, either within or across units, or both.
- It can eliminate (unconscious) biases on the part of the investigator in assigning the treatments.

A sample variable is a static variable, observed as is, in the 'natural' context. Even if observed over time, as in a panel study, such a variable can only exhibit change of uncertain origin. Thus, in sample survey designs, the one variable which may often be used in an attempt to overcome the problem of the direction of causality is time. In most cases, one can assume that events which occur later in time cannot affect earlier events. Economic forecasting is one example where this may not be true. But, one can never be sure to have recorded all of the events which might have had an influence.

Two approaches to collecting chronological information may be distinguished.

(1) In a *prospective* study, a sample of units may be chosen according to the criteria of certain explanatory variables and then followed up in time to see what response is obtained. Here, we find *panel, clinical trial*, and *cohort* studies.

(2) In a *retrospective* study, a sample is chosen according to the present response and the values of previous explanatory and response variables are investigated. In the medical sciences, this is called a *case-control* study.

A prospective study may have either an experimental design with randomization or a sample design. In both cases, we have initial conditions, set experimentally or not, and observe resulting responses. This can be complicated if changing conditions over the course of the study, i.e. evolving explanatory covariables, are taken into account. In contrast, the retrospective study usually requires special analytic procedures. However, it is of little pertinence to repeated measurements.

We conclude that prospective experimental and survey sample designs yield similar types of data structures, with response distributions depending on explanatory variables. The same statistical models can often be applied in the two situations. However, because of randomization, the conclusions, particularly with respect to causality, will be different. Thus, the same broad classification can be applied to both of them.

Let us now consider a different distinction among the general conditions under which repeated responses might be collected. Variables, whether treatment or sample, will generally distinguish among units. But here, we shall concentrate on the relationships among responses on the same unit, starting with the simplest and moving towards the more complex:

- responses are obtained on a unit, without any other variables being available to distinguish among them; we may assume that the responses are obtained simultaneously, although this is not often really true; two subcases occur:
    * the responses are categorical events, which, being indistinguishable, are aggregated as counts of each possible event;
    * responses are measured and left in this raw state;

- responses on a unit are distinguishable by an accompanying set of different treatments and/or by different values of covariates; they can still be assumed to have been obtained simultaneously;
- responses are distinguished by the time or space order in which they are obtained, but by no treatment or covariates;
- responses are distinguished by the time or space order, but treatments and/or covariates also change along the same continuum.

Let us consider each of these possibilities in more detail.

Count data In the simplest case, occurrences of some phenomena on each unit are counted. Because no explanatory variable (time, treatment) distinguishes among these observed events, they can be aggregated as single numbers, the counts. Depending on their type, the data may be modelled by binomial, multinomial, or Poisson distributions. However, most usually, the units are different enough so that the available explanatory variables, describing the differences among units, cannot account for all of the variability. This problem of heterogeneity is commonly known as *extravariation* or *overdispersion* in count data.

Nesting A similar, but slightly more complex, case arises in a number of circumstances. Repeated measurements, which are not simply records of events, are made more or less simultaneously on each subunit. Again, no explanatory variable, such as time or treatment, distinguishes responses within the same unit. But, now all of these observations on a unit cannot be summarized in one number, in the way that the number of events could. Examples include family studies, teratology, and measurements on eyes or teeth. Thus, measurements of height for all children in a family will not be aggregated. We say that responses on the subunits are *nested* within units. All responses on each unit may be relatively homogeneous and should, at least, be about equally related.

The variability of responses on a unit will often be smaller than that across units. If responses on a unit were independent, the covariability among them would be zero. In contrast, here, we shall have a *uniform* intra-class or intra-unit covariability model. All inter-relationships among responses within a unit are equal, but nonzero. In this way, we model directly the stochastic dependence structure of our observations.

This may be clearer if we look more closely at second order relationships, the form of the covariance matrix. This will be block diagonal, with a non-zero 'block' submatrix for responses on each unit. With N observational units and R repetitions per unit, we have $\mathbf{I}_N \otimes \mathbf{\Sigma}$, where

$$\Sigma = \begin{pmatrix} \psi^2 + \delta & \delta & \cdots & \delta \\ \delta & \psi^2 + \delta & \cdots & \delta \\ \vdots & \vdots & \ddots & \vdots \\ \delta & \delta & \cdots & \psi^2 + \delta \end{pmatrix}$$

$$= \psi^2 \mathbf{I}_R + \mathbf{J}_R^T \delta \mathbf{J}_R \qquad (1.1)$$

Here, ψ^2 is the intra-unit variance and δ is both the extra component of variance across units and the common covariance among responses on the same unit, while $\mathbf{I}_R$ and $\mathbf{I}_N$ are $R \times R$ and $N \times N$ identity matrices and $\mathbf{J}_R$ is a $1 \times R$ row vector. This special form of matrix helps to constrain δ to be greater than $-\psi^2/2$, so that the *intra-unit correlation* among pairs of responses on a unit

$$\rho = \frac{\delta}{\psi^2 + \delta} \qquad (1.2)$$

is less than one in absolute value. Note that δ, and ρ, can be negative. The matrix is said to have the property of *compound symmetry*. Because the variance is composed of two (or more, in more complex models) parts, I shall call this a *variance components model*.

Instead of directly modelling the stochastic dependence in this way, another approach, which is useful in some cases, might be to assume that the units were chosen at random from some large population. Then, the parameters in the model which describe the differences among the units are taken to vary over this population, according to some distribution. This is known as a *random effects model*, based on some form of *compound distribution*. The extra variability, or heterogeneity, among units having different values of the explanatory variables has, as its reflection, a relatively closer relationship, or homogeneity, among responses on the same unit or in the same class. Thus, we again obtain a uniform intra-unit relationship. However, here, the responses across units must be heterogeneous as compared to those within, so that the intra-unit correlation must be positive.

Random effects and compound distributions are especially useful for developing models when the second order variances and covariances are not sufficient to describe the stochastic dependence structure. The overdispersion model for count data, mentioned above, is just a special case of such a random effects model, where each class has only one aggregated observation recorded, the count.

Directly modelling this stochastic dependence structure, as in a variance components model, and assuming that some parameter distinguishing the units has a random distribution, as in the random effects model, can both result in essentially the same model. In fact, the random effects model is a more limited case because, as we have seen, there the covariability parameter, say, the covariance, δ, cannot be negative.

Obviously, a study design may also involve several levels of nesting. For example, we might have counts within litters or eyes within families.

Split plots   Often, the simple situation just described for intra-unit co-variability must be modified, because we wish to distinguish among the subunits by means of different treatments or covariates. This is, then, usually called a *split plot* design. The term comes from agriculture, where a number of fields or plots are given different treatments of one factor, but, then, each plot is divided or split into subplots with different treatments of a second factor. Here, the second treatment is an explanatory variable within units. Responses on the same unit or plot may again be expected to be more closely related, so we still have the same form of stochastic dependence structure just described, which, to second order, is uniform correlation. A simpler case occurs if different treatments or covariates do not distinguish among the units, or plots, but only among subplots. This design is called a *randomized block*.

With their nested structure, the split plot designs may have at least two advantages over simpler 'analysis of variance' (ANOVA) designs, such as factorial designs.

(1) If a treatment can be applied to a whole unit (plot) at once, rather than separately to subunits, this may reduce costs.

(2) Because subunits are usually more uniform than different units, parameters measuring comparisons among conditions may be estimated more precisely.

For example, if treatment and control are applied to different subunits, the unit is serving as its own control. These are also characteristics of crossover designs discussed below.

Unfortunately, the type of uniform covariability model used for nested and split plot studies is often applied indiscriminantly to any data which are repeated measurements. It has even been called *the* repeated measurements model. Rather, we shall now begin to see, in more complicated cases, how other, more complex, models of stochastic dependence will usually be needed.

Longitudinal studies   In many *longitudinal* studies, what we called above prospective studies, each unit receives only one treatment (combination) and/or has specific sample variables recorded and then the response is observed over time. Examples include growth studies and longitudinal health studies. Time is an explanatory variable within units. We wish to compare the differences in the way in which the measurements change over time for different units or groups of units. We are, then, comparing growth curves or time series, and looking for differences among them. However, the number of responses is often smaller than it might be for a single time series or growth curve. In such a case, the covariability among responses often can be modelled by a low order Markov process (see Section 2.10 below). The further apart in time are the observations on an unit, the less closely related they are. From the considerations above, we know that this is not a typical ANOVA design, because there is no randomization over time. Split plot analysis cannot be used.

As an example, let us again model the second order stochastic dependence structure directly by looking at the covariance matrix. The values on the minor diagonals, for a unit, decrease the farther they are from the main diagonal, while off diagonal elements between units are zero. The blocks on the diagonal are now

$$
\Sigma = \begin{pmatrix}
\psi^2 + \xi & \xi\rho(1) & \cdots & \xi\rho(R-1) \\
\xi\rho(1) & \psi^2 + \xi & \cdots & \xi\rho(R-2) \\
\vdots & \vdots & \ddots & \vdots \\
\xi\rho(R-1) & \xi\rho(R-2) & \cdots & \psi^2 + \xi
\end{pmatrix}
\tag{1.3}
$$

The matrix has the Toeplitz form, with all values on a given diagonal equal. The parameter, ψ^2, has the same meaning as before, while ξ and $\rho(|j-k|)$ are the variance component and autocorrelation within a unit. The latter is some function which varies with the distance, $|j-k|$, between observations on a unit. In most cases, variability among units will also be present, so that this type of matrix will need to be combined with that for heterogeneity.

As with the difference between variance components and random effects, similar models to that just presented can often be obtained by taking a given response to be conditional on previous responses on the same unit, instead of directly modelling the stochastic dependence. This type of regression situation is called an *autoregression model*. However, with repeated measurements, as opposed to a single time series, we are not constrained to keep the variance constant over time. For example, in many growth curve situations, the variability increases as the response increases. Models can be developed, and estimated, for such cases.

We, thus, may note a number of characteristics of longitudinal repeated measurement studies (Diggle and Donnelly, 1989):

(1) The data consist of a relatively large number of time series, structured by some type of more or less complex experimental or sample design.

(2) Usually, the series are relatively short.

(3) The times of measurement may be unequally spaced, may include missing values, and may be different among units. This may be the result either of design or of accident.

(4) The series will often be nonstationary in mean and/or in higher order structure.

(5) The research worker is usually most interested in the mean response profile, often some sort of *location model* linking the experimental conditions to the observed series of responses. However, a reasonable model for the higher order structure, at least the second order co-variance structure, is important, if only to provide valid and efficient inferences about the mean response profile. In some cases, it may also be of interest in its own right.

When predictions are required, these will not often be for values outside the time horizon of the observations, as is often done, for example, with econometric time series. Rather, the longitudinal data on available units will be used to predict future values of a unit which has a short series, with prediction only up to the end of the time period of the first set of units.

Time-varying covariates In more complex longitudinal studies, time-varying covariates are measured on each unit or each unit receives a sequence of different treatments, with measurement after each treatment. The first occurs in a sampling context and the second in an experimental design.

Let us first look, in more detail, at the experimental context. All units should not receive the treatments in the same order or treatment and time will be confounded. The result is called a *crossover* or *change-over* design. If the treatments are spaced far enough apart, there should, in many cases, be little *carry-over* or *holdover effect* from one to the next. If this *washout time* is sufficient, all responses on the unit should be about equally related. Although one attempts to construct an appropriate design to minimize carry-over effects, usually it is not certain that sufficient washout time was available and these effects will need to be taken into account. Then, time and both past and present treatment are explanatory variables within units. One is a sample variable and the other an experimental variable. Thus, crossover designs are related to the usual longitudinal studies in somewhat the same way as split plot designs are related to the simpler nested studies.

In the strictly sampling context, design controls cannot be applied, and the analysis of time-varying covariates is even more difficult.

We can see that, for time-varying covariates or treatments, in addition to the usual time, often now called *period, effects*, we can also have *sequence effects*. The effect of a given condition may depend on the immediately preceding conditions and responses, as well as on the progression over time. Because of the complexity of directly modelling this in the stochastic dependence structure, it is often handled by conditioning on previous values in the location model for the mean.

As already noted, in most cases of longitudinal studies, variability among units will also be present, so that some kind of uniform intra-unit relationship will have to be included in the model, as well as the time series effects.

We may summarize the relationship among the four types of designs for intra-unit variables in the following table:

| | | Time Variable | |
		No	**Yes**
	No	Counts Simple nesting	Longitudinal
Condition			
Variable	**Yes**	Split Plot	Time-varying covariates Crossover

However, this simple classification of repeated measurement designs immediately invites generalization. Many designs are much more complex, involving, for example, several levels of nesting. Thus, the basic designs outlined are often combined to obtain the desired result. The analysis will often be even more complex due to problems of missing observations on subunits or over time, of unequal spacing of the responses in time, or of different time spacing for different units. This is particularly true when living things, and especially human beings, are being studied. There may be deaths for unrelated reasons, people may move away, treatments may create undesirable side effects, and so on. One important distinction concerns whether missingness depends on the explanatory variables, on the level of responses, or on both (Heyting, Tolboom, and Essers, 1992).

1.4   Types of responses

In the statistical literature, the most common models for repeated responses assume them to be continuous measurements taking any real value. Such models are most commonly based on some variation on the multivariate normal distribution. This reflects the situation in other areas of classical statistics which have traditionally been based on the univariate normal distribution: multiple regression and ANOVA. There are a number of good reasons for this. The multivariate normal distribution is well known and is relatively easy to manipulate. Of great importance are the facts that, as we shall see in the Chapter 2, all stochastic dependence relationships are described by the second order moments and that these correlation relationships can vary independently of the mean. This also implies that both conditional and marginal models are also normal distributions. All of this greatly simplifies the task of modelling repeated measurements.

However, not all measurements can be considered to be able to take any values on the real line. Nor can they be reasonably modelled with a symmetric, 'bell-shaped' distribution. For the reasons just outlined, many of these types of response observations have traditionally been transformed, in one way or another, in attempts to meet the requirements of the normal model.

Categorical and count data are playing increasingly important roles in statistical observations, and, in a number of fields, now have far more importance than normal type responses. We have already encountered count data in the discussion of repeated measurements designs above. They are repeated event data, where nothing distinguishes the events on a unit, so that they are aggregated as counts. On the other hand, categorical data are repeated observations of events occurring to the same unit, but not aggregated because treatments or covariates distinguish among them. However, the events may be aggregated as frequencies across units having exactly the same profiles of explanatory variables.

Thus, care must be taken, because categorical data may take a form similar to counts, that of frequencies of occurrence of each category of event. The same logistic or log linear models are usually applicable to the two. However, the difference, from our perspective, is that, for a count, the response of interest is the frequency of occurrence of one or more events, necessarily on the same unit, while, for categorical data, the response is an indicator of which of a number of events has occurred. Categorical data are only repeated measurements if observed several times on the same unit.

Two examples may make this clearer.

(1) Consider a study of over-weight people. If we take families as the unit and members of the family as the subunit and we enumerate the number of over-weight individuals of each family, without distinguishing among them, we have count data. If we take individual people as our unit, ignoring family membership, we have an indicator categorical variable of whether or not each person is over-weight. Then, we can obtain the frequencies of over- and under-weight people in our sample, along with distinguishing characteristics such as sex and height.

(2) A longitudinal repeated measurements study might observe the number of industrial accidents per week in each factory over a given period of time. This would give series of count data. But one might also study the presence or absence of several types of industrial accidents over time. These would be categorical data. Frequencies, in this latter case, may be obtained by aggregating over factories with the same time trajectory to obtain transition matrices.

As in these examples, one common use of counts is to measure rates (of over-weightness or of accidents).

A second area which has taken a central role in statistical modelling has been the study of duration, especially survival, data. Such observations have two particularities: their values restricted to the positive domain and the problem of censoring. A very wide variety of models is available, with the most popular perhaps being the semi-parametric Cox proportional hazards model. Given the application of such duration models to survival, repeated measurements have not usually been of primary concern. If we measure survival in living beings, there can be no repetition, because death occurs at the end. However, these models have much wider applicability and the possibility of repetitive observations often exists. Consider the classical measurement of periods between successive breakdowns of a machine or the succession of periods of unemployment or of illness of individuals. This is often known as an *event* or *life history* study.

A close relationship exists between durations and counts in longitudinal event history data. This is clear in the counting process approach to duration data, where the number of events is accumulated over a period of time, thus uniting a duration and a count. This relationship will be used extensively in Part IV.

Event history data is peculiar in that the response, whether taken as elapsed duration between events or cumulated counts of events, is a direct function of time. This contrasts with many other longitudinally observed repeated responses, which are simply attributes of the unit, such as blood pressure, measured at various points in time. They will have an evolution, but not necessarily accumulation. Intermediate are the 'growth' type responses, which have a predominant tendency, either to become larger or smaller, usually up to some asymptotic limit, but where successive differences are not necessarily in the same direction: a growing rat may lose weight over one observation period, but events cannot disappear after they have occurred.

With the ready availability of modern computing power, it is no longer acceptable simply to attempt to transform such count or duration data to normality. Suitable models must be developed and applied. In the past decade or two, rapid advance has been made in these fields for univariate data: logistic and log linear models for counts and categorical data, survival models for duration data. Unfortunately, the actual development of such models for multivariate data has begun to fall behind the need, and the ability, to apply them in concrete cases.

In the study of univariate non-normal data, the theory of generalized linear models has played a very important role. For repeated measurements, we are in a more difficult situation, because no simple and general multivariate analogue of the exponential family exists, upon which a multivariate extension of generalized linear models could be developed. Stochastic dependence relationships cannot be summarized in second order correlations. Second and higher moments usually depend on the mean. Conditional and marginal distributions are not usually of the same form as the corresponding non-normal multivariate distribution. For all of these reasons, as we shall see, a number of rather *ad hoc* methods have often been employed in order to obtain suitable models.

Normal, categorical and count, and duration data do not encompass all possible types of repeated responses which might be recorded. However, currently they do represent the vast majority. In addition, they have been the areas of most active development of univariate statistical models. Thus, more tools and more interesting problems are available. By concentrating on models for these types of data, I hope to provide a useful basis for developing models in most repeated measures situations. In those other, still exceptional, cases, perhaps it can be hoped that the models which have been discussed will yield sufficient clues as to how one might proceed.

The reader who is interested in direct applications of models to repeated measurements data may now wish to skip to Chapter 3, only using the unifying principles of modelling in the next chapter as a reference guide. The reader wishing to master the principles of modelling repeated responses should cover the next chapter in detail, before going on to the subsequent illustrations.

2
Fundamentals of modelling

2.1 Statistical models

Statistical modelling is the attempt to describe variability in observed data by mathematical means. Any such model can only be an approximation to reality, but it is an attempt to elucidate the underlying mechanism which generated the data. In a scientific context, the observations must be, at least, imagined to be repeatable, so that a model may be expected to remain suitable under such recurrence. Various compromises must be made in order to obtain a result which is usable. Consider two extreme situations.

(1) If the mathematical equations being used have enough unknown parameters, they can always be made to describe the *observed* data perfectly. This is known as a *saturated* model. Such a model might appear to provide the ideal solution, but consider a second set of data supposedly arising under identical conditions. What chance will such a model have to describe it well? The least variation from the first set of data cannot be taken into account. Thus, a saturated model is not really modelling the physical mechanism producing the data in its random variability.

(2) A second possibility, then, might be to construct a model which is sufficiently simple that it is readily understandable and can be related to the scientific phenomenon under study. However, if it is too simple, a *minimal* model, it will not represent the observations at hand very well, although it will most likely do just as well (or poorly!) for any other similar data set. No systematic differences remain; everything is random.

Thus, the two primary, and conflicting, criteria with which one must struggle in constructing a statistical model are to have sufficient parameters for a reasonably close fit to the observations and sufficient simplicity to describe some understandable underlying mechanism producing the responses. In a strictly statistical sense, the first is measured by some criterion of goodness of fit, such as the log likelihood or deviance. In this text, this will usually be taken, at least where possible, as a measure of distance from a saturated model. The second is measured by the degrees of freedom, which is simply the number of unknown parameters which could still be added to the model, before it became saturated.

These two criteria constitute a first step to being able to describe the variability in a phenomenon under study. They are principles for selecting among models, but tell us nothing about what models to consider in the first place. For this, we must look more closely at what we mean by variability. In usual statistical practice, two basically different types of variability are considered.

Take a set of observations made under conditions which are essentially identical in all that is of interest to the people making the study. Obviously, every factor which might influence the results cannot be taken into account. One might even go so far as to claim that *no* variable which is not of direct interest should be involved, although this is not always possible or desirable. In any case, the responses will have some inherent variability of this type which must be accounted for in the model. We call this *random* or *stochastic* variability, because we do not try directly to *explain* these differences among the responses, but only to take them into account. This variation is introduced into a statistical model as a stochastic component, the probability distribution. One role of experimental design is to attempt to uniformize this uncontrolled variability over units.

Occasionally, the only goal of a study is to look at the form of a probability distribution, to understand what kind of mechanism might have generated the data. More usually, observations are taken under several conditions, and the goal is to determine the influences of these conditions on the responses observed, the *systematic* variability. In fact, what one is interested in now is how the distribution of responses changes under different conditions. This will be illustrated for two simple models in Figures 2.1 and 2.2 below. This systematic component is called the *location model*, because it describes how the position of the distribution is moving with respect to the axis of responses.

The distinction just made between stochastic and systematic components is often relatively arbitrary, and must be skillfully used to construct an appropriate model. A mis-specified location model will induce spurious random variation which the stochastic component must absorb. If observations on critical explanatory variables, which should be in the location model, are impossible to obtain, the same thing will happen. In some cases, such problems can be explicitly allowed for in the stochastic component, but this is not always the ideal situation. Thus, in this sense, it is often wise to overspecify the location model, as some 'best' model, at least in preliminary analysis, in order to obtain as reasonable an idea as possible of the stochastic component.

Consider the classical regression situation where one has independent observations and wishes to model the location as

$$E[Y_i] = c(\mathbf{x}, \boldsymbol{\beta})$$

The goal is to find some suitable function, $c(.)$, such that the residuals are white noise. In normal theory, this corresponds to the usual hypothesis that the variance is constant. Such regression modelling can be extended to independent responses from other distributions, for example as generalized linear models, although then the best definition of the residuals is not without controversy. However, once the hypothesis of independence among the responses must be abandoned, as with repeated measurements, the distinction between the systematic and stochastic components becomes even much less clear. Criteria for choosing a suitable function for the location now depend both on the form of the distribution and on the form of stochastic dependence among responses.

Thus, in much of statistics, only the systematic component is really of interest. Some reasonable stochastic component is assumed so that the desired information about the location model, the effects of the different known conditions of production of the data, can efficiently be obtained from the available data. Any unknown parameters specific to the *form* of the assumed distribution are taken to be nuisance parameters, to be dealt with in the most expeditious way possible. One such much used model is Cox proportional hazards to study differences in survival among groups without imposing a form on the distribution of survival times. As we shall see throughout this book, the situation is not always so simple when repeated measurements are involved. Much care must usually be taken in modelling the form of the distribution.

In this context, a robust model may be defined as one which yields a relatively stable estimation of the location model over a range of possible response distributions. The most well known example of such relaxation of the assumptions of the model is 'least squares' estimation of a regression equation. Two other examples are quasi-likelihood and generalized estimating equations. For repeated measurements studies, such approaches are sometimes possible. However, often, more care must be taken with the stochastic component, and, in many cases, it may not be considered simply as nuisance parameters, but be of direct interest.

Perhaps the most important unifying concept that has been introduced into statistics in recent decades has been that of generalized linear models, based on the exponential family of distributions. A quick look at them, in the next section, will permit us to see, not only the importance of this unification, but also the distinct peculiarities of the normal distribution, which may, so often, lead one astray. We shall, then, go on in Section 2.3 to consider some of the basic concepts for constructing the location model. Finally, in the rest of this chapter, we look at some general aspects of the stochastic component and, particularly, of specific distributional forms which may be important for repeated measurements models. Discussion will centre on univariate statistical models, insofar as they are pertinent to repeated measures, on several ways in which they can be generalized to a multivariate situation, introducing different forms of stochastic dependence, on families of multivariate distributions, and on stochastic processes. The details of specific multivariate models will be the topic of the rest of the book.

2.2   Specificities of the normal distribution

The normal distribution is a member of the exponential family, so that it is interesting to compare it to other members of that family. In this way, we shall be able to see how certain specificities of this distribution help to simplify its use in modelling multivariate data such as repeated measurements. At the same time, it is useful to be familiar with models based on the exponential family, because certain general multivariate models applicable to repeated measurements can be constructed based on it.

2.2.1   THE EXPONENTIAL FAMILY

Consider a set of *independent* random variables, Y_i $(i = 1, ..., N)$, with observed values, y_i. The *canonical form* of the probability (density) function for the random variable, Y_i, with the *location parameter*, θ_i, is

$$f(y_i; \theta_i) = \exp[y_i \theta_i - b(\theta_i) + c(y_i)] \tag{2.1}$$

Note the duality of the value of the random variable, y_i, and the parameter, θ_i. More generally, in a two parameter distribution, with constant *dispersion* or *scale parameter*, ϕ, we may write

$$f(y_i; \theta_i, \phi) = \exp\{[y_i \theta_i - b(\theta_i)]/a_i(\phi) + c(y_i, \phi)\} \tag{2.2}$$

As an illustration, let us consider two common examples, the Poisson distribution

$$\Pr(y_i; \mu_i) = \frac{\mu_i^{y_i} e^{-\mu_i}}{y_i!} \tag{2.3}$$
$$= \exp[y_i \log(\mu_i) - \mu_i - \log(y_i!)]$$

where $\theta_i = \log(\mu_i)$, $b(\theta_i) = \exp(\theta_i)$, $a_i(\phi) = 1$, $c(y_i, \phi) = -\log(y_i!)$, and the normal distribution

$$f(y_i; \mu_i, \sigma^2) = \frac{1}{\sqrt{2\pi\sigma^2}} e^{-\frac{(y_i - \mu_i)^2}{2\sigma^2}} \tag{2.4}$$
$$= \exp\{[y_i \mu_i - \mu_i^2/2]/\sigma^2 - y_i^2/(2\sigma^2) - \log(2\pi\sigma^2)/2\}$$

where $\theta_i = \mu_i$, $b(\theta_i) = \theta_i^2/2$, $a_i(\phi) = \sigma^2$, $c(y_i, \phi) = -[y_i^2/\phi + \log(2\pi\phi)]/2$.

One specificity of members of the exponential family is the importance of the mean and variance in their definition. For this family, with dispersion parameter, the log likelihood function, for one observation, is

$$\log[L(\theta_i, \phi; y_i)] = [y_i\theta_i - b(\theta_i)]/a_i(\phi) + c(y_i, \phi) \qquad (2.5)$$

and its first derivative or the *score*

$$U_i = [y_i - b'(\theta_i)]/a_i(\phi) \qquad (2.6)$$

When, for independent responses, the sum of the scores over all responses is set equal to zero, we obtain the *likelihood* or *estimating equations*. Then, from these, we have

$$E[Y_i] = b'(\theta_i)$$
$$= \mu_i \qquad (2.7)$$

say, and

$$\text{var}[Y_i] = b''(\theta_i)a_i(\phi) \qquad (2.8)$$

Usually,

$$a_i(\phi) = \frac{\phi}{w_i} \qquad (2.9)$$

where w_i are known prior weights. Call $b''(\theta_i) = \tau_i^2$, the *variance function*, a function of μ_i only, so that

$$\text{var}[Y_i] = a_i(\phi)\tau_i^2$$
$$= \frac{\phi\tau_i^2}{w_i} \qquad (2.10)$$

Here, θ_i is the parameter of interest, while ϕ is usually a nuisance parameter. With ϕ known, y_i is the sufficient statistic for θ_i and the distribution is completely specified by the relationship between the mean and the variance.

In the multivariate context of repeated measurements, the variance of the responses, and the covariances among responses on the same unit, play key roles.

2.2.2 VARIANCE FUNCTIONS AND SCALE PARAMETERS

Because the second order properties of distributions are very important for repeated measurements, let us look more closely at some common distributions which are members of this exponential family.

Poisson distribution

$$a_i(\phi) = 1$$
$$b(\theta_i) = \exp[\theta_i]$$
$$\mu_i = \exp[\theta_i] \tag{2.11}$$
$$\tau_i^2 = \exp[\theta_i]$$
$$= \mu_i$$

Binomial distribution

$$a_i(\phi) = 1$$
$$b(\theta_i) = n_i \log(1 + \exp[\theta_i])$$
$$\mu_i = \frac{n_i \exp[\theta_i]}{(1 + \exp[\theta_i])} \tag{2.12}$$
$$\tau_i^2 = \frac{n_i \exp[\theta_i]}{(1 + \exp[\theta_i])^2}$$
$$= \frac{\mu_i(n_i - \mu_i)}{n_i}$$

Negative binomial distribution

$$a_i(\phi) = \frac{1}{\phi}$$
$$b(\theta_i) = -\log(1 - \exp[\phi\theta_i])$$
$$\mu_i = \frac{\phi \exp[\phi\theta_i]}{(1 - \exp[\phi\theta_i])} \tag{2.13}$$
$$\tau_i^2 = \frac{\phi^2 \exp[\phi\theta_i]}{(1 - \exp[\phi\theta_i])^2}$$
$$= \phi\mu_i + \mu_i^2$$

Normal distribution

$$a_i(\phi) = \sigma^2$$
$$b(\theta_i) = \frac{\theta_i^2}{2}$$
$$\mu_i = \theta_i \tag{2.14}$$
$$\tau_i^2 = 1$$

Gamma distribution

$$a_i(\phi) = \frac{1}{\nu}$$
$$b(\theta_i) = -\log(-\theta_i)$$
$$\mu_i = -\frac{1}{\theta_i} \tag{2.15}$$
$$\tau_i^2 = \theta_i^{-2}$$
$$= \mu_i^2$$

Inverse Gaussian distribution

$$a_i(\phi) = \sigma^2$$
$$b(\theta_i) = -(-2\theta_i)^{\frac{1}{2}}$$
$$\mu_i = -2\theta_i^{-\frac{1}{2}} \tag{2.16}$$
$$\tau_i^2 = -2\theta_i^{-\frac{3}{2}}$$
$$= \mu_i^3$$

In this formulation, ϕ is assumed fixed and only μ_i varies. We may note the specificities of the normal distribution:

(1) Only this distribution has a variance which does not depend on the mean. This implies that the covariances will have the same characteristic. Here lies one of the reasons why the multivariate normal distribution is so easy to define and to use.

(2) The function $b(.)$ defines the cumulants of the distribution. Because this function is quadratic for the normal distribution, only the first two moments are nonzero. Thus, the stochastic dependence among normally distributed responses is completely defined by the (second order) covariances, which do not depend on the mean.

Other common distributions which one might want to use in repeated measurements do not have these two simplifying characteristics.

2.2.3 GENERALIZED LINEAR MODELS

The three components Consider a simple linear (least squares) regression. The model is often written

$$y_i = \beta_0 + \beta_1 x_i + \varepsilon_i \qquad \text{with} \qquad \varepsilon_i \sim N(0, \sigma^2) \tag{2.17}$$

but its structure can be more clearly seen from

$$\mu_i = \beta_0 + \beta_1 x_i \tag{2.18}$$

where μ_i is the mean of a normal distribution. Although it is not evident from the simple case of the normal distribution, three distinct components are, in fact, involved here.

(1) The response distribution Y_i $(i = 1, ..., N)$ are independent random variables, with means μ_i and observed values $y_i = \mu_i + \varepsilon_i$, where the ε_i are the residuals or 'errors'. These random variables share the same distribution from the exponential family, often classically called the error structure. This *probability model* for our data is the *random* or *stochastic* component.

(2) The linear predictor A set of P (usually) unknown parameters, $\boldsymbol{\beta}=$ $[\beta_1, \ldots, \beta_P]^T$, and the corresponding set of known explanatory variables $\mathbf{X}_{N \times P} = [\mathbf{x}_1, \ldots, \mathbf{x}_N]^T$, the design or model matrix, are such that

$$\eta_i = \sum_{l=1}^{P} x_{il} \beta_l \tag{2.19}$$

or

$$\mathbf{H} = \mathbf{X}\boldsymbol{\beta} \tag{2.20}$$

define the *linear predictor*. This describes how the location of the response distribution changes with the explanatory variables. This *mathematical* or *location model* for our data is the *systematic component*. In more general cases of interest, outside the GLM context, the structure describing change in location will be nonlinear. Such may be the case, for example, with growth curves.

To clarify our ideas before continuing, let us look again at the simple linear normal regression model. This may be represented as in Figure 2.1. Next, consider the similar case for the binomial distribution, simple linear logistic regression, plotted in Figure 2.2. The first thing we notice is that the latter linear regression does not produce a straight line; it has the logistic form. As well, the *shape* of the distribution changes with changes in the location, because the variance depends on the mean in this distribution. Obviously, our linear predictor or location model cannot be connected to the binomial probabilities in the same simple way as the identical linear structure is to the means of the normal distribution. This leads us to the third point.

(3) The link function The relationship between the mean of the ith response and its linear predictor is given by the *link function* $g_i(.)$:

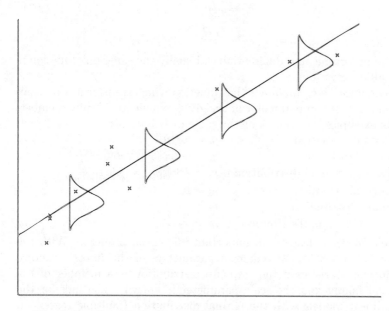

Fig. 2.1. Graphical representation of a simple linear normal regression.

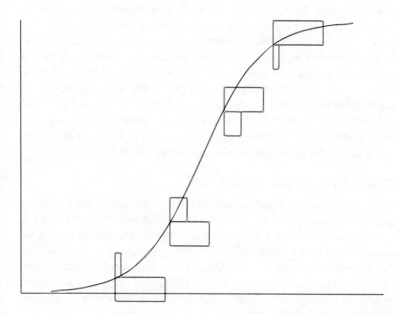

Fig. 2.2. Graphical representation of a simple linear logistic regression.

$$\eta_i = g_i(\mu_i) \tag{2.21}$$
$$= \mathbf{x}_i^T \boldsymbol{\beta}$$

It must be monotonic and differentiable. Usually the same link function is used for all observations.

The *canonical link function* is that function which transforms the mean to a canonical location parameter of the given exponential family member. Again, as examples, let us look at the same distributions:

Poisson distribution $\qquad\qquad \eta_i = \log[\mu_i]$

Binomial distribution $\qquad\qquad \eta_i = \log[\frac{\pi_i}{1-\pi_i}] = \log[\frac{\mu_i}{n_i-\mu_i}]$

Negative binomial distribution $\quad \eta_i = \frac{1}{\phi}\log[\pi_i] = \frac{1}{\phi}\log[\frac{\mu_i}{\phi+\mu_i}]$

Normal distribution $\qquad\qquad\quad \eta_i = \mu_i$

Gamma distribution $\qquad\qquad\quad \eta_i = \frac{1}{\mu_i}$

Inverse Gaussian distribution $\quad\; \eta_i = \frac{1}{\mu_i^2}$

These link functions lead to an important inferential property. With the canonical link function, all unknown parameters of the *linear* structure have sufficient statistics if the response distribution is a member of the exponential family and the scale parameter is known. The link for the common linear models with the normal distribution (multiple regression and ANOVA) is especially simple, because it is the identity. The other most commonly used member of this family is the log linear model for categorical data, and its special case, the logistic model. There, as its name suggests, the log of the mean is equated to the linear predictor.

Generalized linear models may be fitted using a generalization of 'least squares' called iterated weighted least squares (IWLS). This has been implemented in such programs as GLIM, GENSTAT, XLisp-Stat, and S Plus.

When we combine the probability model with the location model, we obtain a full statistical model to describe our data (Lindsey, 1974b). The reader not familiar with the theory of generalized linear models may consult Dobson (1990) and McCullagh and Nelder (1989) for more details on constructing and analysing them in the univariate situation.

2.2.4 FURTHER GENERALIZATIONS

One extension to the GLM family permits us to include a large number of distributions which can be important in repeated measurements. If the desired distribution would be a member of the exponential family except for one parameter, fitting the model is still relatively simple. For example, the Weibull distribution is such an extension of the exponential distribution. In such cases, a program implementing IWLS allows us to fit the desired model by adding a simple additional iterative step, on the extra parameter, to the estimation procedure.

In the same way that what is commonly called 'least squares' estimation is maximum likelihood estimation of normal distribution models, without the assumption of normality, so, more generally, 'quasi-likelihood' estimation (Wedderburn, 1974) is the corresponding extension of generalized linear models. In fact, what we are doing is only specifying the relationship between the mean and the variance, ignoring higher order relationships. The idea of such an approach is to take advantage of the robustness of the estimation of these models when one is not sure that the model assumptions are applicable to the data at hand.

Occasionally, it may be necessary to go even further in loosening the assumptions of a standard model, such as a GLM, than just assuming quasi-likelihood. Estimates for a model are generally obtained by setting the first partial derivatives of the log likelihood function, the score, equal to zero and solving the resulting system of, usually nonlinear, equations. These are the *likelihood* or *estimating equations* and the resulting estimates, maximum likelihood estimates. For the GLM family, IWLS is the most common procedure for solving them. With independent observations, this involves iteratively weighting by a diagonal variance matrix. However, for stochastically dependent responses, this matrix will no longer be diagonal; the covariances are nonzero. For the normal distribution, we obtain a multivariate normal, because of the unique properties outlined above. But, if we substitute such a general covariance matrix into the likelihood equations of another member of the GLM family, we generally do not obtain a result from which we can back up by integration to obtain a true likelihood function, and hence a statistical model. Such equations are called *generalized estimating equations*. They may often provide useful and interpretable results, by analogy with the multivariate normal distribution, but have no direct physical interpretation in terms of generation of the responses from a statistical model. They are described further in Section 2.9 below.

As it stands, the discussion in this section has limited direct usefulness for repeated measurements, because it has primarily been dealing with the univariate case. These definitions for the exponential family and generalized linear models point out the difficulties which must be overcome in order to construct models for repeated measures, because they are the simplest class of univariate models available.

2.3 The location model

As we saw above for the exponential family and shall see more generally in the next section, probability distributions usually have one parameter in their specification which determines their location on an axis whose domain is the variable which is being described. As in generalized linear models, the parameter is very often the mean or expected value of the random variable, $E[Y] = \mu$, or some transformation of it. Then, the goal is to describe how this parameter changes under the various observable conditions, the systematic component of the model.

In classical normal theory linear models, the mean parameter of the normal distribution is equated to a linear predictor describing changing conditions; the link function is the identity. The model is especially simple because the form of the normal distribution, its dispersion or variance, need not change as the mean varies. This situation is not typical, because the linear predictor will generally be equated to some *function* of the mean, the link function, and the form of the distribution will almost always depend on its mean. Such relationships for a number of distributions were described in Section 2.2.2.

Thus, in Equation (2.21) above, we saw that the link function relates to the location parameter

$$\eta_i = g_i(\mu_i)$$
$$= \mathbf{x}_i^T \boldsymbol{\beta}$$

and in Equation (2.10), that the variance is a function of the mean

$$\text{var}(Y_i) = a_i(\phi)\tau_i^2$$
$$= \frac{\phi \tau_i^2}{w_i}$$

for generalized linear models, where the distribution is a member of the exponential family. For other distributions, the same ideas are applicable, although the analytic results are not so simple.

A linear model is one in which the unknown parameters to be estimated are incorporated into the location model in a linear fashion. On the other hand, the variables may be involved in a nonlinear way, such as being squared, having a logarithm, and so on. This has been important in classical theory, and in generalized linear models, for strictly computational reasons. Estimates can be obtained by linear algebra, the manipulation of vectors and matrices. The introduction of a link function permits nonlinearity in the relationship between the mean and the linear predictor. However, this comes at the expense of requiring iterations of the linear algebra, for example, by the use of the Newton–Raphson method or IWLS. Other optimization methods, such as the simplex or the calculation of numerical gradients, are also used. In repeated measurements, such as growth curves, forms of nonlinearity in the parameters are often required, directly in the location model, but this is primarily a numerical, and not strictly a statistical, problem. However, linear models still play an important role, and it is useful to look at them a little bit more closely.

2.3.1   TYPES OF EXPLANATORY VARIABLES

In Chapter 1, we distinguished experimental variables or treatments and sample variables, or covariates, by the way in which the data are collected.

Here, we shall classify variables according to their internal structure. Thus, two types of explanatory variables can usually be distinguished.

(1) Quantitative, usually continuous, variables are measurable in some way. Classically, regression analysis has been used to handle them. A simple model with identity link would be written

$$\mu_i = \beta_0 + \beta_1 x_i$$

(2) Qualitative or factor variables simply index nominally different conditions. Analysis of variance is the classical approach to them. Again, a simple model would be

$$\mu_i = \mu + \alpha_i$$

However, as is well known, both can be handled in exactly the same way. A factor variable is fitted as regression on a series of binary indicator or 'dummy' variables indexing the conditions. And both can be combined into the same model, known classically as analysis of covariance.

Quantitative variables can be transformed into factors by appropriately cutting them into a number of discrete categories. Conversely, if the labels of a factor variable are measures, a second way of representing it, instead of as a series of binary indicators, is as a polynomial of appropriate degree. The two are equivalent, but provide alternative interpretations of the data. Often, it is useful to use *orthogonal polynomials*, because they allow independent interpretation of each order of effect (linear, quadratic, cubic, ...).

When polynomials are used, they may be thought of as a Taylor or Maclaurin series approximation

$$
\begin{aligned}
\eta_i &= g_i(\mu_i) \\
&= c(0) + c'(0)x_1 + c''(0)x_1^2/2 + \cdots \\
&= \beta_0 + \beta_1 x_1 + \beta_2 x_1^2 + \cdots
\end{aligned}
\tag{2.22}
$$

to some unknown nonlinear functional form, $c(x_1)$, of the explanatory variable, x_1. Thus, they should only be considered when such a function is unknown theoretically. And care should be taken not to use polynomials of too high an order, usually not more than quadratic, because the model will be inherently unstable in replications of the data.

Explanatory variables are said to *interact* if the effect of one on the location of the response distribution depends on the value of a second. If not, they are said to be additive. Interactions can be of any order, involving more than two variables, although, especially in the factor variable case, they rapidly become difficult to interpret. In almost all cases, if an interaction of some order is retained in a model, all lower order interactions encompassed by the variables involved, including all such main effects, must also be included for the model to make sense. These are known as *hierarchical models*.

In the continuous case, inclusion of interactions often implies that the appropriate powers of the same variables should also be included. Thus, a model with a two-way interaction between two continuous variables should usually also include the squares of the two variables as well. A three-way interaction would mean pulling in all two-way interactions, all quadratics and all cubics for the three variables. One reason that this is important is so that the series approximations to unknown variables, described above, make sense.

In the context of repeated measurements, continuous explanatory variables will be most important in explaining how repeated responses vary, for example, over time. Qualitative variables will usually distinguish different treatments or conditions under which the series of responses were obtained.

2.3.2 WILKINSON–ROGERS NOTATION

When a large number of variables are present in a linear model, and, especially in the presence of many interactions, the traditional way of presenting the model, as above, although mathematically precise, can be cumbersome and unclear. Wilkinson and Rogers (1973) introduced an alternative notation, which not only has the advantage of being very clear to use in many situations, but is also directly transferable to an electronic computing environment.

The Wilkinson and Rogers notation uses the variable names, instead of the corresponding parameters, in the formula. It uses a number of operators:

- addition of effects: +
- simple interaction: ·
- crossing or hierarchical interaction: *
- nesting: /
- deletion: −

The addition and deletion operators can be both unary and binary. All of the others are binary. There are also a number of more complex operators which will not be discussed here. Because the system is meant to be interactive and sequential, the unary operators function with respect to the immediately preceding model, adding or deleting one or more terms.

The binary addition operator connects all terms in the model, in much the same way as it does in the classical formulae involving the parameters. Thus, a model with three main effect terms would be represented as

$$A + B + C$$

It can also connect more complex terms made up of interactions.

The simple interaction operator defines relationships among variables which are interactions. A two-way factorial design would be represented as

$$A + B + A \cdot B$$

More complex models are readily developed.

The analogous models for continuous variables can be written in the same way, although powers can be used as a short form. The equivalent expression to the two-way factorial just given would be

$$A + B + A^2 + B^2 + A \cdot B$$

Note that this contains more terms than the previous expression. Not all computer implementations of this notation are able to handle such products of continuous variables.

So far, the notation provides little advantage over the traditional equations. In our context, the remaining two operators provide the real benefits of simplicity and clarity. Consider again the two-way factorial model. This may also be written

$$A * B \equiv A + B + A \cdot B$$

The hierarchical interaction operator is primarily important with factor variables. It automatically includes all of the necessary lower order terms in the model. In more complex situations, this results in greatly simplified formulae. Take a three-way factorial design, but without the three-way interaction. This may be written

$$A * B * C - A \cdot B \cdot C \equiv A + B + C + A \cdot B + A \cdot C + B \cdot C$$

One context where such notation is especially useful is in the specification of log linear and logistic models for cross-classified frequency or contingency tables, where many levels of interactions may be involved.

The nesting operator, as its name suggests, is used for variables which only have meaning within each category of some other variable. In such cases, the marginal term for the nested variable is irrelevant.

$$A/B \equiv A + A \cdot B$$

By the efficient use of brackets, even more complex expressions can be succinctly written:

$$(A + B) * C \equiv A + B + C + A \cdot C + B \cdot C$$

$$(A * B)/C \equiv A + B + A \cdot B + A \cdot B \cdot C$$

For clarity, the actual names of the variables, or suitable abbreviations, can be placed in the formulae. It should be clear, however, that this notation is particularly suited to qualitative or factor variables and that it is not applicable to nonlinear models.

2.4   Univariate probability distributions

One of the fundamental hypotheses of much of standard statistical modelling is the independence among observations. This permits the use of univariate probability distributions which can be simply multiplied together to obtain the joint probability of the set of responses and, hence, the likelihood function. Because the same unit is being repeatedly observed in repeated measurements studies, the problem is not so simple here. Theoretically, all models should be based on appropriate multivariate distributions for the responses on a unit. Unfortunately, most such distributions available have been developed, not for modelling observable physical phenomena, but for making inferences around the normal distribution models. They are of little use in our context. In the last few years, a number of suitable models have begun to appear, but this is still one area of statistics which is notably underdeveloped.

The simplicity of the normal distribution, with its identity link, its variance not depending on the mean, and its zero higher order moments, means that the multivariate normal model is also particularly simple. We shall be using it extensively in Part II of this book. However, extension of such methods to other distributions, by analogy, must be made with great care because of this deceptive simplicity. In Sections 2.6 and 2.7, we shall look at some of these possibilities, but it is first, perhaps, useful to pass quickly in review some of the existing univariate distributions which might be useful in repeated measurements. When available, other appropriate non-normal multivariate distributions will be introduced in their proper place later in the book.

Although certain specific characteristics of some members of the exponential family of distributions were given in Section 2.2.2, they will also be included here, in a different form, for completeness. As for the rest of the book, the distributions listed here may be classified as being appropriate for normal models, counts, and categorical data, or duration data.

2.4.1   NORMAL DISTRIBUTIONS

Normal distribution   The random variable in a normal distribution can take any value on the real line. Such data may be generated when a very large number of unknown influences combine additively to produce the observed values.

$$f(y; \mu, \sigma^2) = \frac{1}{\sqrt{2\pi\sigma^2}} e^{-\frac{(y-\mu)^2}{2\sigma^2}} \tag{2.23}$$

The distribution is symmetric and is characterized by the variance parameter not depending on the mean.

(Exponential family)

Log normal distribution   The random variable in a log normal distribution can take any value on the positive real line. Such data may be generated when a very large number of unknown influences combine multiplicatively to produce the observed values.

$$f(y; \mu, \sigma^2) = \frac{1}{y\sqrt{2\pi\sigma^2}} e^{-\frac{(\log[y]-\mu)^2}{2\sigma^2}} \tag{2.24}$$

The distribution is asymmetric, but otherwise has many characteristics similar to those for the normal distribution. Indeed, log normal models can be fitted to data using all of the classical procedures simply by taking logarithms of the responses. This distribution is also suitable as a basis for duration models, in Section 2.4.3 below.

(Exponential family; accelerated failure time model)

Power-transformed normal distribution   This model, introduced by Box and Cox (1964), attempts to transform data to normality by

$$y^{(v)} = \frac{(y + v_2)^{v_1} - 1}{v_1} \tag{2.25}$$

which, when $v_1 \to 0$, gives the log normal distribution.

$$f(y; \mu, \sigma^2, v_1, v_2) = \frac{1}{\sqrt{2\pi\sigma^2}} e^{-\frac{(y^{(v)} - \mu)^2}{2\sigma^2}} J(v; y) \tag{2.26}$$

where $J(v; y)$ is the Jacobian of the transformation. Although such a model may give reasonable empirical results, it should be avoided if a more suitable theoretical model can be found. There are several reasons.

(1) it will not usually provide very useful information on an underlying data generation mechanism,

(2) the resulting transformed response will often be such that the location model is difficult to interpret,

(3) a model for one set of data may often not be found applicable to subsequent sets.

(Exponential family for v fixed)

Inverse Gaussian distribution   Suppose that we have Brownian motion (see Section 2.10 below) with positive drift, ν and variance per unit time, κ. Then, the inverse Gaussian distribution arises as the duration to the hitting time, i.e. until the first passage time to a fixed barrier at v.

$$f(y; v, \nu, \kappa) = \frac{v}{\sqrt{2\pi\kappa y^3}} e^{-\frac{(v-\nu y)^2}{2y\kappa}} \tag{2.27}$$

Because only two of the parameters are identifiable, we reparameterize as $\mu = v/\nu$ and $\sigma^2 = \kappa v/\nu^3$. This is another distribution whose main application is to durations.

(Exponential family)

2.4.2   CATEGORICAL AND COUNT DISTRIBUTIONS

Poisson distribution   The random variable from a Poisson distribution can take any non-negative integral value. This distribution may most usefully be thought of as describing the number of events which have occurred at random to a unit in a given period of time, or in a given space.

$$\Pr(y; \mu) = \frac{\mu^y e^{-\mu}}{y!} \tag{2.28}$$

Because $E[Y] = \text{var}[Y]$ for this distribution, their ratio, $E[Y]/\text{var}[Y]$, provides a measure of departure from randomness. Values greater than one indicate uniformity, those less than one clumping.

This model is very important in stochastic processes in time or space. It applies to the number of events occurring in successive small intervals, where occurrences in these intervals are independent. It is called a Poisson process (see Section 2.10 below) and corresponds to the intervals between events having an exponential distribution. A third application is to cross-classified frequency data, because, if the total number of observations is fixed, it is identical to a multinomial distribution.

If the random variable is allowed to take non-negative real values, it is called the *Eulerian distribution*.

(Exponential family)

Multinomial distribution   The multinomial is a distribution able to describe many sets of data, because it only assumes independence among the responses and because all empirical observations must actually be made within discrete intervals of measurement. It is simple in that it has few theoretical assumptions, but complex in that it has many parameters. The variable consists of the possible observable categories, although this is ambiguous, because the distribution actually involves the frequency of occurrence of each category in multiple observations.

$$\Pr(y_k) = \pi_k \quad \forall k \tag{2.29}$$

$$\Pr(n_1, \ldots, n_K; \pi_1, \ldots, \pi_K) = \left(\frac{\sum_k n_k}{n_1 \ldots n_K} \right) \prod_{k=1}^{K} \pi_k^{n_k} \qquad (2.30)$$

where $\sum_k \pi_k = 1$. The multinomial distribution forms the basis for log linear models for categorical data. It can also be used as the basis for comparing all other univariate distributions with independent responses (Lindsey, 1974a; Lindsey and Mersch, 1992).

(Exponential family)

Binomial distribution The binomial distribution is an important special case of the multinomial, when there are only two categories.

$$\Pr(n_1, n_2; \pi_1) = \left(\frac{n_1 + n_2}{n_1} \right) \pi_1^{n_1} (1 - \pi_1)^{n_2} \qquad (2.31)$$

In modern statistics, this distribution has come to play a role at least as important as the normal distribution.

One possible interpretation of this distribution is to suppose that some unobserved, underlying variable exists with one of a normal, a logistic, or a Weibull distribution. If its value for a unit is below a critical level, one of two possible observable events occurs, otherwise the second. Then, such binary events may be modelled by a binary regression with, respectively, a probit, logistic, or complementary log log link.

(Exponential family)

Negative binomial distribution The random variable of the negative binomial or Polya distribution is the count of the number of failures until a fixed number of successes is achieved. It can, thus, be used as a discrete duration distribution.

$$\Pr(y; n, v) = \frac{\Gamma(y + n)}{y! \Gamma(n)} \left(\frac{1}{1 + v} \right)^n \left(\frac{v}{1 + v} \right)^y \qquad (2.32)$$

where $\pi = v/(1+v)$ and $\Gamma(\cdot)$ is the gamma function. Once the observations have been made, the likelihood is identical to that for a binomial model. An important special case, when $n = 1$, is the geometric distribution.

The number of successes, n, can also be taken as unknown and estimated as a parameter, in this case, not necessarily being an integer. As we shall see, this yields a generalization of the Poisson distribution for overdispersion, where the mean does not equal the variance. In this context, it is derived by compounding the Poisson distribution with the gamma (see Section 2.6 below).

(Exponential family for n fixed)

Beta-binomial distribution The beta-binomial distribution may be used when binomial counts are under- or overdispersed, i.e. when the mean-variance relationship, given in Section 2.2.2, does not hold.

$$\Pr(n_1, n_2; \kappa, \upsilon) = \binom{n_1 + n_2}{n_1} \frac{B(\kappa + n_1, \upsilon + n_2)}{B(\kappa, \upsilon)} \tag{2.33}$$

where $B(\cdot, \cdot)$ is the beta function. It can be derived by compounding the binomial distribution with the beta distribution (which is not presented here, because it is not used, in its own right, as a basis for models; however, see Section 5.2.1). A generalization is the beta-multinomial distribution.

(Not in the exponential family)

2.4.3 DURATION DISTRIBUTIONS

Duration distributions are often called survival or reliability distributions, from their use in medicine and engineering, but they have much wider application. Responses which are durations have several important characteristics which distinguish them from most others.

(1) They must be non-negative and are usually positive.

(2) They are usually positively skewed.

(3) Because time and money of the observer are limited and units may be susceptible to disappear from observation, observations may be *censored*: only part of some durations may be recorded, although this still provides useful information.

Two functions, related to the probability distribution, are

- the *survival function*, which is the probability of surviving, or, more generally, of a period lasting until at least a given time,

$$\begin{aligned} S(y) &= \Pr(Y > y) \\ &= 1 - F(y) \end{aligned} \tag{2.34}$$

- the *risk*, the *hazard*, the *failure rate*, or the *mortality rate*, which is the instantaneous probability that the duration will terminate at that point, conditional on it having continued until then, or the *intensity* with which termination occurs:

$$\omega(y) = \frac{f(y)}{S(y)} \tag{2.35}$$

Any one of the three functions defines a model completely.

Two special cases are of particular interest. If the hazard function can be written

$$\omega(y_i; \mathbf{x}_i) = \omega_0(y_i) g^{-1}(\eta_i) \tag{2.36}$$

where $\omega_0(y_i)$ is a baseline hazard function when $\mathbf{x}_i = 0$, $g(\cdot)$ is the link function, and η_i the linear predictor, we have a *proportional hazards model*. Usually, a log link is used. Hazards for all units are proportional, depending on the explanatory variables. If the base function, $\omega_0(y_i)$, is left unspecified, we have a semi-parametric model, known as the Cox proportional hazards model (Cox, 1972a).

If the hazard function can be written

$$\omega(y_i; \mathbf{x}_i) = \omega_0(y_i e^{-\eta_i}) e^{-\eta_i} \tag{2.37}$$

it is called an *accelerated life* or *failure time model*. The effect of explanatory variables is to accelerate or decelerate the time to completion of the duration. It yields a simple regression model for the log duration.

Exponential distribution   If the phenomenon involves no aging effect, the exponential distribution will be applicable.

$$f(y; \theta) = \theta e^{-\theta y} \tag{2.38}$$

The failure rate is constant, equal to θ, which is also the reciprocal of the mean duration. In a stochastic process, this distribution, for the intervals between events, corresponds to a Poisson process, with intensity θ, for the counts of the numbers of events in fixed intervals.

(Exponential family; proportional hazards model; accelerated failure time model)

Pareto distribution   If the log durations have an exponential distribution, the durations will have the Pareto distribution, although a second parameter, κ, the minimum, is usually included.

$$f(y; \theta, \kappa) = \frac{\kappa\theta}{y^{\theta+1}} \tag{2.39}$$

The hazard function is inversely proportional to the duration,

$$h(y; \theta) = \frac{\theta}{y} \tag{2.40}$$

Because of the transformation, this distribution does not have the same simple properties as the exponential.

(Exponential family)

Weibull distribution   For the Weibull distribution, the durations arise from a 'weakest link' situation, whereby one of a variety of (unrecorded) causes may bring about the event ending the duration. The first one to occur, the weakest, determines the duration.

$$f(y; \mu, \kappa) = \kappa \mu^{-\kappa} y^{\kappa-1} e^{-(y/\mu)^{\kappa}} \qquad (2.41)$$

The parameter, μ is the mean duration, the location parameter. The second parameter, κ, is a shape parameter. It can be interpreted as a power transformation of an exponentially distributed variable, so that, when $\kappa = 1$, we have the exponential distribution. The hazard function is

$$h(y; \mu, \kappa) = \kappa \mu^{-\kappa} y^{\kappa-1} \qquad (2.42)$$

It is monotone increasing for $\kappa > 1$ and decreasing for $\kappa < 1$.

(Exponential family for κ fixed; proportional hazards model; accelerated failure time model)

Extreme value distribution   The extreme value or Gompertz distribution, as its name implies, is used for extreme phenomena.

$$f(y; \mu, \kappa) = \kappa \mu^{-\kappa} e^{\kappa y - (e^y/\mu)^{\kappa}} \qquad (2.43)$$

The hazard function is

$$h(y; \mu, \kappa) = \kappa \mu^{-\kappa} e^{\kappa y} \qquad (2.44)$$

It is monotone increasing for $\kappa > 0$ and decreasing for $\kappa < 0$. If the log duration has an extreme value distribution, then the duration has a Weibull distribution. A generalization, with one additional parameter, is called the Gompertz–Makeham distribution.

(Exponential family for κ fixed; proportional hazards model)

Gamma distribution   With the gamma distribution, the durations arise as the total of a sequence of exponentially distributed intervals.

$$f(y; \mu, \phi) = \frac{\phi^\phi y^{\phi-1} e^{-\frac{y\phi}{\mu}}}{\mu^\phi \Gamma(\phi)} \qquad (2.45)$$

Then, if ϕ is an integer, it is the number of exponential intervals making up the duration. For $\phi = 1$, it reduces to the exponential distribution. Whereas the Weibull distribution might be thought of as a number of processes operating in parallel, with the weakest determining the duration time, the gamma would have them in series, with the duration being their sum. The hazard function is

$$h(y; \mu, \phi) = \frac{\phi^\phi y^{\phi-1} e^{-\frac{y\phi}{\mu}}}{\mu^\phi [\Gamma(\phi) - \Gamma(\frac{y\phi}{\mu}, \phi)]} \qquad (2.46)$$

where $\Gamma(\cdot,\cdot)$ is the incomplete gamma function. It decreases to a constant value $1/\mu$ for $\phi < 1$ and increases to this same constant for $\phi > 1$.

(Exponential family; accelerated failure time model)

Logistic distribution   Like the normal distribution, the logistic has a random variable with values anywhere on the real line.

$$f(y; \theta, \kappa) = \frac{\kappa e^{-\theta - \kappa y}}{(1 + e^{-\theta - \kappa y})^2} \qquad (2.47)$$

Except in the tails, it is usually indistinguishable from the normal distribution.

The log logistic distribution, obtained by replacing y by $\log y$,

$$f(y; \theta, \kappa) = \frac{\kappa e^{-\theta} y^{-\kappa}}{y(1 + e^{-\theta} y^{-\kappa})^2} \qquad (2.48)$$

which allows only non-negative values, is usually more suitable for duration data. The location parameter is θ, while κ is a shape parameter. The hazard function is

$$h(y; \theta, \kappa) = \frac{\kappa}{y(1 + e^{-\theta} y^{-\kappa})} \qquad (2.49)$$

It increases to a maximum, then decreases. This distribution is very similar in form to the log normal, and can usually only be distinguished from it if a large number of observations is available in the tail.

(Not in the exponential family; accelerated failure time model)

Not all of the univariate distributions just listed will be used in the remainder of the book. However, they have been provided here for convenience, because the methods of model construction for repeated measures, which will be described, are applicable to all of them, and many more. In Sections 2.6 and 2.7, I shall discuss a number of general ways in which univariate distributions can be converted to multivariate forms useful for our purposes.

2.5   Families of Multivariate Distributions

As we have seen, one reason that the multivariate normal distribution is relatively simple to understand and use is because all of the stochastic dependence structure among responses is contained in the covariances or correlations. The first basic problem in constructing any other multivariate distribution is deciding how to define this dependence. Holland and Wang (1987b) derive a local dependence function for bivariate distributions, based on a cross-product ratio. For discrete distributions, it is a difference equation, the *local log odds ratio*,

$$\gamma(y_1, y_2) = \log \left[\frac{\Pr(y_1, y_2) \Pr(y_1 + 1, y_2 + 1)}{\Pr(y_1 + 1, y_2) \Pr(y_1, y_2 + 1)} \right] \tag{2.50}$$

and for continuous distributions a differential equation

$$\gamma(y_1, y_2) = \frac{\partial^2}{\partial y_1 \partial y_2} \log[f(y_1, y_2)] \tag{2.51}$$

Generalization to multivariate distributions does not involve simply writing down one higher order difference or differential equation. Instead, one equation is required for each possible combination of two or more responses. Thus, a trivariate distribution would require three second order and one third order equation to determine completely its stochastic dependence structure. Then, the complete local stochastic dependence structure, along with the marginal distributions, is sufficient to define a multivariate distribution.

Although it is theoretically possible to construct multivariate distributions based on given local stochastic dependence structures, the results are usually intractable. The local dependence function is more useful in studying the stochastic dependence in multivariate distributions constructed by more indirect approaches.

Useful multivariate distributions can be derived in a number of ways. Hougaard (1987) gives a general discussion for duration distributions. For example, useful general families of multivariate distributions may be defined in terms of their univariate marginal distributions or their univariate conditional distributions. Unfortunately, it is rarely possible to have the same univariate marginal and conditional distribution. If one is chosen to have a known form, the other will almost always be analytically very complex. We shall look at the two cases in turn in this section. In the following sections, we consider some general procedures for the construction of multivariate models as compound distributions and by conditioning.

2.5.1 MULTIVARIATE MARGINAL DISTRIBUTIONS

One possible approach to the construction of an appropriate multivariate distribution is by the specification of its univariate marginals. This is not sufficient to define completely the distribution and many supplementary conditions have been proposed, usually for specific marginals. Here, we shall look at one general method.

For simplicity, consider the bivariate case. Suppose that the desired univariate marginal distributions are given by their cumulative distribution functions, $F_1(y_1)$ and $F_2(y_2)$. Then, a bivariate distribution with these marginals can be formed as

$$F_{12}(y_1, y_2) = c^{-1}\{c[F_1(y_1)] + c[F_2(y_2)]\} \tag{2.52}$$

where $c(\cdot)$ is a function from $[0,1]$ to $[0,\infty]$ with negative first derivative and positive second derivative, and $c(1) = 0$. Examples of such functions include

$$
c(u) = \begin{cases} u^{1-\nu} - 1 & \nu > 1 \\ -\log(u) & \nu = 0 \\ 1 - u^{1-\nu} & 0 < \nu < 1 \end{cases}
$$
$$
c(u) = -\log\left[\frac{1-\nu^u}{1-\nu}\right] \qquad \nu > 1
$$
$$
c(u) = |\log(u)|^{\nu} \qquad \nu > 1
$$
$$
c(u) = \frac{1}{1-\nu}\log\left[\frac{1+\nu(u-1)}{u}\right] \qquad -1 \le \nu \le 1
$$
$$
c(u) = \frac{\log(u)[\log(u)-2\nu]}{2\nu^2} \qquad \nu > 0
$$

(2.53)

Genest and MacKay (1986a and b) call these copulas.

Usually, one will be interested in a multivariate distribution where all of the marginals have the same form. Then, we see from the way in which they are constructed using Equation (2.52), that these multivariate distributions will have a high degree of symmetry. For this reason, they are of most interest in repeated measurements studies for modelling heterogeneity, because this implies a symmetry among the responses on a unit.

A subfamily of the copulas, described by Marshall and Olkin (1988) and Oakes (1989), is that for which $c^{-1}(\cdot)$ is a Laplace transform, yielding compound distributions which will be discussed further below in Section 2.6. These have been used for modelling dependency or frailty in duration data, for example, by Clayton (1978), Clayton and Cuzick (1985), Lindley and Singpurwalla (1986), and Oakes (1982, 1986, 1989).

If we use the first function given above, $c^{-1}(\cdot)$ is the Laplace transform of a gamma distribution, with index $1/(\nu - 1)$. The joint probability function is

$$
F_{12}(y_1, y_2) = \{[F_1(y_1)]^{1-\nu} + [F_2(y_2)]^{1-\nu} - 1\}^{-\frac{1}{\nu-1}} \qquad (2.54)
$$

with marginals $F_k(y_k)$. Independence occurs for $\nu = 1$, i.e. for $c(u) = -\log(u)$, because then $F_{12}(y_1, y_2) = F_1(y_1)F_2(y_2)$. This is the compound or frailty distribution of Clayton (1978) and Oakes (1982).

2.5.2 EXPONENTIAL FAMILY CONDITIONALS DISTRIBUTIONS

A second approach to specifying a suitable multivariate distribution involves definition of the univariate conditional distributions. General results are available when these are members of the exponential family. Recall, from Equation (2.1) above, that distributions in the exponential family have the form

$$
f(y_i; \theta_i) = \exp[y_i\theta_i - b(\theta_i) + c(y_i)]
$$

Suppose, now, that we wish to create a multivariate distribution whereby the conditional distributions have this form (Arnold and Strauss, 1991; Arnold, Castillo, and Sarabia, 1992).

Consider the simple bivariate case, without explanatory variables, where k indexes the two responses. The distribution will be

$$f(y_{i1}, y_{i2}; \theta_{i1}, \theta_{i2}, \rho_i) = \exp[y_{i1}\theta_{i1} + y_{i2}\theta_{i2} + \rho_i y_{i1} y_{i2}$$
$$+ c_1(y_{i1}) + c_2(y_{i2}) - d(\theta_{i1}, \theta_{i2}, \rho_i)] \quad (2.55)$$

where $d(\theta_{i1}, \theta_{i2}, \rho_i)$ replaces $b(\theta_i)$ as the normalizing constant. The parameter, ρ_i, measures the stochastic dependence between the two responses; when it is zero they are independent. Thus, the local stochastic dependence structure is constant over all values of the responses.

An example is a bivariate distribution with Poisson conditional distributions. This has the form

$$f(y_{i1}, y_{i2}; \theta_{i1}, \theta_{i2}, \rho_i) = \exp[y_{i1}\theta_{i1} + y_{i2}\theta_{i2} + \rho_i y_{i1} y_{i2}$$
$$- \log(y_{i1}!) - \log(y_{i2}!) - d(\theta_{i1}, \theta_{i2}, \rho_i)]$$
$$= \frac{\mu_{i1}^{y_{i1}} \mu_{i2}^{y_{i2}} \nu_i^{y_{i1} y_{i2}}}{y_{i1}! y_{i2}! e^{d(\theta_{i1}, \theta_{i2}, \rho_i)}} \quad (2.56)$$

where $\theta_{ik} = \log(\mu_{ik})$, $\rho_i = \log(\nu_i)$, and $0 \leq \nu_i \leq 1$ with independence for the value $\nu_i = 1$. The conditional distribution of y_{i1}, given y_{i2}, is Poisson with mean, $\mu_{i1}\nu_i^{y_{i2}}$, and symmetrically for y_{i2}.

These simple results may easily be extended to the more complex cases

- more than two responses
- explanatory variables for the location, whereby, θ_i becomes a vector, so that the θ_{ik} also become vectors, and ρ_i a matrix,
- different distributions for the various conditionals.

However, even in the simple cases, the results are not easy to use because $d(\theta_{i1}, \theta_{i2}, \rho_i)$ cannot usually be written down explicitly, making estimation of the parameters very difficult.

This family is symmetrical in the way conditioning is applied. For this reason, it is of limited usefulness in repeated measurements. Responses on a unit will usually be ordered in time or space and we shall want a hierarchical series of conditional distributions, following this order. We look at this case in Section 2.7 below.

2.5.3   EXPONENTIAL DISPERSION MODELS

Multivariate models may also be constructed directly through the exponential dispersion family (Jørgensen, 1992), denoted by $\mathrm{ED}(\boldsymbol{\mu}, \sigma^2)$:

$$f(\mathbf{y}; \boldsymbol{\theta}) = \exp\{\phi[\mathbf{y}^T \boldsymbol{\theta} - b(\boldsymbol{\theta})] + c(\phi, \mathbf{y})\}$$

where $\mu = E[\mathbf{Y}]$ and $\sigma^2 = 1/\phi$. The bivariate Poisson distribution given above is one example. This family has the convolution property,

$$\mathrm{ED}^*(\boldsymbol{\theta}, \phi_1) * \ldots * \mathrm{ED}^*(\boldsymbol{\theta}, \phi_R) = \mathrm{ED}^*(\boldsymbol{\theta}, \phi_1 + \cdots + \phi_R)$$

where $\mathbf{Z} = \phi\mathbf{Y} \sim \mathrm{ED}^*(\boldsymbol{\theta}, \phi)$ when $\mathbf{Y} \sim \mathrm{ED}(\boldsymbol{\mu}, \sigma^2)$, so that it can be used to model a process with *stationary, independent increments*.

For example, a gamma process is defined by $Y \sim \mathrm{Ga}(\mu, \phi^{-1})$, with $Z = \phi^{-1}Y \sim \mathrm{Ga}(\phi\mu, \phi^{-1})$, so that

$$\Delta Z_i \sim \mathrm{Ga}[\phi\mu(t_i - t_{i-1}), \{\phi(t_i - t_{i-1})\}^{-1}]$$

or $Z(t) \sim \mathrm{Ga}[\phi\mu t, (\phi t)^{-1}]$

For the normal distribution, we have Brownian motion, a Wiener process, and for the Poisson distribution, a Poisson process, which will be discussed further in Section 2.10, and used extensively in subsequent chapters. Less common models include, for example, Bernoulli, negative binomial or Polya, and inverse Gaussian processes. When the distribution is infinitely divisible, the process is in continuous time. Thus, this provides us with one general basis for model construction of stochastic processes within the exponential family.

2.6   Compound distributions

The ideal way to construct a model for the dependence among responses is directly in terms of the stochastic dependence structure, as for the variance components model of Section 1.3. Often, for non-normal distributions, this is difficult, or even intractable. One possible indirect way around this problem, for a constant dependence among responses on a unit, is to imagine a random effect, yielding a *compound distribution*.

As we have seen, the *location* parameter of a univariate distribution describes certain basic characteristics of the population to which it pertains, those related to the mean of the variable. A location *model* is used to describe how these characteristics vary *systematically*, in observable ways, in subsets of the population, as conditions change. As already discussed in Section 2.1 above, all such conditions are never observable, and, indeed, are never pertinent for a given problem at hand. Thus, a location parameter, itself, might sometimes be thought to have some random distribution across the population, independent of the distribution of the responses. The result is called a compound distribution.

In itself, this idea is of little use, because the random fluctuation of the location parameter will tend to absorb the systematic variation of the location model, which is of interest. There is no longer an unambiguous distinction between the stochastic and systematic components. A 'location' parameter is being used to create stochastic dependence. However, in

repeated measurement studies, this approach can often serve our purpose
because of the two types of relationships between the responses present.
Responses on the same unit can be expected to be more closely related
than those across units. Indeed, the latter are assumed to be independent, while the former are not. Then, responses on the same unit might
be expected to have the same location parameter, conditional on any unit-
specific variables, which may change over the responses on that unit. On
the other hand, if the units are selected randomly from a larger population, the value of the location parameter can vary randomly across units.
This distribution for the location parameter may account for the random
variability or heterogeneity among individuals, which is not relevant to the
study at hand, and for which, in any case, no explanatory variables are
available.

The random variation may be introduced in two ways. The location
parameter in the probability model may be given some distribution and the
location model, describing systematic changes, introduced into the resulting
compound distribution. This is called a *population-averaged* model (Zeger,
Liang, and Albert, 1988), because covariates describing differences among
the responses specific to a unit cannot be introduced. Or the location model
may first be included in the probability model and, then, some parameter
in it given a distribution. This is called a *subject-specific* model, where such
covariates can be used. In either case, we obtain a joint distribution of the
responses and the random parameter.

The (marginal) distribution of the responses, known as a *compound
distribution* can, then, be obtained by integration. If $\Pr(y_{ijk}|\lambda_{ij})$ is the
probability (density for continuous distributions) of the responses, given
the random parameter, and $p_j(\lambda_{ij})$ is the *compounding distribution* of the
parameter, then

$$\Pr(y_{ij1}\ldots y_{ijR}) = \int \left[\prod_{k=1}^{R} \Pr(y_{ijk}|\lambda_{ij})\right] p_j(\lambda_{ij})d\lambda_{ij} \qquad (2.57)$$

where i is the unit, j indexes conditions across units, and k those within
units. Because the 'location parameter', λ_{ij}, which is given the random
distribution, disappears with the integration, it can, in fact, be a purely
imaginary construct. Thus, we obtain a joint multivariate distribution
of all responses having fixed location parameter, those on the same unit.
However, except in very special cases, this distribution cannot be obtained
in closed form, so that numerical integration must be applied, if models
based on it are to be used.

One open question has not yet been faced. What is the choice of
the compounding distribution, $p_j(\lambda_{ij})$, for the parameter? Hopefully, this
might be done on theoretical grounds, in the context of a specific scientific

study. Often, this is not possible. Then, two basic types of choices may be distinguished on purely mathematical and statistical grounds.

2.6.1 CONJUGATE DISTRIBUTIONS

Let us look, first, at cases where we can obtain a closed form for our multivariate distribution. Here, the distribution of the location parameter is known as the *conjugate*. For members of the exponential family, it can be written down in general (see Diaconis and Ylvisaker, 1979, and Morris, 1983). Because the mean parameter can often take on only a restricted range of values, this must be taken into account in the choice of compounding distributions.

Giving a distribution to the mean, when it is constant for all observations on a unit (a population-averaged model), is equivalent to giving the distribution to the canonical parameter. Recall, from Equation (2.1), that the distributions which are members of this family have the form

$$f(y; \phi | \lambda) = \exp\{[y\theta - b(\theta)]/a(\phi) + c(y, \phi)\}$$
$$= \exp\{[yg(\lambda) - b(g(\lambda))]/a(\phi) + c(y, \phi)\}$$

where $g(\cdot)$ is the canonical link function and λ is now the random mean parameter. Then, the conjugate distribution will be

$$p(\lambda; \kappa, \upsilon) = \exp\{\kappa g(\lambda) - b(g(\lambda))/\upsilon + d[\kappa, \upsilon]\} \tag{2.58}$$

where $d[\cdot, \cdot]$ is a term not involving λ. The resulting closed-form marginal distribution, for one response, is

$$f(y; \kappa, \upsilon, \phi) = \exp\{d[\kappa, \upsilon] + c(y, \phi)$$
$$- d[\kappa + y/a(\phi), \upsilon a(\phi)/(\upsilon + a(\phi))]\} \tag{2.59}$$

The expected value of λ can be found from the conjugate distribution, as a function of κ and υ. Then, a function of this, usually the same link function as would be used for the original exponential family distribution, is equated to the location model, in the usual way.

Let us consider two examples.

Poisson distribution Recall from Equation (2.3) that this distribution has the form

$$\Pr(y | \lambda) = \exp[y \log(\lambda) - \lambda - \log(y!)]$$

The conjugate distribution is, then,

$$p(\lambda; \kappa, \upsilon) = \exp\{\kappa \log(\lambda) - \lambda/\upsilon + d[\kappa, \upsilon]\} \tag{2.60}$$

so that λ will have a gamma distribution, with

$$d[\kappa, v] = -(\kappa + 1)\log(v) - \log[\Gamma(\kappa + 1)] \qquad (2.61)$$

The resulting marginal distribution is

$$\begin{aligned}
\Pr(y|\kappa, v) &= \exp\{d(\kappa, v) - \log(y!) - d[\kappa + y, v/(v + 1)]\} \\
&= \frac{\Gamma(\kappa + y + 1)v^y}{\Gamma(\kappa + 1)y!(v + 1)^{\kappa+y+1}} \qquad (2.62)
\end{aligned}$$

where $\kappa = n - 1$ in the negative binomial distribution of Equation (2.32). Because $E[Y] = \kappa v$, a log linear model, with log link, can be constructed in terms of this.

Normal distribution From Equation (2.4), this distribution is given by

$$f(y; \psi^2|\lambda) = \exp\{[y\lambda - \lambda^2/2]/\psi^2 - y^2/(2\psi^2) - \log(2\pi\psi^2)/2\}$$

The conjugate distribution is

$$p(\lambda; \kappa, v) = \exp\{\kappa\lambda - \lambda^2/(2v) + d[\kappa, v]\} \qquad (2.63)$$

which is, itself, a normal distribution for λ, with mean κv, variance v, and

$$d[\kappa, v] = -\frac{\kappa^2 v}{2} - \frac{\log(2\pi v)}{2} \qquad (2.64)$$

The resulting marginal distribution is also normal

$$\begin{aligned}
f(y; \kappa, v, \psi^2) &= \exp\{d[\kappa, v] - y^2/(2\psi^2) - \log(2\pi\psi^2)/2 \\
&\qquad - d[\kappa + y/\psi^2, v\psi^2/(v + \psi^2)]\} \\
&= \exp\left\{ \frac{\kappa v y}{\psi^2 + v} - \frac{\kappa^2 v^2}{2(\psi^2 + v)} \right. \\
&\qquad \left. - \frac{y^2}{2(\psi^2 + v)} - \frac{\log[2\pi(\psi^2 + v)]}{2} \right\} \qquad (2.65)
\end{aligned}$$

with mean κv and variance $\psi^2 + v$, so that a linear location model, with identity link, can be used.

These two examples were given for only one observation, for simplicity. When more than one observation has the same value for the location parameter, we obtain a multivariate distribution. For example, the resulting multivariate normal distribution has a covariance matrix with variance $\psi^2 + v$ on the diagonal and covariance v off diagonal. In Equation (1.1) of Section 1.3 above, we saw that this is the appropriate form of covariance matrix for nested and split plot data, with $v = \delta$. Because ψ^2 and v are both components of the total variance, this was called a variance components model; it is developed further in Section 3.2 below.

In the marginal distribution, two parameters, κ and v, replace the mean parameter, λ. One function of them becomes the new location parameter, and another is a shape parameter. Once this distribution has been derived, the shape parameter may often be allowed to take on a larger range of values than was possible in the compounding distribution. For example, in the compounding normal distribution, v is the variance and must be positive. However, in the resulting marginal distribution, $\psi^2 + v$ is the variance, so that we only require $\psi^2 + v > 0$ and v could be negative. This is a useful conclusion, because v is also the covariance among responses on a unit. The same sort of argument can be used with other distributions (see, for example, Prentice, 1986).

In this approach, the location model was introduced after obtaining the marginal distribution for Y. Thus, it describes changes in the location among groups of observations having the same mean. Within the group, which, for repeated measurements, is the unit, there is a multivariate distribution with common stochastic dependence structure among all responses. The location model describes differences in means across units, but cannot take into account differences in mean among responses on a unit, because there are assumed to be none. We have seen that this is called a population-averaged approach, because it describes averages over a set of responses, here, those on a unit.

2.6.2   RANDOM EFFECTS

Suppose, now, that we first construct a location model for our responses, before giving a distribution to one of its parameters. In this way, we can incorporate changing conditions for the responses on a unit within the location model, giving a subject-specific approach, because the model can contain information on the different responses of a subject or unit.

Let us add, to our location model, a parameter to index the unit explicitly:

$$\eta_i = g_i(\mu_i)$$
$$= \lambda_i + \mathbf{x}_i^T \boldsymbol{\beta} \tag{2.66}$$

Some of the variables in the vector, $\mathbf{x}_i$, may describe changing conditions for the unit, while other refer to conditions common to groups of units.

Now, as for the previous approach, we suppose that this parameter, λ_i, is constant for a unit, but varies randomly across units. Note that, in contrast to Section 2.6.1, here the location is not assumed to remain constant for all responses on a unit; only the intercept does. When we integrate to obtain the marginal distribution of the responses, we do not now, in general, find a simple closed form, no matter what distribution we might choose for the random parameter. The integral of the product of conditional probabilities is usually intractable and estimation must rely on numerical methods. But, we do obtain a multivariate distribution.

Because the random parameter can usually take any value on the real line, very often, the compounding distribution is taken to be normal. If numerical integration must be performed, there is frequently little good technical reason for making such a choice. Any distribution is as easy to use as another, and very arbitrary shapes of distributions can be introduced by judiciously choosing the quadrature points and corresponding weights.

Again, one special case should be noted in passing. The normal distribution with identity link, with normal compounding distribution, yields a closed form solution for the marginal multivariate distribution. As might be expected, it is, again, multivariate normal, and is the classical random effects or mixed model. More surprising, perhaps, is the fact that this model is identical to that found above for the conjugate case of variance components. Thus, for the normal distribution, the same form of model has both population-averaged and subject-specific interpretations.

2.6.3 RANDOM COEFFICIENTS

In Equation (2.22) of Section 2.3.1, we saw how polynomials are often used to approximate unknown functions in a location model. Now, suppose that the location of the distribution varies as a function, $c(x_1, x_2, \ldots)$, of several continuous variables and, for simplicity, that we wish to approximate it, at least locally, by a linear relationship. In addition, we only (can) observe one of them, x_1. For example, suppose that we study the growth of a plant as a function of the amount of nitrogen fertilizer made available. Such a response relationship will depend on temperature and rainfall, among other things, but such measurements may not be available. Thus, the response relationship will be varying randomly with changes in these unknown variables.

Then, our model becomes

$$\begin{aligned}
\eta_i &= g_i(\mu_i) \\
&= c(0, x_2, \ldots) + c'(0, x_2, \ldots)x_1 \\
&= \beta_0(x_2, \ldots) + \beta_1(x_2, \ldots)x_1
\end{aligned} \tag{2.67}$$

where the derivative is with respect to x_1. Here, the parameters to be estimated are a function of the unobserved variables and, hence, vary across the units. This is a clear case of *heterogeneity* of the population, because all of the observed units cannot be assumed identical for the unobserved variables. The model is called a *random coefficients model*. In survival studies, the distribution of the unobserved variable is known as the *frailty*. Although heterogeneity can be of importance, in itself, for repeated measurements, as elsewhere, it has special significance here. This is so because the models used for it can be applied to the heterogeneity of observations across units as compared to the repeated responses on a unit.

Thus, a further extension of the random effects model involves giving distributions to some or all of the parameters in the location model.

2.7 Conditional models

In models using compound distributions, the responses on a unit are all equally interdependent. Modelling such stochastic dependence will very often be necessary in repeated measures because of the heterogeneity of responses across units as compared to the relative homogeneity within a unit. However, this is usually not the only type of stochastic dependence which may be present. Suppose that the observations form part of a longitudinal study on each unit. Then, one could expect that responses closer together in time would be more closely related. The same would usually be true of spatially related observations.

We have some stochastic mechanism producing the responses. At a given point in time, an observation is recorded. If there is time dependence, the response at the next time point will depend on the value already produced, and observed. Thus, at each point in time, the probability of a given response being generated will depend on or be conditional on the values generated previously.

In Section 2.5.2 above, we looked at a family of conditional distributions which was symmetrical in that each response, given all of the others, had a specific distribution. This might be appropriate for certain kinds of spatial dependence. In contrast, for time dependence, we require a hierarchical series of conditional distributions, following the order of the observed responses. Although apparently univariate conditional probabilities are used,

$$\Pr(y_t|y_1,\ldots,y_{t-1}) = \frac{\Pr(y_1,\ldots,y_t)}{\Pr(y_1,\ldots,y_{t-1})} \qquad (2.68)$$

a multivariate distribution of the responses on a unit is implicitly involved.

To proceed, one may choose a given form either for the conditional distribution, on the left hand side of Equation (2.68), or for the multivariate distribution, on the right side. In general, the conditional distribution will be different from the multivariate and marginal distributions, because the ratio of two multivariate distributions does not yield a conditional distribution of the same form. The one or the other will most often be intractable. As usual, the normal distribution is an exception, because then all three are normal.

The hierarchical relationship among the responses on a unit implies, by recursion, that the conditional probabilities are independent:

$$\Pr(y_1, \ldots, y_t) = \Pr(y_1) \Pr(y_2|y_1) \cdots \Pr(y_t|y_1, \ldots, y_{t-1}) \qquad (2.69)$$

so that the likelihood is composed of a product of terms, and univariate analysis may be used. The existence of this relationship means that multivariate distributions with known conditional form are much easier to fit than those with known marginal form. However, this is only true for a strict one-way ordering, such as with time. Multivariate models for spatial phenomena cannot be decomposed in this conditional way and, hence, usually require much more complex procedures.

The formulation of Equation (2.69), however, highlights a potential problem. The unconditional distribution of the first response is required. The solution will depend upon how the initial conditions are conceived to have come about.

Usually, the dependence does not extend very far in time. This will be especially true for repeated measures, where the series on a unit will often be very short. Then, the random variables, Y_t and Y_{t-k} may be assumed (conditionally) independent for $k > M$, where Y_{t-k} is said to have a *lag k* with respect to Y_t. This is the property of a *Markov process* of order M (see Section 2.10).

Because few useful non-normal multivariate distributions are available, suitable models can often only be obtained by direct construction of the conditional distribution. The resulting multivariate distributions in the ratio will, in general, have an intractable form. Moreover, such models need not condition on all of the information in the previous responses. The question is to decide how each response depends on the previous ones.

One may wish to imitate normal-theory autoregression. But there, both the marginal and conditional distributions are also normal, the AR(1) autoregression coefficient is also the autocorrelation, defining the autocovariances, and successive responses are linearly related. Not all of these characteristics can be simultaneously preserved in generalizations. Thus, perhaps the simplest possibility is to use the earlier responses as explanatory variables, as the name, autoregression, implies. But, this is not the only generalization, as we shall see in the following subsections.

2.7.1 AUTOREGRESSION

Thus, one possible approach is to define the conditional distribution in terms of the location model. In other words, the latter can also incorporate previously generated values of the response, in addition to the other explanatory variables. Thus, the location model could be

$$\eta_{it} = g_i(\mu_{it})$$

$$= \sum_{h=1}^{M} \rho_{ih} y_{i,t-h} + \boldsymbol{\beta}_i^T \mathbf{z}_{it} \tag{2.70}$$

This model, with the response entered with M lags, is one way to define an *autoregressive process* of order M, or an AR(M). Note that, unless the link is the identity, this construction does not give a linear dependence between the mean and previous observed values.

Things are not always necessarily as simple as writing down Equation (2.70). What we have done is to relate some *function* of the mean directly to the previously realized values. Once again, the simplicity of the normal distribution may have led us astray. There, that (link) function is the identity, so that the mean would depend directly on previous responses. In other cases, it may make more sense to transform the responses in the same way as the mean:

$$\eta_{it} = g_i(\mu_{it})$$
$$= \sum_{h=1}^{M} \rho_{ih} g_i(y_{i,t-h}) + \boldsymbol{\beta}_i^T \mathbf{z}_{it} \tag{2.71}$$

However, just as there is no absolute law saying that the canonical link must be used, so there is no reason why the previous responses should always be transformed in the same way as the mean. Where possible, the choices should be made for theoretical scientific reasons. There are even cases where it is impossible to link-transform the lagged responses. Two which readily come to mind are binary data and Poisson data containing zero counts. In neither case could the canonical link be used for the responses, even though it poses little problem for the mean, which does not take on these extreme values.

A second useful modification of Equation (2.70) is to make the present mean response depend, not on the previous response, but on the difference between the (link-transformed) previous response and its linear predictor

$$\eta_{it} = g_i(\mu_{it})$$
$$= \sum_{h=1}^{M} \rho_{ih} [g_i(y_{i,t-h}) - \boldsymbol{\beta}_i^T \mathbf{z}_{i,t-h}] + \boldsymbol{\beta}_i^T \mathbf{z}_{it} \tag{2.72}$$

The increased complexity in this model is only apparent, because dependence among responses is now restricted to a pure stochastic component, the difference between the previous (link-transformed) response and its location model. This is more clearly seen if we rewrite Equation (2.72) in terms of these differences:

$$\eta_{it} - \boldsymbol{\beta}_i^T \mathbf{z}_{it} = g_i(\mu_{it}) - \boldsymbol{\beta}_i^T \mathbf{z}_{it}$$

$$= \sum_{h=1}^{M} \rho_{ih}[g_i(y_{i,t-h}) - \boldsymbol{\beta}_i^T \mathbf{z}_{i,t-h}]$$

Successive differences are related by the ρ_{ih}.

The transformed observed values, $g_i(y_{i,t-h})$, are being used to predict the new transformed expected value, $g_i(\mu_{it})$. Thus, if $|\rho_{ih}| > 1$, the (link-transformed) mean in Equation (2.71) or the successive differences in Equation (2.72) may be growing or oscillating without bounds. One way to overcome this problem, at least in simple cases, as with only one lag ($M = 1$), would be to use differences of successive link-transformed responses. Thus, for example, in a growth study, one would use increases in size at each period, instead of the absolute size each time. For count data, this might imply modelling successive ratios of counts. In this way, we model stationary, independent increments (Section 2.5.3 above and Section 2.10 below). Need it also be added that it is usually a more realistic approach to use nonlinear predictors instead of $\boldsymbol{\beta}^T \mathbf{z}_t$?

As we shall see in more detail in Chapter 4 (see, also, Section 2.10 below), the normal distribution is a special, simple case again here. The autoregression model can also be written as a multivariate normal distribution for the data, where the covariances have the form which was presented in Equation (1.3) of Section 1.3. The values on each minor diagonal are symmetrically equal and usually decrease with the distance from the main diagonal, i.e. with distance in time. The way in which they decrease depends on the number of lags in the model. Thus, normal autoregression has both a conditional and a direct multivariate interpretation. This results from the identity link and from the simple variance function, and because the stochastic dependencies involve no higher moments and because the conditional and multivariate distributions are of the same form. Other distributions usually do not have such simple relationships.

Thus, conditioning on past values in this simple linear fashion cannot create a conditional model for the complete past history of the process, except in special cases, such as the normal distribution. Only in such cases will Equation (2.70) establish simple correlations among responses, the covariance structure. More generally, the form of this structure and of higher order relationships will depend on the distribution and on the link function; neither the marginal nor the conditional distribution of the responses necessarily will have a tractable form.

Just as with compound distributions in Section 2.6 above, the distinction between systematic and stochastic components is ambiguous. The parameter, ρ_{ih}, appears to be a location parameter, but, in fact, it describes the stochastic dependence among responses.

2.7.2 THE CONDITIONAL EXPONENTIAL FAMILY

In the exponential family, a special form of conditioning is often possible and useful. Thus, the conditional exponential family of Markov processes (Feigin, 1981) is defined by

$$f(y_t; \theta) = \exp[\theta d(y_t, y_{t-1}) - b(\theta, y_{t-1}) + c(y_t, y_{t-1})] \qquad (2.73)$$

We have

$$E[d(Y_t, Y_{t-1})] = \frac{\partial b(\theta, y_{t-1})}{\partial \theta} \qquad (2.74)$$

If $b(\theta, y_{t-1})$ factors:

$$b(\theta, y_{t-1}) = b_1(\theta) b_2(y_{t-1}) \qquad (2.75)$$

this is called a conditionally additive exponential family, because the cumulant has the form of an additive process.

Consider, as an example, a gamma distribution for $Y_t | y_{t-1}$, with parameters, $\rho = \mu_t / \phi_t$ and $\phi_t = y_{t-1}$, from Equation (2.45):

$$f(y_t | y_{t-1}; \rho) = \frac{y_t^{y_{t-1}-1} e^{-\frac{y_t}{\rho}}}{\rho^{y_{t-1}} \Gamma(y_{t-1})} \qquad (2.76)$$

Then,

$$\mu_t = E[Y_t | y_{t-1}]$$
$$= \rho y_{t-1} \qquad (2.77)$$

Because μ_t and ϕ_t must both be greater than zero, the autocorrelation in this model is always positive.

This is a second generalization of the usual normal-theory autoregression model. It is a multivariate distribution for which the univariate conditional distribution is in the exponential family (gamma in our example). However, it can be further generalized by taking a member of the exponential dispersion family (Jørgensen, 1986, 1987, 1992), as described in Section 2.5.3 above.

2.7.3 BINOMIAL THINNING

The previous two approaches defined specified conditional distributions. The derivation of an appropriate multivariate autoregressive-type distribution which has given marginal distributions is more complex. It is also rather artificial, because it attempts to keep the linear model for the mean, characteristic of the normal distribution. Consider first the case where the marginals follow some exponential family distribution with a continuous response variable having mean μ. Because variables from most such distributions have only positive values, this will place restrictions on the parameter values.

Take a $0-1$ binary series, c_t, with fixed binomial probability, ρ, of 1 at each time point and, independently, a second series, d_t, having the desired marginal distribution, both unobservable. Then, let

$$y_t = \rho y_{t-1} + (1 - c_t)d_t \qquad (2.78)$$

Note that, when $c_t = 1$, the relationship between consecutive responses is deterministic. Gaver and Lewis (1980) develop this approach for the gamma distribution. In the simpler case of a marginal exponential distribution, with mean, $\mathrm{E}[Y_t] = \mathrm{E}[D_t] = \mu$, this yields a series of periods between events, an EAR(1), whereby a large value, when $c_t = 0$, is followed by a run of decreasing values, with geometrically distributed lengths, for which $c_t = 1$.

This model can be modified such that

$$y_t = \rho^* d_{t-1} + (1 - c_t)y_{t-1} \qquad (2.79)$$

where $\Pr(C_t = 1) = \rho^*$, now called a TEAR(1). This has nondeterministic runs of rising values of geometrically distributed lengths while $c_t = 0$, followed by a sharp fall to a short duration when $c_t = 1$. The two can be combined to produce the NEAR(1) model (Lawrance and Lewis, 1981a).

The approach of Equation (2.78) must be slightly modified when the response is discrete. For example, in the Poisson case, ρy_{t-1} may be defined as a sum of y_{t-1} binary variables, each with probability, ρ. Then, $(1-c_t)d_t$ must be a Poisson variable with mean, $(1-\rho)\mu$. This eliminates the deterministic runs of the EAR(1) and yields a Markov chain. McKenzie (1986, 1988b) gives results for the negative binomial and Poisson distributions.

These models can be generalized to higher order processes, and to models allowing negative autocorrelation (Lawrance and Lewis, 1981b), but, even in the simple cases, they generally have intractable likelihood functions. Only in special circumstances will such models be found to correspond to some physical mechanism under study.

2.8   The dynamic generalized linear model

Another approach to longitudinal data, similar in some ways both to autoregression and to random effects, and in fact able to encompasses both, is to have the coefficients evolve over time according to a Markov process. This is a *dynamic generalized linear model*, which is usually estimated by a procedure called the Kalman filter. Although originally proposed as the dynamic linear model for normal data, it has been extended to other distributions, notably by West, Harrison, and Migon (1985), Kitagawa (1987), Fahrmeir (1989), Harvey (1989) and Harvey and Fernandes (1989).

The location model, now called the *observation* or *measurement equation*, is

$$g_i(\mu_{it}) = \boldsymbol{\lambda}_{it}^T \mathbf{v}_{it} \qquad (2.80)$$

where $\boldsymbol{\lambda}_{it}$ is a random vector of coefficients, defining the *state* of unit i at time t, with a distribution conditional on the previous responses and on $\mathbf{v}_{it}$. In contrast to the random effects model, here coefficients can vary (over time) on the same unit, as well as across units. The state is simply the minimum past and present information necessary to predict a future response. Because the set of all possible states is a mathematical construct, this is often, confusingly, called *the* state space model. However, any stochastic process (see Section 2.10 below) has a state space.

The state of the system is, then, taken to evolve over time according to a *state transition equation*

$$E[\boldsymbol{\lambda}_{it}] = \mathbf{T}_{it} \boldsymbol{\lambda}_{i,t-1} \qquad (2.81)$$

where $\mathbf{T}_{it}$ is the first order Markovian *state transition matrix* (see Section 2.10 below). Note that $\boldsymbol{\lambda}_{it}$ may contain values before time t as well as present values. The distributions of Y_{it} and $\boldsymbol{\lambda}_{it}$ are assumed to be independent. Then, the multivariate probability is given by the recursive relationship of Equation (2.69) above.

The dynamic generalized linear model for an autoregression of order M has measurement and state equations

$$g_i(\mu_{it}) = [1, 0, \ldots]\boldsymbol{\lambda}_{it} \qquad (2.82)$$

$$E\left[\begin{pmatrix} \lambda_{it} \\ \lambda_{i,t-1} \\ \vdots \\ \lambda_{i,t-M+1} \end{pmatrix}\right] = \begin{pmatrix} \rho_{i1} & \cdots & \rho_{i,M-1} & \rho_{iM} \\ 1 & \cdots & 0 & 0 \\ \vdots & \ddots & \vdots & \vdots \\ 0 & \cdots & 1 & 0 \end{pmatrix} \begin{pmatrix} \lambda_{i,t-1} \\ \lambda_{i,t-2} \\ \vdots \\ \lambda_{i,t-M} \end{pmatrix}$$

For a random effects model, the equations are

$$g_i(\mu_{it}) = \mu + \lambda_{it} \qquad (2.83)$$
$$E[\lambda_{it}] = 0$$

a special case of Equation (2.66) above. Simple models, such as these, can be combined in any desired way.

West, Harrison, and Migon (1985), West and Harrison (1989), Harvey (1989), and Harvey and Fernandes (1989) use the canonical link for all response distributions, so that the distribution of the random coefficients can be conjugate. Kitagawa (1987) approximates the probability by a piece-wise linear function, a form of numerical integration.

Filtering means estimating the current state given responses up to the present. The Kalman filter is a sequential or recursive procedure, yielding new distributions at each time point. Using Bayes' theorem, we have

$$p(\boldsymbol{\lambda}_{it}|\mathcal{F}_{it}) = \frac{\Pr(y_{it}|\boldsymbol{\lambda}_{it}, \mathcal{F}_{i,t-1})p(\boldsymbol{\lambda}_{it}|\mathcal{F}_{i,t-1})}{\Pr(y_{it}|\mathcal{F}_{i,t-1})} \qquad (2.84)$$

where $\mathcal{F}_{it}$ denotes the history of responses for unit i up to and including time t, i.e. the vector of responses $(y_{i1}, \ldots, y_{it})^T$, with all pertinent relationships among them. In Equation (2.84), $p(\boldsymbol{\lambda}_{it}|\mathcal{F}_{it})$ is called the *filtering* or *observation update* and

$$\Pr(y_{it}|\mathcal{F}_{i,t-1}) = \int_{-\infty}^{\infty} \Pr(y_{it}|\boldsymbol{\lambda}_{it}, \mathcal{F}_{i,t-1})p(\boldsymbol{\lambda}_{it}|\mathcal{F}_{i,t-1})d\boldsymbol{\lambda}_{it} \qquad (2.85)$$

while $\Pr(y_{it}|\boldsymbol{\lambda}_{it})$ is the usual distribution, if there were no random coefficients, defined by the observation equation.

The *one-step-ahead prediction* or *time update*,

$$p(\boldsymbol{\lambda}_{it}|\mathcal{F}_{i,t-1}) = \int_{-\infty}^{\infty} p(\boldsymbol{\lambda}_{it}|\boldsymbol{\lambda}_{i,t-1})p(\boldsymbol{\lambda}_{i,t-1}|\mathcal{F}_{i,t-1})d\boldsymbol{\lambda}_{i,t-1} \qquad (2.86)$$

is defined by the transition equation. Both of these integrals are usually complicated when the distributions are not normal and the link is not the identity. We shall be interested in the conditional distribution, $\Pr(y_{it}|\mathcal{F}_{i,t-1})$, to calculate the likelihood function.

Most often, such dynamic models are used in forecasting. However, here, they are particularly important as a unifying framework for many of the models of repeated measurements, as well as a means of calculating the likelihood function and estimating the parameters in difficult cases (see, for example, de Jong, 1988). Thus, two advantages of this Kalman filter-type approach, even in the case of a model based on the normal distribution, are that it can be extended to handle unequally spaced time intervals and missing observations and that it encompasses many useful models as special cases.

2.9   Generalized estimating equations

A third approach to converting univariate distributions for use in a multivariate situation must also be considered. It involves working with the estimation procedure, instead of directly with the models. Second order stochastic dependence, in the form of a covariance matrix, is introduced rather artificially. For this reason, in most cases, this approach does not yield a true statistical model, based on a probability distribution.

Recall, from Equation (2.6) that the score or estimating equations for independent observations from the exponential family with dispersion parameter, obtained by differentiating the log likelihood function with respect to the location parameter, are, for one observation,

$$U_i = [y_i - b'(\theta_i)]/a_i(\phi)$$

and that $b'(\theta_i) = \mu_i$. Setting the sum of scores for all observations equal to zero and solving yields the maximum likelihood estimates of the parameters. To obtain estimates of the P parameters, β_l, in a location model, $\eta_i = g(\mu_i) = \boldsymbol{\beta}^T \mathbf{z}_i$, we substitute to get the appropriate estimating equations,

$$\sum_{i=1}^{N} \sum_{k=1}^{R} \left[(y_{ik} - \mu_{ik}) \frac{\partial \theta_{ik}}{\partial \mu_{ik}} \frac{\partial \mu_{ik}}{\partial \eta_{ik}} z_{ilk} \right] = 0 \tag{2.87}$$

With the R *independent* observations on unit i, this may be written, in matrix notation,

$$\sum_{i=1}^{N} \mathbf{Z}_i \mathbf{D}_i \mathbf{d}_i = 0 \tag{2.88}$$

where $\mathbf{Z}_i$ is the $P \times R$ design matrix for the unit,

$$\mathbf{D}_i = \operatorname{diag} \left[\frac{\partial \theta_{ik}}{\partial \mu_{ik}} \frac{\partial \mu_{ik}}{\partial \eta_{ik}} \right] \tag{2.89}$$

is an $R \times R$ matrix and $\mathbf{d}_i = \mathbf{y}_i - \boldsymbol{\mu}_i$ is a $R \times 1$ vector. The fact that $\mathbf{D}_i$ is diagonal indicates that the observations on the unit are being modelled as independent.

Pairwise stochastic dependence can be introduced into the model by making $\mathbf{D}_i$ nondiagonal, as originally proposed by Gilmour, Anderson, and Rae (1985). Liang and Zeger (1986) call these generalized estimating equations. Unfortunately, once this stochastic dependence is introduced, it is, in general, no longer possible to integrate back to obtain a likelihood, or a probability-based model (McCullagh and Nelder, 1989, pp. 333-336).

The remaining question is how to structure a nondiagonal matrix, $\mathbf{D}_i$. Because we have pairwise stochastic dependence, it would be nice if we could construct a matrix which could be interpreted in terms of correlations. Recall again, from Equation (2.10), that

$$\operatorname{var}(Y_{ik}) = a_{ik}(\phi) \tau_{ik}^2$$

$$= \frac{\phi \tau_{ik}^2}{w_{ik}}$$

Let $\mathbf{U}_i = \mathrm{diag}(\phi \tau_{ik}^2 / w_{ik})$. Then, the required $R \times R$ correlation matrix, $\mathbf{R}_i$ can be obtained from the covariance matrix defined by

$$\mathbf{\Sigma}_i = \mathbf{U}_i^{\frac{1}{2}} \mathbf{R}_i \mathbf{U}_i^{\frac{1}{2}} \qquad (2.90)$$

The estimating equations become

$$\sum_{i=1}^N \mathbf{C}_i \mathbf{\Sigma}_i^{-1} \mathbf{d}_i = 0 \qquad (2.91)$$

where

$$\mathbf{C}_i = \left[\frac{\partial \mu_{ik}}{\partial \beta_l} \right]$$

$$= \mathbf{Z}_i \mathrm{diag} \left[\frac{\partial \mu_{ik}}{\partial \eta_{ik}} \right] \qquad (2.92)$$

Because

$$b''(\theta_{ik}) = \frac{\partial \mu_{ik}}{\partial \theta_{ik}}$$

$$= \frac{\tau_{ik}^2}{w_{ik}} \qquad (2.93)$$

when $\mathbf{R}_i$ is the identity matrix, we recover the original independence model, Equation (2.88). This matrix can now be chosen with a certain number of unknown parameters so as to model the appropriate heterogeneity and/or time dependent structure, such as uniform intra-unit correlation or longitudinal autocorrelation.

Such 'models' can be constructed at both the population-averaged and subject-specific levels. Their major drawback is the lack of direct interpretation in terms of a concrete physical mechanism generating the data, because, in general, they have no probabilistic basis. As well, if several different stochastic structures, $\mathbf{R}_i$, are being considered, it is difficult to compare them, since no deviances, but only Wald tests, are available (see Section 2.11 below).

2.10   Stochastic processes

A stochastic process involves a random variable, say Y_t, which is varying over time (or space). The main properties distinguishing among such processes are

(1) Its *state space*: the set of all its possible observed values, i.e. of all possible responses. This is a mathematical concept, not to be confused with the physical space over which the variable may be changing. State spaces are either discrete (finite or countably infinite) or continuous.

(2) The index, t, which also may be either discrete or continuous, usually referring to time.

(3) The nature of the dependence among the random variables, Y_t.

A process is said to be *strictly stationary* if all sequences of observations of equal length have identical multivariate distributions

$$\Pr(Y_1 = y_1, \ldots, Y_R = y_R) = \Pr(Y_{t+1} = y_1, \ldots, Y_{t+R} = y_R) \quad \forall t, R \tag{2.94}$$

It is *second order stationary* if

$$E[Y_t] = \mu \quad \forall t$$
$$E[(Y_t - \mu)(Y_{t+h} - \mu)] = \gamma(h) \quad \forall t, h \tag{2.95}$$

Because a multivariate normal distribution is completely defined by its first two moments, if the process is normal and second order stationary, it is strictly stationary. This is not generally true of other distributions. In this text, stationarity will always be strict.

Much of classical time series analysis is involved with transforming data to obtain stationarity. In the repeated measures context, with several series, this is rarely necessary and should be avoided.

Especially important models are those with a process having *stationary, independent increments* (Doob, 1953, pp. 96–98, 391–424), because this allows one to use standard univariate techniques. In general, such a process is not stationary. Common examples include the Poisson process and Brownian motion, both described below, but, as mentioned in Section 2.5.3, such processes can be constructed for other common distributions (see Jørgensen, 1992).

Because many repeated measurements are longitudinal observations, stochastic processes in time play an important role. They may be modelled in either discrete or continuous time, although they are always necessarily measured in discrete time. A brief description of a number of the common ones will be given in this section.

Point process Consider an event which can occur at irregularly spaced intervals in time, known as *waiting times*, such that no two events occur simultaneously. The sequence of events is known as a *point process*. The number of events in given time intervals, the waiting time between successive events, or, more generally, the intensity of occurrence of events may be of interest. Other variables may be measured each time an event occurs, yielding a *marked point process*.

Counting process   The cumulated number, N_t, of events up to time, t, in a point process is known as a *counting process* with *intensity*, $\omega(t|\mathcal{F}_{t-})$, defined by

$$\omega(t|\mathcal{F}_{t-})dt = \Pr(\text{an event in } (t, t+dt)|\mathcal{F}_{t-})$$
$$= \Pr(dN_t = 1|\mathcal{F}_{t-}) \tag{2.96}$$

where $\mathcal{F}_{t-}$ is the complete history up to, but not including, t, called the *filtration*. As we saw in Section 2.4.3, for duration or survival models, the intensity is also known as the risk, hazard rate, failure rate, or mortality rate.

If $E[M_t] < \infty$ and

$$E[M_{t+k} - M_t|\mathcal{F}_{t-}] = 0 \qquad \forall t \quad 0 < k < \infty \tag{2.97}$$

M_t is called a *martingale*. For a counting process, $M_t = N_t - \int_0^t \omega(t|\mathcal{F}_{t-})dt$ fulfils this condition. A *stopping* or *Markov time* is a time which depends only on $\mathcal{F}_{t-}$, and not on events after t.

Important simplifications occur when the intensity depends only on the complete history through N_t: $\omega(t|N_t)$. Special cases, some of which are discussed further below, include

- the ordinary *Poisson process*, with

$$\omega(t|N_t) = \omega \tag{2.98}$$

  the intensity is always the same (the only counting process with stationary independent increments);
- the *nonhomogeneous Poisson process*, with

$$\omega(t|N_t) = \omega(t) \tag{2.99}$$

  the intensity is a function of time;
- the *pure birth* or *Yule process* with

$$\omega(t|N_t) = N_t\omega \tag{2.100}$$

  the intensity is proportional to the number of previous events;
- the *nonhomogeneous birth process*, with

$$\omega(t|N_t) = N_t\omega(t) \tag{2.101}$$

  the intensity, proportional to the number of previous events, is also a function of time;

- the *renewal process*, with

$$w(t|N_t) = w(t - t_{N_t}) \qquad (2.102)$$

  the intensity depends on the time since the last event, starting over after each event;
- the *semi-Markov* or *Markov renewal process*, with

$$w(t|N_t) = w_t(t - t_{N_t}) \qquad (2.103)$$

  the form of the intensity function is changing with time, as well as depending on the time since the last event. This usually occurs because the process changes state at each event.

Those processes with an intensity depending on time are nonstationary. As we see, this dependence may be on the elapsed time or on the number of previous events. In more complex cases, it may also depend on other time-varying covariates.

Poisson process In a point process, suppose that

(1) the probability of an event in $(t, t + \Delta t)$ is $w\Delta t + o(\Delta t)$
(2) the occurrence of events in $(t, t + \Delta t)$ is independent of what happens before t
(3) the probability of more than one event in $(t, t + \Delta t)$ is $o(\Delta t)$

Then, the number of events in $(0, t)$, N_t, has a Poisson distribution with mean, $\mu = E(N_t) = wt$ and the point process is known as a *Poisson process* with intensity or rate w (see Section 2.4.2 above). This process has stationary independent increments.

If the intensity is a function of time, $w(t)$, this is a nonhomogeneous Poisson process. If $w(t)$ is stochastic, it is a *doubly stochastic* or *Cox process* (Cox, 1955).

Renewal process When the waiting times between events in a point process are distributed independently, they make up a *renewal process*. If we have a Poisson process, the intervals have an exponential distribution. More generally, they can have any duration distribution, for example, from Section 2.4.3 above. Because of the independence, standard survival data techniques can be used for modelling.

Survival distribution If the first occurrence of an event in a point process is absorbing, the waiting time to this event is said to have a *survival distribution*. This is a 'renewal' process which is not renewed. In the social sciences, these are sometimes called single spell or one episode data. If several different types of events can end the interval, we have competing risks.

Markov process   Any series of observations where

$$\Pr(y_k|y_1, \ldots, y_{k-1}) = \Pr(y_k|y_{k-1}) \tag{2.104}$$

so that each response only depends on the immediately preceding one, is known as a *Markov process*. More generally, this is a Markov process of order one, so that, if a response only depends on the M preceding responses, it is said to be a process of order M. It is usually necessary to assume that a stochastic process has some finite order, M (less than the number of observations over time), for otherwise its multivariate distribution continues to change with each additional observation.

If the variable for a Markov process can only take discrete values or states, it is known as a *Markov chain*. Usually, there are a finite number of possible discrete states and observations are made at equally spaced discrete intervals. Then, the probabilities of change of state from one period to the next can be represented by a square *transition matrix* of non-negative values, with the entries of each line summing to one. A point process can be modelled as a special case with only two states.

When the response variable for a Markov process is continuous, we have a diffusion process.

Random walk   Suppose binary events, with values, or state space, -1 and 1, and corresponding probabilities, π and $1 - \pi$, can occur at times $t = 1, \ldots, R$. Then, the cumulated total, y_t, constitutes a special case of a Markov chain, called a (discrete-valued) *random walk*. It is also a simple example of a martingale.

One may be interested in the time to the first return to the origin, when $y_t = 0$ again, or to the first passage through K, when $y_t = K$ for the first time, the *hitting time*. If the initial value, y_0 is greater than zero and there is an upper limit, K, this is the classical ruin problem for the time until return to the origin or first passage through K.

The random walk has *absorbing barriers* if it stops upon arrival at zero or K and *reflecting barriers* if it returns in the opposite direction. The transition matrix for a random walk has the probabilities, π and $1 - \pi$, on the first two minor diagonals, one on either side of the main diagonal, and zero elsewhere, except for a one on the first and last lines (its position determined by whether the barrier is absorbing or reflecting). If $\pi \neq 0.5$, the random walk has *drift*.

A random walk can be generalized so that jumps greater than one are possible, with probabilities depending only on the distance. Then, the transition matrix has all elements on a given (minor) diagonal equal. It is symmetric if there is no drift. A random walk can be further generalized to jumps which are a continuous variable, with mean zero, a special case of a Markov process.

Autoregression   A Markov process, observed on equally spaced discrete time intervals, where the continuous response has a normal distribution with constant variance, is called an *autoregression*, in a more restricted but more usual sense than that of Section 2.7.1 above. Again, this can be of order M, signified by AR(M). An AR(M) can be written

$$\mu_t = \alpha_0 + \sum_{h=1}^{M} \rho_h y_{t-h} \qquad (2.105)$$

If the initial response is assumed to have a stationary distribution and the roots of the corresponding characteristic equation

$$m^M - \sum_{h=1}^{M} \rho_h m^{M-h} = 0 \qquad (2.106)$$

are not all less than one in absolute value, the series is nonstationary (Priestley, 1981, pp. 116–135). Another possibility is that the initial value be fixed, instead of having a stationary distribution, i.e. to condition on it.

If $\rho_1 = 1$ for an AR(1), the first differences, $\Delta y_t = y_t - y_{t-1}$, are a continuous-valued random walk, with drift, if $\alpha_0 \neq 0$ and, also, a 'renewal' process, with 'intervals' having a normal distribution. If time is continuous, we have a continuous AR or a CAR, necessary for data with unequally spaced or missing observations.

Brownian motion or Wiener process   If a continuous-valued random walk, observed in continuous time, has increments which are independently normally distributed, with the variance proportional to the time intervals,

$$\Delta y_t \sim N(\mu \Delta t, \sigma^2 \Delta t) \qquad (2.107)$$

it is called *Brownian motion* or a *Wiener process*. This process has stationary independent increments. If $\mu \neq 0$, the process has drift. With positive drift and absorbing barriers, the time to absorption, called the hitting time, has an inverse Gaussian distribution (see Section 2.4.3 above).

2.11   Inference

To conclude this chapter, it may be appropriate here to say a word about the methods of inference to be used. This book is about the construction of models and their application in the repeated measurements context. In modern statistics, with several competing approaches to making inferences, the lowest common denominator is the (log) likelihood ratio comparing competing models. Minus two times this value, called the deviance, will be used for drawing the required inferences. GLIM users should, however, note that deviance there only refers to that part of the log likelihood containing the linear (location) parameters; any scale parameter (variance) is concentrated out. Thus, deviance, as used here, is only identical to the GLIM terminology when the model has no (estimated) scale parameter, such as in binomial and Poisson models.

The likelihood function defines a set of probability-based models which one is entertaining as having possibly generated the observed data. If this function is so complex as to be intractable, then there is a good chance that it cannot provide useful and interpretable information on the data generating mechanism. Generalized estimating equations (Section 2.9) are another extreme, where, as we shall see, it is not clear what the competing 'models' can tell us about the phenomena of interest. Thus, the basic assumption underlying this approach is the Fisherian one that a model is more plausible or likely if it makes the observed data more probable.

However, this approach, by itself, does not take into account the complexity of the model, as usually measured by the degrees of freedom. Although it is not clear that these incommensurable quantities, fit measured by the likelihood, and complexity can be sensibly compared statistically, as opposed to scientifically, the common approach is to calibrate the deviance, at least approximately, by the appropriate value from a Chi-squared distribution. When significant differences among models are mentioned, this will usually be the meaning. Occasionally, the quicker, but much more approximate, procedure of comparing the parameter value to its standard error, a Wald test, will be used. However, this can be unreliable, and even very misleading, especially for non-normal models, unless the sample size is very large. In addition, it is not invariant to parameter transformation. Exceptionally, when classical normal theory ANOVA or regression models are used, the standard F test will be given (which is, in fact, equivalent to inspecting the deviance).

One disadvantage of the strictly likelihood approach occurs when a number of 'nuisance' parameters are present. In such cases, better estimators may be obtained by marginal or conditional likelihood estimation (Kalbfleisch and Sprott, 1970). However, this is only really feasible within the generalized linear model context, where sufficient statistics for the parameters are available. One example, which we shall see below, is the normal distribution variance components and random effects models, where such conditional or marginal estimation is often called restricted maximum likelihood (REML). Another is the Rasch model for categorical data.

Dobson (1990) provides an excellent introduction to statistical modelling, in the generalized linear model context. It is surprisingly difficult to find material on fitting stochastic models to data. Most texts concentrate either on the behaviour of stochastic models under specific conditions, with illustrative applications, or on the estimation of some asymptotic statistics, such as means or variances. The reader may like to consult a standard general introduction to stochastic processes such as Ross (1989) or Karlin and Taylor (1975), and, the more statistical Cox and Lewis (1966).

Part II

Normal distribution models

3
Heterogeneous populations

3.1 The multivariate normal distribution

In this part, we shall be dealing with responses which take the form of quantitative measurements having any value on the real line. In addition, we shall be assuming that the distribution of these observations has the symmetrical form of a normal distribution (or can be transformed to it, as in the log normal distribution). Classically, these have been the assumptions most often used for the analysis of repeated measurements data, so that a rich variety of models is available and their properties are well known.

If the repeated data are ordered, in time, in space, or by dose, for example, one of the first things that should be done whenever starting the process of model construction for a set of repeated measurements data, whether thought to be normally distributed or not, is to plot the data. A second is to calculate and examine the empirical (sample) covariance matrix. This will usually be done for various possible location models which are of interest, to see what influence they may have on any patterns in the matrix.

Because repeated measurements are inherently stochastically dependent responses, we shall be concerned, here, with the multivariate normal distribution. Such a distribution, with an unspecified covariance structure, has frequently been used for repeated measurements. In some cases, such generality is necessary, when the stochastic dependence relationships among the responses have no specific structure. It might also be justified as a sort of semi-parametric model, when one is only interested in the mean relationships. Although this may have the advantage of a minimum of hypotheses, it usually involves reduced efficiency and validity in inferences about the parameters of interest (Altham, 1984). With R repeated observations per unit, $R(R+1)/2$ parameters must be estimated in the covariance matrix. In any case, it will obviously not be possible to fit such a model if R is too large relative to the number of units, N. Thus, we shall be primarily interested in models where the stochastic dependence structure, as specified by the covariance matrix, is relatively simple, describing parsimoniously the relationships involved. However, we shall first look at the more general case in this section.

Thus, in general, we have a multivariate normal distribution, independent for each of the N units,

$$\mathbf{Y} \sim \mathrm{MVN}(\mathbf{M}, \mathbf{I}_N \otimes \boldsymbol{\Sigma}) \tag{3.1}$$

with $\mathbf{Y}$ the response matrix and $\mathbf{M}$ the matrix of means, both $N \times R$ matrices, while $\boldsymbol{\Sigma}$, the covariance matrix, is $R \times R$. Suppose, first, that we do not have treatments or covariates varying with the responses on a unit, and, thus, distinguishing among them. We are in the standard multivariate ANOVA (MANOVA) situation, with the explanatory variables only distinguishing among (groups of) units. Our location model to describe the differences among units is

$$\mathrm{E}[\mathbf{Y}] = \mathbf{M} \tag{3.2}$$
$$= \mathbf{XB} \tag{3.3}$$

where $\mathbf{X}$, the design matrix, is $N \times C$ describing the C conditions of the experimental or sample survey design and $\mathbf{B}$ is the $C \times R$ matrix of unknown parameters. So far in this model, no assumptions have been made about the covariance matrix. Because the structure of $\boldsymbol{\Sigma}$ is left unspecified, we have a total of $R(R+1)/2 + C \times R$ unknown parameters to estimate. Such a model allows no way of relating together the repeated responses on a unit through a location model. Their only relationship is stochastically, through the covariance matrix.

Potthoff and Roy (1964) suggested a generalization of the MANOVA model, which overcomes this problem, so that it can be used in growth curve studies and related problems. Attention concentrates on the location model for the mean responses. Their model may be written in the form

$$\mathrm{E}[\mathbf{Y}] = \mathbf{XBZ} \tag{3.4}$$

Here, $\mathbf{Y}$ and $\mathbf{X}$ are the response and design matrices, as above, while $\mathbf{B}$ is now a $C \times P$ parameter matrix and $\mathbf{Z}$ is a $P \times R$ matrix of covariates changing with the responses on a unit. It often contains $P - 1$ orthogonal polynomials describing how the response curve varies over time for a unit. For example, in a growth experiment to study the difference in effect of several types of food on animals, $\mathbf{X}$ would describe these treatment differences. Then, if $\mathbf{Z}$ were a suitable polynomial in time, $\mathbf{B}$ would give the average growth profile for each food type.

The structure of $\boldsymbol{\Sigma}$ is left unspecified in this model. In fact, in their original presentation, Potthoff and Roy assumed that $\boldsymbol{\Sigma}$ was unknown, but had to supply an independent estimate of it. However, it can be estimated by maximum likelihood. On the other hand, the model does cover a number of simpler cases of location models. For example, with only one initial

Table 3.1. Distances (mm.) from the centre of the pituitary to the pteryomaxillary fissure in girls and boys aged 8, 10, 12, and 14. (Potthoff and Roy, 1964)

		Age		
	8	10	12	14
Girls	21	20	21.5	23
	21	21.5	24	25.5
	20.5	24	24.5	26
	23.5	24.5	25	26.5
	21.5	23	22.5	23.5
	20	21	21	22.5
	21.5	22.5	23	25
	23	23	23.5	24
	20	21	22	21.5
	16.5	19	19	19.5
	24.5	25	28	28
Boys	26	25	29	31
	21.5	22.5	23	26.5
	23	22.5	24	27.5
	25.5	27.5	26.5	27
	20	23.5	22.5	26
	24.5	25.5	27	28.5
	22	22	24.5	26.5
	24	21.5	24.5	25.5
	23	20.5	31	26
	27.5	28	31	31.5
	23	23	23.5	25
	21.5	23.5	24	28
	17	24.5	26	29.5
	22.5	25.5	25.5	26
	23	24.5	26	30
	22	21.5	23.5	25

condition, $\mathbf{X}$ is an $N \times 1$ vector of ones. It can also be generalized to a multivariate response at each time point. However, it is only easily fitted if the design is balanced and there are no missing values.

We shall apply this model to the data which Potthoff and Roy originally used. These are measurements of the distance in millimetres from the centre of the pituitary to the pteryomaxillary fissure for boys and girls measured at four different ages (Table 3.1). Here, $N = 27$, $R = 4$, and $C = 2$. Our matrix, $\mathbf{X}$, has a first column of ones and a second column with zero for girls and one for boys. $\mathbf{Z}$ has the general form

$$\begin{pmatrix} 1 & 1 & 1 & 1 \\ -3 & -1 & 1 & 3 \\ 1 & -1 & -1 & 1 \\ -1 & 3 & -3 & 1 \end{pmatrix}$$

with $P = 4$. Because the location model is saturated, fitting these linear, quadratic, and cubic orthogonal polynomials when there are four measurements in time is equivalent to fitting a different parameter at each time.

Let us first fit the full Potthoff and Roy model, with differences among ages and between sexes. This is a saturated model to which we can compare the others. The estimated covariance matrix is given by

$$\hat{\Sigma} = \mathbf{Y}^T(\mathbf{I} - \mathbf{X}(\mathbf{X}^T\mathbf{X})^{-1}\mathbf{X}^T)\mathbf{Y}/N \qquad (3.5)$$

This matrix, and the corresponding correlation matrix (below the diagonal), are

$$\begin{pmatrix} 5.014 & 2.516 & 3.621 & 2.510 \\ 0.571 & 3.875 & 2.710 & 3.071 \\ 0.661 & 0.563 & 5.978 & 3.825 \\ 0.522 & 0.726 & 0.728 & 4.616 \end{pmatrix}$$

Because we are not using the orthogonal polynomials, but the equivalent model with a different mean at each age, $\mathbf{Z}$ is the identity matrix and the parameter estimates, given by

$$\hat{\mathbf{B}} = (\mathbf{X}^T\mathbf{X})^{-1}\mathbf{X}^T\mathbf{Y} \qquad (3.6)$$

are

$$\begin{pmatrix} 21.182 & 22.227 & 23.091 & 24.091 \\ 1.693 & 1.585 & 2.628 & 3.378 \end{pmatrix}$$

The first line gives the average distance for each age for the girls and the second line the corresponding differences for the boys. The two profiles are plotted in Figure 3.1.

The standard errors of the parameter estimates for this model can be obtained from their covariance matrix

$$((\mathbf{X}^T \otimes \mathbf{I}_R)(\hat{\Sigma} \otimes \mathbf{I}_N)^{-1}(\mathbf{X} \otimes \mathbf{I}_R))^{-1} \qquad (3.7)$$

which gives

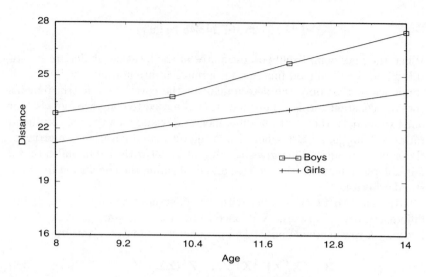

Fig. 3.1. Average distances from the centre of the pituitary to the pteryomax-illary fissure in boys and girls aged eight to fourteen, from Table 3.1.

$$\begin{pmatrix} 0.291 & 0.090 & 0.066 & 0.095 & -0.206 & -0.006 & 0.078 & 0.001 \\ 0.090 & 0.291 & 0.092 & 0.066 & 0.009 & -0.206 & -0.007 & 0.081 \\ 0.066 & 0.092 & 0.290 & 0.089 & 0.078 & 0.004 & -0.205 & -0.005 \\ 0.095 & 0.066 & 0.089 & 0.282 & -0.013 & 0.076 & 0.006 & -0.198 \\ -0.206 & 0.009 & 0.078 & -0.013 & 0.348 & 0.025 & -0.102 & 0.042 \\ -0.006 & -0.206 & 0.004 & 0.076 & 0.025 & 0.344 & 0.034 & -0.102 \\ 0.078 & -0.007 & -0.205 & 0.006 & -0.102 & 0.034 & 0.348 & 0.036 \\ 0.001 & 0.080 & -0.005 & -0.198 & 0.042 & -0.102 & 0.036 & 0.343 \end{pmatrix}$$

The standard errors are the square roots of the diagonal elements

$$\begin{pmatrix} 0.539 & 0.539 & 0.539 & 0.531 \\ 0.590 & 0.587 & 0.590 & 0.585 \end{pmatrix}$$

indicating that the sex differences are significantly different from zero.

A simpler model is one which has no differences between sexes. The only change to our previous formulation is that the second column of **X** disappears so that it is a vector of ones. The covariance/correlation matrix is now

$$\begin{pmatrix} 5.706 & 3.164 & 4.695 & 3.890 \\ 0.626 & 4.482 & 3.716 & 4.364 \\ 0.711 & 0.635 & 7.645 & 5.968 \\ 0.600 & 0.759 & 0.795 & 7.371 \end{pmatrix}$$

The parameters are here the averages over sex

$$(22.185 \ 23.167 \ 24.648 \ 26.093)$$

When the maximum likelihood estimates of the location model are substituted into the likelihood function for a multivariate normal distribution, it simplifies so that only the determinant of the covariance matrix remains (besides constant terms). This may, then, be used to compare models. For our two models, the resulting difference in deviances or the log likelihood ratio, is $N \log(|\hat{\mathbf{\Sigma}}_0|/|\hat{\mathbf{\Sigma}}|)$, where $\hat{\mathbf{\Sigma}}_0$ is the estimated covariance matrix for the simpler model. This gives a value of 13.69 with 4 d.f., indicating a fair difference between the models, and confirming the conclusion from the standard errors.

We can also fit the model with no differences among the ages, but differences between sexes. $\mathbf{X}$ is once again a 27×2 matrix, while $\mathbf{Z}$ now becomes a row vector of ones. The parameter estimates, given by

$$\hat{\mathbf{B}} = (\mathbf{X}^T\mathbf{X})^{-1}\mathbf{X}^T\mathbf{Y}\hat{\mathbf{\Sigma}}^{-1}\mathbf{Z}^T(\mathbf{Z}\hat{\mathbf{\Sigma}}^{-1}\mathbf{Z}^T)^{-1} \qquad (3.8)$$

are

$$\begin{pmatrix} 22.357 \\ 2.045 \end{pmatrix}$$

The first value is the average distance, over ages, for girls, and the second is the difference from this for boys. The covariance matrix, now given by

$$\hat{\mathbf{\Sigma}}_0 = (\mathbf{Y} - \mathbf{X}\hat{\mathbf{B}}\mathbf{Z})^T(\mathbf{Y} - \mathbf{X}\hat{\mathbf{B}}\mathbf{Z})/N \qquad (3.9)$$

is (with, as usual, correlations below the diagonal)

$$\begin{pmatrix} 6.960 & 3.112 & 2.078 & -1.096 \\ 0.583 & 4.088 & 2.211 & 1.907 \\ 0.293 & 0.407 & 7.223 & 6.734 \\ -0.123 & 0.279 & 0.742 & 11.412 \end{pmatrix}$$

We see that ignoring the age differences has induced negative correlation.

The deviance, compared to the full model, is 49.38 with 6 d.f., indicating even greater significance of the difference among ages than those between sexes. The standard errors of the parameter estimates for this model can be obtained from their covariance matrix,

$$((\mathbf{X}^T \otimes \mathbf{Z})(\hat{\mathbf{\Sigma}}_0 \otimes \mathbf{I}_N)^{-1}(\mathbf{X} \otimes \mathbf{Z}^T))^{-1} \qquad (3.10)$$

which gives

$$\begin{pmatrix} 0.149 & -0.067 \\ -0.067 & 0.164 \end{pmatrix}$$

with standard errors

$$\begin{pmatrix} 0.386 \\ 0.405 \end{pmatrix}$$

So far, the full model is the only one which fits satisfactorily. Let us, then, try a linear trend in age. For this, we use the first two rows of the matrix, $\mathbf{Z}$, of orthogonal polynomials, given above. The parameter estimates are now

$$\begin{pmatrix} 22.665 & 0.476 \\ 2.272 & 0.350 \end{pmatrix}$$

The first line is the intercept and the linear slope for the girls; the second line gives the differences of each of these for the boys as compared to the girls. As we know, distance is increasing with age, giving the positive slopes. It is starting higher and increasing faster for the boys than for the girls. The covariance/correlation matrix is

$$\begin{pmatrix} 5.119 & 2.441 & 3.611 & 2.522 \\ 0.544 & 3.928 & 2.718 & 3.062 \\ 0.653 & 0.561 & 5.980 & 3.824 \\ 0.519 & 0.719 & 0.728 & 4.618 \end{pmatrix}$$

and the deviance, as compared to the full model, is 2.97 with 4 d.f., indicating that this model is acceptable. It is not necessary to fit a model with a quadratic term. Remember that the model with both quadratic and cubic terms is identical to the full model.

As we have seen by looking at the covariance/correlation matrices, this generalized MANOVA model allows the stochastic dependencies among the repeated measurements to take on any arbitrary structure. In the rest of Part II, we shall study certain specific simpler forms of structure for this matrix. Specifically, in this chapter, we shall look at cases where the stochastic dependence among all responses on a unit is identical, the variance components model and its extensions. In the simplest cases, this is the same as saying that some units give consistently higher and some lower responses under the same conditions. It is the simplest form of structure to be applied to the covariance matrix, that of compound symmetry (Section 1.3).

3.2 Variance components

Up until now, we have stochastic dependence among responses on the same unit, but this dependence does not take into account 'distance' among these observations. One possible simplification which can be made with respect to the completely unknown covariance matrix is to model it directly in such a way that all responses on a unit are the same distance apart. This model has a constant covariance among all pairs of responses on a unit. All such responses are equally related, but those across different units are independent. This is just one special case of the multivariate Potthoff and Roy model.

As we saw in Equation (1.1) of Section 1.3 above, the constant covariance model has a covariance matrix on a unit which can be written in the form

$$\mathbf{\Sigma} = \psi^2 \mathbf{I}_R + \mathbf{J}_R^T \delta \mathbf{J}_R$$

Thus, we model the variance of any response as being made up of two parts, the variation in responses on the same unit, ψ^2, and a component which is the additional variance across units. The latter also corresponds to the constant covariance, δ, among responses on the same unit. Recall that δ is a *component* of the variance, not a variance, and a covariance, so that it need not be positive. When the design of the data is reasonably balanced, estimates of these two parameters can be obtained from a classical univariate ANOVA table. I shall present this easily applicable method in the present section.

The variance components approach is based on the partition of variation among responses into two basic types. One of these is the variation across units, usually measured with respect to sums or averages over all responses within each unit. The other is variation among the responses on each unit. The variability across units is assumed constant or homogeneous under all treatment conditions, in the same way that variability within units is assumed to be the same for all units. The response variability across units is usually larger than that within units. Then, if the experiment is reasonably balanced, a univariate model may be analysed by decomposing the variance into components. Such balance will probably never be attained in a sample survey design. The reader not familiar with such models is referred to a standard text, such as Searle (1971a).

In the simplest case, where no variables distinguish among responses on a unit, we have a simple nested design. Estimation of the intra-class correlation in such a design is used in many fields: in epidemiology, to measure the degree of family resemblance; in genetics, to estimate the heritability of traits; in psychology, to evaluate the reliability of tests or assessors. In the slightly more complex situation, where the subunits have different treatments, we have the typical split plot situation. Some treatment(s) distinguishes the responses on the same unit, which are assumed more or less simultaneous in time. Common applications include, in agriculture, field experiments on productivity of crops under various conditions and, in medicine, teratological studies of drug effects on litters of animals.

Consider the classical tuberculin assay of Fisher (1949), given here in

Table 3.2. Skin thickening responses (mm.) of cows in a biological assay of tuberculins in a Latin square. (Fisher, 1949)

Site	Cow Group			
	I	II	III	IV
1	454 (A)	249 (B)	349 (C)	249 (D)
2	408 (B)	322 (A)	312 (D)	347 (C)
3	523 (C)	268 (D)	411 (A)	285 (B)
4	364 (D)	283 (C)	266 (B)	290 (A)

A: standard double treatment.

B: standard single treatment.

C: Weybridge single treatment.

D: Weybridge half treatment.

Table 3.3. ANOVA table for the tuberculin assay data of Table 3.2.

	SS	d.f.	MSS	F
Cow Group	0.4723	3	0.1574	
Site	0.0833	3	0.0275	38.05
Treatment	0.1760	3	0.0587	81.17
Residual	0.0043	6	0.0007	

Table 3.2. The response of interest is the amount of thickening of the skin, in millimetres, observable in a set number of hours after injection of the tuberculin. Although 120 cows were used, only totals for groups of 30 cows with the same treatment are given in the table. These four groups of 30 cows only differed in which of the various tuberculins was applied to each of the four sites on the neck. This assignment of the tuberculin types to the different sites followed a Latin square design. Because the responses for the four sites come from the same cow, they will be related. The unit or plot is the cow and the subunit or subplot is the site on the cow.

A standard analysis of variance, using logarithms of the response values, i.e. a log normal distribution, gives the results in Table 3.3. Here, we must interpret the sums of squares differently from the way that it is done in the usual fixed effects ANOVA case. We must decompose them into components of variance because of the interdependence among sites on the same cow. The residual line in the table gives the estimate of the variance, ψ^2, within cows. Because each value in the table is the sum for 30 cows, this value must be multiplied by 30 so as to refer to individual cows. Thus, our estimate of this variance is 0.0217.

The cow group line in the table is a combination, $\psi^2 + R\delta$, of this variance and the additional variation, δ, among cows. Because measurement in a cow group is an average of $R = 4$ sites, the latter variation is accumulated that number of times. We can obtain an estimate of this component by subtraction, giving a value of 1.175 ($= (0.1574 \times 30 - 0.0217)/4$). The total variance for a response is the sum of these two values: 1.197. Thus, we have estimated the *stochastic* parameters from a 'best' or full model including site, treatment, and cow group.

We obtain the following covariance matrix:

$$\begin{pmatrix} 1.197 & 1.175 & 1.175 & 1.175 \\ 1.175 & 1.197 & 1.175 & 1.175 \\ 1.175 & 1.175 & 1.197 & 1.175 \\ 1.175 & 1.175 & 1.175 & 1.197 \end{pmatrix}$$

The simple compound symmetry of this matrix contrasts with the complexity of those in the previous section. If we divide the matrix by the value on the diagonal, by Equation (1.2), we obtain the corresponding correlation matrix: the correlation among responses on sites on the same animal is estimated as 0.98. This reflects the fact, seen from the difference in size between the two variance components, that variability among cows is much larger than that among sites on a cow.

With these data, we cannot estimate the full model of the previous section, with unconstrained covariance matrix, because we do not have available the replications of the site-treatment combinations. Thus, we cannot verify if our model for compound symmetry is acceptable by comparing it to that saturated model.

The *location* parameters, and their standard errors, are estimated by fitting a model with only the two effects of interest, site and treatment. We average over cow groups. The parameter vectors for these effects, with the first category of each variable as baseline, are estimated as

$$\begin{pmatrix} 0.000 & -0.204 & 0.007 & -0.208 \end{pmatrix}$$

for the four treatment contrasts, showing that the smaller doses (B and D) have much less effect, and

$$\begin{pmatrix} 0.000 & 0.093 & 0.128 & -0.053 \end{pmatrix}$$

for the four sites, showing that sites two and three react more strongly to the tuberculin. Standard F tests, given in Table 3.3, show that both sets of contrasts are significantly different from zero.

In summary, the two components of variance in the covariance matrix are estimated by fitting a complete location model, including parameters for differences among units, and decomposing the variance, while the location parameters of interest are estimated from the simpler location model which only contains them. In this example, in fact, the estimates from the full model are identical to those for the simpler model, because of the orthogonal design. However, the estimates of their standard errors are much

Table 3.4. Energy loss in transformers (units unknown). (Coons, 1957)

	Location			
	Edge		Centre	
	Tension			
	Low	High	Low	High
	75	70	122	114
Glass	69	51	91	125
Coating	66	68	102	84
	56	50	–	72
	42	59	39	82
Carlite	75	66	50	46
Coating	–	106	74	77
	62	102	34	56

too small (0.019, as opposed to 0.163 if the simpler model is used).

This procedure for obtaining estimates of the parameters of the multivariate normal model, by decomposing the variance into components, does not yield the usual maximum likelihood estimates. Instead, it gives *marginal* or *conditional* maximum likelihood estimates, in this context often called restricted maximum likelihood (REML) estimates. The difference is simply that sums of squared deviations are divided by the appropriate degrees of freedom instead of by N.

Because the variance component estimates are obtained by subtraction, they may be negative. This is in agreement with the model we are fitting, because δ, although a component of variance, is also a covariance. A negative value indicates negative correlation among responses on subunits (Nelder, 1954), resulting from greater variability within than among units.

The univariate analysis of variance for the multivariate variance components model may still be used if there is some unbalance resulting from values missing in the data. An additional indicator covariate, with zero for all observations except that missing, which has a value of one, may be fitted for each missing observation (Berk, 1987, originally proposed by Bartlett, 1937). Equivalently, if a weighting facility is available in the program used to perform the analysis, the missing values are weighted out.

A new example may be used to illustrate this procedure. Consider data on energy loss by coils in transformers, cut from 8 different sheets of steel, in Table 3.4, from Coons (1957). Four coils were made from each sheet, the subplots, coming from different locations (edge or centre) and subject to different tensions (low or high). Four sheets received glass coatings and four received carlite coatings. Two observations are missing. The (location) parameter values of interest, obtained from the simpler model, are

Table 3.5.   ANOVA table for the transformer data of Table 3.4.

	SS	d.f.	MSS	F
Coating	2001.0	1	2001.0	2.97
Coating/Sheets	4043.0	6	673.8	
Spring	1106.4	3	368.8	1.53
Location	529.8	1	529.8	2.20
Tension	541.2	1	541.2	2.25
Interaction	35.2	1	3.2	0.15
Coating/Spring	6065.5	3	2021.8	8.41
Residual	3845.1	16	240.3	

$$(0.000 \ -6.833)$$

for the differences between coatings,

$$(0.000 \ -6.75 \ 38.50 \ 32.25)$$

for the differences in the combination of spring locations and tensions, and

$$(0.000 \ 30.33 \ -48.92 \ -26.67)$$

for the interaction of spring and coating. F tests from the ANOVA in Table 3.5 show that only the interaction of coating with spring location and tension is significantly different from zero, so that we should look separately at the effects of location and tension for each coating.

Again, the variance estimates are obtained from the full model. The residual line in the ANOVA table gives the estimate of the variance, ψ^2, within sheets: 240.3. The coating/sheets line in the table is a combination, $\psi^2 + R\delta$, of this variance and the covariance, δ, among sites. We obtain the estimate of this covariance by subtraction, giving 108.35 ($= (673.8 - 240.3)/4$). The total variance for an observation is the sum of these two values: 348.65. Then, the correlation among observations on the same sheet is estimated as 0.31 ($=108.35/348.65$). The estimated covariance matrix for observations on a sheet again has the compound symmetry form:

$$\begin{pmatrix} 348.65 & 108.35 & 108.35 & 108.35 \\ 108.35 & 348.65 & 108.35 & 108.35 \\ 108.35 & 108.35 & 348.65 & 108.35 \\ 108.35 & 108.35 & 108.35 & 348.65 \end{pmatrix}$$

In this example, it is possible to compare this with the sample covariance matrix of the saturated model:

$$\begin{pmatrix} 105.954 & 49.604 & 38.286 & -17.265 \\ 49.604 & 262.440 & 104.863 & 10.690 \\ 38.286 & 104.863 & 206.389 & 24.289 \\ -17.265 & 10.690 & 24.289 & 341.190 \end{pmatrix}$$

We see that the two matrices are not at all similar. The difference in deviance between the two stochastic structures is 15.42 with 8 d.f., indicating that compound symmetry is a questionable model.

3.3 Random effects models

A more indirect approach than modelling the stochastic dependence structure directly in the covariance matrix is to construct a random effects model (Section 2.6.2). This emphasizes the variability of responses across units, instead of the homogeneity of responses on a unit.

Suppose that we have a seemingly standard no-interaction two-way analysis of variance model,

$$\mathrm{E}[Y_{ik}|\lambda_i] = \mu + \lambda_i + \beta_k \tag{3.11}$$

where λ_i would describe the differences among the units and β_k the set of response conditions on each unit. For simplicity, we have a randomized block design with no treatment differences distinguishing among the units, but only different conditions within units.

Now suppose, in addition, that the units are a random sample from some larger population. Then, the λ_i, which describe the average differences in responses among the units, could be taken to have some random distribution over the population, instead of being fixed by the experimental design. Usually, this distribution is taken to be identically normal, $\mathrm{N}(0, \delta)$, independently of the conditional distribution of $Y_{ik}|\lambda_i$ which is $\mathrm{N}(0, \psi^2)$, so that responses on a unit are conditionally independent. In matrix form, the linear model can be written

$$\mathrm{E}[\mathbf{Y}|\boldsymbol{\Lambda}] = \mathbf{J}_N^T \mathbf{B} \mathbf{Z} + \boldsymbol{\Lambda} \mathbf{J}_R \tag{3.12}$$

Here, as usual, $\mathbf{J}_N$ and $\mathbf{J}_R$ are row vectors of ones, $\mathbf{B}$ a $1 \times P$ parameter vector, and $\mathbf{Z}$ has a first line of ones for the mean and $P-1$ subsequent lines zeroes and ones coding the contrasts for the response conditions within a unit, so that μ is included in the vector $\mathbf{B}$. $\boldsymbol{\Lambda}$ is the column vector of N random parameters. This is called a *mixed* or *random effects model*. The unconditional, marginal expectation of the response, after integration, is

$$\mathrm{E}[\mathbf{Y}] = \mathbf{J}_N^T \mathbf{B} \mathbf{Z} \tag{3.13}$$

so that $\boldsymbol{\Lambda}$ is an imaginary or latent parameter which does not even exist in the final model, only in its derivation.

Table 3.6. Visual acuity of 7 subjects looking through 4 powers of lenses with each eye, measured as time lag in msec. (Crowder and Hand, 1990, p. 30)

Power							
1	3	6	10	1	3	6	10
Left Eye				Right Eye			
116	119	116	124	120	117	114	122
110	110	114	115	106	112	110	110
117	118	120	120	120	120	120	124
112	116	115	113	115	116	116	119
113	114	114	118	114	117	116	112
119	115	94	116	100	99	94	97
110	110	105	118	105	105	115	115

Such a random effects model implies a constant correlation among all responses on a unit, which is what we require. In fact, this model has the same location model and stochastic structure as the components of variance model of the previous section, and can be handled in much the same way. The only difference is that δ is now a variance, that of λ_i, and, hence, must be non-negative. Although both models have the same multivariate normal distribution, the random effects model is less general, and hence less useful, than the components of variance approach, because negative intra-unit correlations are excluded. In some situations, the uniform covariance must theoretically be positive, but, in general, the covariance matrix need only be positive definite. A negative covariance indicates that there is more variability among responses within each unit than across units. This can arise when a study is based on very similar units, but can also be generated by specific mechanisms, for example, of repulsion among responses on a unit.

Let us consider an example where we have no treatment variables on the units, but two levels of nesting. Seven subjects were tested for visual acuity of each eye, from Crowder and Hand (1990, p. 30), as given in Table 3.6. The response is the time lag between a light flash and electric response at the back of the cortex. The eye is the subunit or first level of nesting. On each eye, four tests are applied, with different powers of lenses. Thus, we have four repeated measurements on each subunit, or eight repeated measurements on each unit.

Here, our model is more complex than Equation (3.12). Let us assume that the subjects are a random selection from a larger population. This is the random effects model described above. It contains a component of variance for the correlation between eyes. Because it does not make much sense to suppose that the two eyes are a random selection from a larger population of eyes of the subject, we might think of the correlation among measurements on the same eye as a component of variance, as in the previous section.

Our model is

$$E[Y_{ihk}|\lambda_{1i}, \lambda_{2ih}] = \mu + \alpha_i + \beta_h + \lambda_{1i} + \zeta_k + \lambda_{2ih} + \nu_{hk} \quad (3.14)$$

with subjects indexed by $i = 1,\ldots,7$, eyes by $h = 1,2$, and powers by $k = 1,\ldots,4$, where $\lambda_{1i} \sim N(0, \delta_E)$, $\lambda_{2ih} \sim N(0, \delta_P)$, and $Y_{ihk}|\lambda_{1i}, \lambda_{2ih} \sim N(0, \psi^2)$. In the Wilkinson and Rogers (1973) notation of Section 2.3.2, it is

$$SUBJ * EYE + POW \cdot (1 + SUBJ + EYE)$$

Then, the marginal expectation is

$$\mu_{ihk} = E[Y_{ihk}]$$
$$= \mu + \alpha_i + \beta_h + \zeta_k + \nu_{hk} \quad (3.15)$$

or

$$SUBJ + POW * EYE$$

We have an 8×8 covariance matrix for each individual subject. It has a double level of symmetry, from the nesting.

$$\begin{pmatrix} \psi^2 + \delta_P + \delta_E & \cdots & \delta_P + \delta_E & \delta_E & \cdots & \delta_E \\ \vdots & \ddots & \vdots & \vdots & \ddots & \vdots \\ \delta_P + \delta_E & \cdots & \psi^2 + \delta_P + \delta_E & \delta_E & \cdots & \delta_E \\ \delta_E & \cdots & \delta_E & \psi^2 + \delta_P + \delta_E & \cdots & \delta_P + \delta_E \\ \vdots & \ddots & \vdots & \vdots & \ddots & \vdots \\ \delta_E & \cdots & \delta_E & \delta_P + \delta_E & \cdots & \psi^2 + \delta_P + \delta_E \end{pmatrix}$$

The upper left and lower right quadrants of the matrix give the relationships among powers within the left and right eyes, while the other two give the relationship between eyes. We see that the variance for the random parameter for eyes, δ_E, provides the extra component of variance across subjects, while the variance for powers, δ_P, provides that across eyes.

The ANOVA is given in Table 3.7. The residual gives the estimate of $\hat{\psi}^2 = 12.8$. The subject/power line is $\hat{\psi}^2 + 4\hat{\delta}_P$, giving an estimate of $\hat{\delta}_P = 2.00$. The subject/eye line is $\hat{\psi}^2 + 8\hat{\delta}_E$, giving an estimate of $\hat{\delta}_E = 5.85$. Thus, the correlation between the eyes of a subject is 0.24 $(= 5.85/(12.8 + 2.00 + 5.85))$ and that among responses on an eye 0.38 $(= (2.00 + 5.85)/(12.8 + 2.00 + 5.85))$. These correlations are relatively

Table 3.7. ANOVA table for the visual acuity data of Table 3.6.

	SS	d.f.	MSS	F
Subject	1379.0	6	289.5	
Eye	46.5	1	46.5	0.78
Subject/Eye	357.4	6	59.6	
Power	140.7	3	46.9	2.24
Subject/Power	375.9	18	20.9	
Eye/Power	40.6	3	13.5	1.05
Residual	231.0	18	12.8	

small because there is little variation in response among the subjects in this study. With only seven subjects, we do not have enough observations to check the hypothesis of compound symmetry.

The F values in the ANOVA table do not indicate any significant differences, either between eyes or among power tests.

In situations where the possibility of negative intra-unit correlations can be excluded, the random effects model is identical to the variance components approach. Then, the choice for the investigator is between two interpretations of one and the same model.

3.4 Random coefficients models

A more complex extension of the previous models is the random coefficients model, originally proposed by Elston and Grizzle (1962). Suppose that the responses follow a model of the form

$$E[\mathbf{Y}|\mathbf{\Lambda}] = \mathbf{XBZ} + \mathbf{\Lambda V} \tag{3.16}$$

where

$$\mathbf{Y}|\mathbf{\Lambda} \sim \mathrm{MVN}(\mathbf{XBZ} + \mathbf{\Lambda V}, \mathbf{I}_N \otimes \mathbf{\Psi}) \tag{3.17}$$

and, independently,

$$\mathbf{\Lambda} \sim \mathrm{MVN}(\mathbf{0}, \mathbf{I}_N \otimes \mathbf{\Delta}) \tag{3.18}$$

so that

$$\mathbf{Y} \sim \mathrm{MVN}(\mathbf{XBZ}, \mathbf{I}_N \otimes (\mathbf{\Psi} + \mathbf{V}^T \mathbf{\Delta V})) \tag{3.19}$$

Here, $\mathbf{Y}$ is an $N \times R$ matrix, $\mathbf{X}$ $N \times C$, $\mathbf{B}$ $C \times P$, $\mathbf{Z}$ $P \times R$, $\mathbf{\Lambda}$ $N \times S$, $\mathbf{V}$ $S \times R$, $\mathbf{\Delta}$ $S \times S$, and $\mathbf{\Psi}$ $R \times R$, where there are N individuals, C design conditions across units, P design conditions within units, R repetitions on a unit, and S random coefficients per unit.

Usually, the responses within units are taken to be conditionally independent, so that $\mathbf{\Psi} = \psi^2 \mathbf{I}_R$. Generally, $\mathbf{Z}$ and $\mathbf{V}$ are different, although, $\mathbf{V}$

may just contain a subset of the rows of $\mathbf{Z}$, so that $S < P$. As elsewhere in this chapter, the notation does not allow for unequal numbers of responses on different units, but can easily be generalized.

The innovation of this model, with respect to the random effects model, is that the parameters describing the change in response over repetitions on a unit now can have random distributions. Formerly, only the mean response was random, so that $\mathbf{V}$ was a vector of ones. Now, the corresponding matrices, $\mathbf{Z}$ and $\mathbf{V}$, will usually take some (orthogonal) polynomial form.

As we saw in Section 2.6.3, random coefficients can arise when important explanatory variables are missing. In such cases, all coefficients should be random, so that $S = P$. When $S < P$, the selection might be made in order to obtain a reasonable structure for the covariance matrix. However, this is often difficult to justify theoretically. Why should some coefficients of the location model for the mean within units be random and others not?

This model encompasses a number of special cases. In their original presentation, Elston and Grizzle have both $\boldsymbol{\Psi}$ and $\boldsymbol{\Delta}$ as a diagonal matrices. In other words, both the conditional responses and the random coefficients are uncorrelated. Secondly, as we just saw, the random effects model is obtained when $\mathbf{V}$ is just a vector of ones. Another special case occurs if we set $\boldsymbol{\Delta} = 0$ with $\boldsymbol{\Psi}$ completely arbitrary or $\boldsymbol{\Psi} = \mathbf{0}$ and let $\boldsymbol{\Delta}$ be completely arbitrary with $\mathbf{V} = \mathbf{I}_R$. We, then, have the original multivariate growth curve model of Potthoff and Roy, presented in Section 3.1 above. Laird and Ware (1982) develop the more general form, with $\mathbf{U} = \mathbf{I}_N$, but $\boldsymbol{\Delta}$ nondiagonal. The case where $\boldsymbol{\Psi}$ is not diagonal will be the subject of the next chapter.

In the simple case where the covariance matrices are diagonal, we can fit the models as in the previous sections, using only univariate techniques. Let us look at the same data which Elston and Grizzle originally used, from a dental study on the ramus height, in mm., measured in a cohort of boys four times over a period of a year and a half, and reproduced in Table 3.8. In contrast to the data sets of the previous two sections, here we have a time variable distinguishing responses on the same unit. The responses are plotted in Figure 3.2. Several of the boys appear to have rapid growth in one of the six month periods, while the rest are fairly linear.

Once again, the covariance matrix is estimated by decomposing the variance of the full model, while the parameters are estimated from the simpler model. We shall fit a straight line location model ($P = 2$) to the changes in ramus height over age. This gives

$$\mathrm{E}[Y_t] = 50.08 + 0.467 z_{2t}$$

where z_{2t} is an element of the linear orthogonal polynomial in time, $(-3, -1, 1, 3)$, the second line of $\mathbf{Z}$, while $\mathbf{X}$ is an $N \times 1$ column vector of ones. The ANOVA is given in Table 3.9; we see that the linear slope is very

Table 3.8.   Ramus heights (in mm.) for 20 boys measured at 4 time points. (Elston and Grizzle, 1962)

	Age		
8	8.5	9	9.5
47.8	48.8	49.0	49.7
46.4	47.3	47.7	48.4
46.3	46.8	47.8	48.5
45.1	45.3	46.1	47.2
47.6	48.5	48.9	49.3
52.5	53.2	53.3	53.7
51.2	53.0	54.3	54.5
49.8	50.0	50.3	52.7
48.1	50.8	52.3	54.4
45.0	47.0	47.3	48.3
51.2	51.4	51.6	51.9
48.5	49.2	53.0	55.5
52.1	52.8	53.7	55.0
48.2	48.9	49.3	49.8
49.6	50.4	51.2	51.8
50.7	51.7	52.7	53.3
47.2	47.7	48.4	49.5
53.3	54.6	55.1	55.3
46.2	47.5	48.1	48.4
46.3	47.6	51.3	51.8

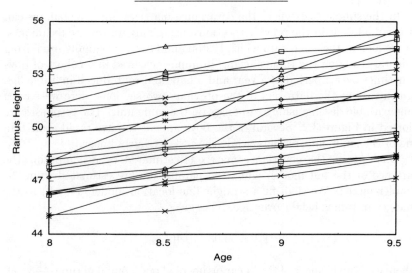

Fig. 3.2.   Growth in ramus heights for 20 boys from age 8 to $9\frac{1}{2}$ from Table 3.8.

Table 3.9. ANOVA for the ramus height data of Table 3.8.

	SS	d.f.	MSS	F
Subjects	476.10	19	25.06	
Linear Time	87.05	1	87.05	51.21
Subjects/Linear Time	32.26	19	1.70	
Residual	7.74	40	0.19	

significantly different from zero.

From the ANOVA table, the estimate $\hat{\psi}^2 = 0.193$. The line for subjects gives $\psi^2 + R\delta_0$, from which, by subtraction, $\hat{\delta}_0 = 6.22$. The line for the interaction gives $\psi^2 + \delta_1 \sum x_t^2$, where x_t are the orthogonal polynomials. Again, by subtraction, we have $\hat{\delta}_1 = 0.0752$. We can now construct our estimated covariance matrix for the responses by matrix multiplication, $\Psi + V^T \Delta V$. Here, $\Psi = \psi^2 I_4$, $\Delta = \text{diag}(\delta_0, \delta_1)$ and the two lines of V are ones and the orthogonal polynomial. We obtain

$$\begin{pmatrix} 7.087 & 6.442 & 5.991 & 5.540 \\ 6.442 & 6.485 & 6.141 & 5.991 \\ 5.991 & 6.141 & 6.485 & 6.442 \\ 5.540 & 5.991 & 6.442 & 7.087 \end{pmatrix}$$

The values decrease the further they are from the main diagonal; responses further apart in time are less closely related. Let us compare this with the unconstrained sample covariance matrix with the same linear location model for the mean

$$\begin{pmatrix} 6.014 & 5.880 & 5.488 & 5.271 \\ 5.880 & 6.129 & 5.848 & 5.627 \\ 5.488 & 5.848 & 6.575 & 6.599 \\ 5.271 & 5.627 & 6.599 & 7.092 \end{pmatrix}$$

Here, the values are generally increasing along each diagonal element. This is not reproduced in our model matrix. However, the deviance to compare the two models is 12.09 with 7 d.f., indicating little difference between them. That comparing this linear regression location model to that with a different mean at each time point has a deviance of 0.21 with 2 d.f., confirming that a linear regression is sufficient.

If a higher degree polynomial were needed, two addition lines would be added to the ANOVA table for each term and calculations analogous to those performed above done for each of them.

At this point, it may be useful to compare this random coefficients model with the simpler random effects and independence models (using the linear regression for the mean). Both of these may be obtained from our ANOVA table.

The estimate of ψ^2 in the random effects model is the sum of the interaction and residual lines: $0.678 \, (= (32.26 + 7.738)/59)$, while that of δ is again obtained by subtraction: $6.096 \, (= (25.06 - 0.678)/4)$. Thus, we have the matrix

$$\begin{pmatrix} 6.773 & 6.096 & 6.096 & 6.096 \\ 6.096 & 6.773 & 6.096 & 6.096 \\ 6.096 & 6.096 & 6.773 & 6.096 \\ 6.096 & 6.096 & 6.096 & 6.773 \end{pmatrix}$$

The deviance with respect to the random coefficients model is 31.80 with 1 d.f. indicating a much poorer fit.

The variance, $\psi^2 = \sigma^2$, for the independence model is obtained from the sum of the three lines, excluding the linear time: $6.62 \, (= (476.1 + 32.26 + 7.738)/78)$. The corresponding matrix has this value on the diagonal and zero elsewhere. The deviance with respect to the random effects model is 110.07 with 1 d.f., so that it is even much poorer than the latter model.

Our random coefficients model has produced a covariance matrix with decreasing stochastic dependence as distance in time between responses increases. This will be discussed further at the end of this section. However, we should already note that, in this longitudinal study, instead of using a random effects model for these data, it would be preferable to model directly the decreasing stochastic dependence of responses on a boy with increasing distance in time. Such models will be presented in the next chapter.

When the design is unbalanced, due to missing values or unequally spaced observations, the simple univariate analysis can no longer be used. A number of different approaches to estimation of the parameters have been advocated: direct maximization of the likelihood function (Jennrich and Schluchter, 1986; Lindstrom and Bates, 1988), the EM algorithm (Laird and Ware, 1982; Laird, Lange, and Stram, 1987), and the state space model, using the Kalman filter (Jones and Ackerson, 1990). The first and last are usually faster and more flexible. In addition, they are applicable for models in which the time series aspect of changing responses is modelled, as in the next chapter. The direct method is best when there are many units and few responses per unit, while the state space approach is faster in the opposite situation.

As an example of unbalanced data, consider a study to evaluate the *in vivo* ultrafiltration characteristics of 41 hollow fibre dialyzers (Vonesh and Carter, 1987, reproduced in Appendix A1). Each of three centres used a different type of dialysate delivery system to monitor transmembrane pressure (TMP). Because the three to five pressures used to evaluate each dialyzer varied in value for different machines, we have both unequal numbers of and unequally spaced observations. The response is the ultrafiltration rate

Table 3.10. Deviances for the ultrafiltration data in Appendix A1.

	Indep.	Rand. Effects	Rand. Coeff.	Corr. Rand. Coeff.	d.f.
No Effects	2184.75	2180.75	1729.88	1673.58	159
Linear	1553.10	1459.13	1402.79	1400.84	158
Nonlinear	1552.55	1458.80	1401.32	1399.93	156
Parallel	1479.54	1439.16	1390.63	1390.24	156
Interaction	1462.05	1399.42	1377.93	1371.02	154

(UFR) for given TMP. The goal of the study was to estimate the linear relationship between the two, which can be clearly seen in the plots, given in Figure 3.3.

Although the responses are obtained over time, we do not know if TMP values were applied in increasing order. If so, they are confounded with time and we do not have a crossover design. However, because the machines should not have a memory (unless a machine damaged at one trial affects subsequent results), this should not be important: observations closer together in time on the same machine should not necessarily be more closely related than those far apart, except insofar as they depend on the value of TMP used. Although the variances, with increasing pressure, might not be constant, there is little indication of this in Figure 3.3. Thus, a compound symmetry model should be appropriate. Then, all 'subplots' are the identical object, given different treatments of TMP. The analysis of deviance is given in Table 3.10; each column has one less degree of freedom than the one to its left.

The parameters estimated for the model with different linear regressions for the three centres and a nondiagonal random coefficients covariance matrix are

$$\hat{\mathbf{B}} = \begin{pmatrix} 1174. & 4.411 \\ -85.40 & -0.2856 \\ -82.42 & -0.3418 \end{pmatrix}$$

$$\hat{\mathbf{\Delta}} = \begin{pmatrix} 1440. & -0.522 \\ -0.522 & 0.032 \end{pmatrix}$$

with the estimate of ψ^2 equal to 704.413. The values of TMP have been centred at their mean, 305.7, for the calculations. The negative covariance between the two coefficients indicates that, when the slope is greater, the intercept is less, and vice versa. Thus, the lines are rotated randomly around the central values of TMP.

From Table 3.10, the difference in deviance for the uncorrelated, as opposed to correlated, random coefficients is 6.91 with 1 d.f. for the interaction

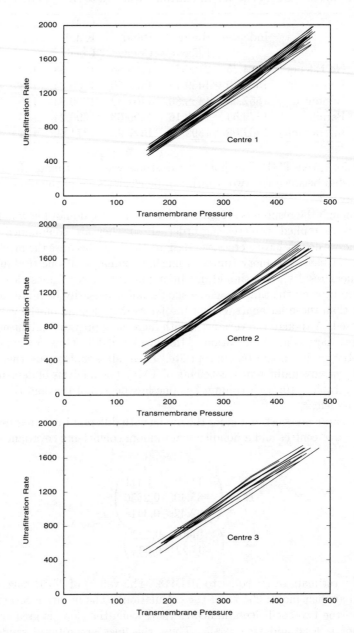

Fig. 3.3. Plots of ultrafiltration rates against transmembrane pressure, from data in Appendix A1.

model; for the random effects model, with the same location parameters, it is 28.40 with 2 d.f., indicating the rejection of both of them. When the interaction correlated random coefficients model is compared to the model with identical linear lines for all centres, the change in deviance is 29.82 with 4 d.f. and to that for parallel lines, 19.22 with 2 d.f. On the other hand, the data do not appear to demonstrate a nonlinear change in mean UFR with TMP. Thus, all simpler models are unacceptable.

The parameter estimates proposed by Vonesh and Carter, using TMP with origin at zero pressure, were obtained by the method of moments instead of maximum likelihood. Nevertheless, the deviance is very similar, indicating that the likelihood surface is rather flat for this model. The location model is similar in the two cases.

Let us now look more closely at the stochastic component. Because each dialyzer has measurements taken at different values of TMP, it is not possible to present a general covariance matrix for these data. However, we can, at least approximately, see what it looks like at certain fixed values of TMP. Let us take the values 160, 260, 350, and 450 as representative. Then, our covariance matrix, using Equation (3.19), is estimated as

$$
\begin{pmatrix}
2958.8 & 1744.9 & 1286.4 & 776.9 \\
1744.9 & 2255.2 & 1376.1 & 1182.1 \\
1286.4 & 1376.1 & 2161.3 & 1546.7 \\
776.9 & 1182.1 & 1546.7 & 2656.3
\end{pmatrix}
$$

where TMP is centred with origin at 305.7, while that of Vonesh and Carter is

$$
\begin{pmatrix}
1992.0 & 1268.9 & 1242.8 & 1213.9 \\
1268.9 & 2245.1 & 1804.8 & 2086.8 \\
1242.8 & 1804.8 & 3004.7 & 2872.5 \\
1213.9 & 2086.8 & 2872.5 & 4439.6
\end{pmatrix}
$$

where TMP has origin at zero pressure. Our model has the covariance matrix for the responses with smaller variances in the middle of the diagonal and larger on the two ends. Theirs has the variances increasing along the diagonal. This difference arises from the different origin of TMP, as well as from the difference in estimation procedures. Let us now investigate this further.

For most data sets, the profile likelihood function of the covariance matrix for a random coefficients model will be rather flat. However, in certain cases, it is possible that one of the models may fit better, when the only difference between them arises from difference in the selection of the origin for the explanatory variable(s). This is a problem which Elston (1964) warns about.

Consider an example where **V** contains a first order orthogonal polynomial for 3 points for an ordered explanatory variable changing with response on each unit, such as TMP above. Let $\mathbf{\Psi} = \mathrm{diag}(\psi^2)$, and $\mathbf{\Delta} = \mathrm{diag}(\delta_0, \delta_1)$. Then, the covariance structure of the responses is

$$
\begin{pmatrix}
\psi^2 + \delta_0 + \delta_1 & \delta_0 & \delta_0 - \delta_1 \\
\delta_0 & \psi^2 + \delta_0 & \delta_0 \\
\delta_0 - \delta_1 & \delta_0 & \psi^2 + \delta_0 + \delta_1
\end{pmatrix}
\tag{3.20}
$$

Note that the variances on the diagonal are not constant and that the greater is the difference between values of the explanatory variable, the smaller is the covariance. This is the form of matrix we have found in our last two examples.

When the random coefficients are uncorrelated, as in this example, the form of the covariance matrix depends on the parameterization of the location model for the coefficients ($\mathbf{V}$). For example, changing the orthogonal polynomials to polynomials in the original observations will completely alter the form of the covariance structure among responses. As Elston states, this makes the model generally unrealistic. A second problem is that we may virtually be restricted to modelling the change in response within a unit as a polynomial in the explanatory variable, if we wish to obtain a reasonable covariance matrix. Thus, great care must be taken if the random coefficients are uncorrelated.

These problems add to the weakness of the random effects model that negative covariances are excluded. Here, such negative values are no longer completely impossible, as seen in the examples just given. However, the range of values of covariances is unnecessarily restricted when the δ_i are taken to be the variances of some of the location parameters. Estimation by univariate ANOVA methods, as in the first example of this section, overcomes this problem at the expense of violating the hypothesis about the distributions of the parameters (because their variances could be negative). But, globally, it gives a somewhat more reasonable multivariate model, as was the case before with the variance components approach.

On the other hand, modelling a random effect for a coefficient may sometimes have physical meaning and direct usefulness. Consider repeated measurements for a growth curve, where a suitable location model is simple linear regression. Comparison of variability in slopes of the line within different treatment groups may be of interest. This can be obtained from estimates of a suitable model for $\mathbf{\Delta}$. However, more complex random structures will only occur in specialized situations (for a genetics example, see Karlin and Taylor, 1981, pp. 184–188). When we add to this the specification and estimation problems of random nonlinear coefficients (Vonesh, 1992), it will usually be preferable to model the covariance matrix directly.

Thus, one curious aspect of the random coefficients model is that it might even be applied to cases where the explanatory variable is time, as we did for the example of ramus heights above. The covariance structure of the random coefficients model is sufficiently flexible that it can accommodate the kinds of relationships often found in responses over time. When V includes a polynomial in time, a covariance structure with decreasing correlation among responses further apart in time may result. Thus, in a time series setting, Equation (3.20) would indicate a nonstationary model. Although this particular structure does not correspond to that usually found in growth studies, where the variance often increases over time, that could be obtained by taking time with origin zero, instead of using orthogonal polynomials. However, it is almost always preferable to model stochastic dependence over time directly in Ψ. Not only will this provide a more easily interpretable model, but often will require fewer parameters than a nondiagonal Δ.

The flexibility of random coefficients models may make them useful for certain data with peculiar covariance structures. But, due to the necessary arbitrariness in choosing which coefficients are random, they usually do not provide a satisfactory description of the mechanism which might have generated the data. If some such structure is known to exist in the covariance matrix, it is most often preferable to model it directly, as will be done in the next chapter.

4
Longitudinal studies

4.1 Autoregression models

In the previous chapter, we saw how to allow for heterogeneity, the generally observed phenomenon that responses from the same unit are usually uniformly more closely related than responses across units. Some units systematically give higher responses than others under apparently identical conditions. Although this is an important type of relationship among repeated measurement observations, it is certainly not the only one. Many repeated measurements studies involve observation of responses over time (or space) and often the evolution of responses is of special importance. Because the same unit is producing several successive responses, those which are closer together will often tend to be more closely related. In such cases, these relationships must also be included in the model. Let us first look at such time series effects in isolation before combining them with heterogeneity in Section 4.2 below.

One simple case can be eliminated immediately. If only two repeated observations are made per unit, heterogeneity and time series effects are not distinguishable. Often, the research worker is interested in a difference of mean, before and after, and models of the previous chapter can be used.

We shall begin by only considering repeated measures which are observed at discrete, equally-spaced time intervals, with no missing values. As well, the stochastic component of each time series will, at first, be assumed stationary. After that, we shall relax the assumptions about how the observations are spaced and the form of the time dependence, as well as generalizing the model to allow for heterogeneity.

4.1.1 SERIAL CORRELATION

The repeated measurements in a study rarely form a long series, so that elaborate time series modelling is usually either not necessary or impossible. Often, a simple autoregression model is sufficient. With short series, a first order model, or AR(1), may fit the data well. We shall restrict ourselves to discussion of this model, although the generalization to an AR(M) is straightforward.

An AR(1) model implies that each response depends directly only on the immediately preceding one. Let us explicitly model the stochastic dependence structure by defining the covariance matrix, although it will not be obvious from this approach where the term, autoregression, comes from. We express the repeated measures as a multivariate normal distribution. Then, the distribution for the R responses on a unit is

$$\mathbf{Y}_i \sim \mathrm{MVN}(\mathbf{B}_i^T \mathbf{Z}_i, \boldsymbol{\Sigma}) \tag{4.1}$$

where $\mathbf{B}_i^T \mathbf{Z}_i$ is the appropriate location model, depending only on the current values of $\mathbf{z}_{it}$. Let ρ be the autocorrelation between successive responses. Then, the covariance matrix will be

$$\boldsymbol{\Sigma} = \frac{\xi}{1 - \rho^2} \begin{pmatrix} 1 & \rho & \cdots & \rho^{R-2} & \rho^{R-1} \\ \rho & 1 & \cdots & \rho^{R-3} & \rho^{R-2} \\ \vdots & \vdots & \ddots & \vdots & \vdots \\ \rho^{R-2} & \rho^{R-3} & \cdots & 1 & \rho \\ \rho^{R-1} & \rho^{R-2} & \cdots & \rho & 1 \end{pmatrix} \tag{4.2}$$

where the variance, ξ, has been standardized by $1 - \rho^2$ for computational convenience. The matrix will hold for all values of R. It is symmetric because the response is equally closely related to one before as to one after. Thus, it is not ordered only in one direction, as one might expect with responses in time, but is reversible, making it also applicable to bidirectional change, such as responses in space. For $|\rho| < 1$, the stochastic dependence among repeated responses is decreasing geometrically with distance between them, in the desired Toeplitz form. The parameter, ρ, is the first order autocorrelation in this AR(1) model. We must have $|\rho| < 1$ for another reason, stationarity (Section 2.10); otherwise, $\boldsymbol{\Sigma}$ will not be positive definite and we cannot fit the model.

Although it may not be obvious from Equations (4.1) and (4.2), we are making a special assumption about the distribution of the first response on each unit; this distribution is taken to be the marginal distribution,

$$Y_{i1} \sim \mathrm{N}\left(\boldsymbol{\beta}_i^T \mathbf{z}_{i1}, \frac{\xi}{1 - \rho^2}\right) \tag{4.3}$$

which is stationary because it does not depend on previous values. Here, we see clearly that $|\rho| < 1$ for a stationary distribution, because otherwise the variance would be negative. Such an assumption is usually most satisfactory under experimental conditions, where the units are first observed at some presumably stable baseline level before starting treatments. Below, we shall look at another, conditional, approach which assumes that the initial response is fixed and, hence, does not impose stationarity. Other, more complex, assumptions are also possible. In all cases, the decision about how to handle these initial conditions of the series will be critical, especially because of the short length of the series.

One interesting aspect of this model is that the inverse of the covariance matrix

$$\mathbf{\Sigma}^{-1} = \frac{1}{\xi} \begin{pmatrix} 1 & -\rho & \cdots & 0 & 0 \\ -\rho & 1+\rho^2 & \cdots & 0 & 0 \\ \vdots & \vdots & \ddots & \vdots & \vdots \\ 0 & 0 & \cdots & 1+\rho^2 & -\rho \\ 0 & 0 & \cdots & -\rho & 1 \end{pmatrix} \qquad (4.4)$$

has a very simple Cholesky decomposition, $\mathbf{\Sigma}^{-1} = \mathbf{A}^T \mathbf{A}$, where

$$\mathbf{A} = \frac{1}{\sqrt{\xi}} \begin{pmatrix} \sqrt{1-\rho^2} & 0 & \cdots & 0 & 0 \\ -\rho & 1 & \cdots & 0 & 0 \\ \vdots & \vdots & \ddots & \vdots & \vdots \\ 0 & 0 & \cdots & 1 & 0 \\ 0 & 0 & \cdots & -\rho & 1 \end{pmatrix} \qquad (4.5)$$

More generally, the number of minor diagonals in $\mathbf{A}$ corresponds to the order of the autoregression.

If ρ is known or an estimate can be obtained, one way to fit the model is to multiply the response vector and all of the explanatory variables by $\mathbf{A}$ and then use standard multiple regression techniques (Watson, 1955). In other words, the appropriate filtering operation, a weighted combination of present and past values, is applied to all of the variables before fitting the multiple regression.

For an AR(1), one way to estimate ρ is by directly calculating the first order autocorrelation,

$$\tilde{\rho} = \frac{\sum (y_t - \mu)(y_{t-1} - \mu)}{\sum (y_t - \mu)^2} \qquad (4.6)$$

As an extension of this, for an AR(M), the autoregression coefficients can be obtained by solving the Yule-Walker equations,

$$\mathbf{R}\tilde{\rho} = \mathbf{r} \qquad (4.7)$$

where $\mathbf{R}$ is a $M \times M$ matrix with ones on the main diagonal and estimates of the autocorrelations of order m on the mth off-diagonal, and $\mathbf{r}$ is a vector of the estimated autocorrelations up to order M.

Another way to fit the serial correlation model is by using the previous, or lagged, response, and the corresponding lagged explanatory variables, as extra explanatory variables in a univariate multiple regression. The fact that the response at time t is being regressed on previous responses gives this model its name, an autoregression. The location model may be written

$$E[Y_{it}|y_{i,t-1}] = \rho(y_{i,t-1} - \boldsymbol{\beta}_i^T \mathbf{z}_{i,t-1}) + \boldsymbol{\beta}_i^T \mathbf{z}_{it} \qquad (4.8)$$

where ρ is now called the autoregression coefficient and $\boldsymbol{\beta}_i^T \mathbf{z}_{it}$ is the unconditional mean response at time t for unit i.

If we rewrite Equation (4.8) as

$$E[Y_{it}|y_{i,t-1}] - \rho y_{i,t-1} = \boldsymbol{\beta}_i^T (\mathbf{z}_{it} - \rho \mathbf{z}_{i,t-1}) \qquad (4.9)$$

we again see the other way of interpreting this model. As with premultiplying the variables by the Cholesky decomposition above, the filtered differences in the explanatory variables are being used to explain the corresponding filtered differences in responses.

In the general form of this model, the explanatory variables, $\mathbf{z}_{it}$ are varying over time. They define the *state* of the unit at each time point, as described by Anderson and Hsiao (1982). We have used only the current state to determine the mean in Equation (4.1) or, equivalently, included the previous state in Equation (4.8), and, thus, removed that (location) part of the previous response which was due to that state. In this way, we are left with only a 'pure' autoregression dependence or serial correlation, plus the random variability of the normal distribution. Thus, the response fluctuates around equilibrium level defined by the explanatory variables, via $\boldsymbol{\beta}^T \mathbf{z}_{it}$. It is the difference between the mean response at t and this quantity which follows a stationary autoregression process. If, at time t, $\mathbf{z}_{it}$ changes momentarily from its usual pattern, and then goes back to its former level, the distribution at time t is affected but not that at time $t+1$ or later.

We have here a standard univariate normal distribution model, so that the variance, ξ, is assumed constant. This is also a condition for second order stationarity. Nevertheless, the responses, y_{it}, themselves, may, in fact, be first order nonstationary, because the mean can change over time, although the differences between responses and their means

$$Y_{it}^* = Y_{it} - \boldsymbol{\beta}_i^T \mathbf{z}_{it} \qquad (4.10)$$

are stationary. However, this autoregression model is not a *simple* multiple linear regression; $\boldsymbol{\beta}_i$ appears twice in Equation (4.8), making it nonlinear. If multiple regression is to be used, some iterative scheme must be applied.

Because the first observed response on each unit has no precedent, with this approach, we can no longer keep it in the model as a response; we are assuming that it is just a fixed value. With short series, this can be of considerable importance. Moreover, it means that we lose a fair proportion of the responses on each unit. This is known as a conditional model, because each response is conditional on the previous value. Because of the fixed initial conditions, it is not reversible, as was the multivariate model above, and is, thus, not applicable to spatial phenomena.

This complete conditional model with fixed initial value, corresponding to Equation (4.8), can be derived from Equations (4.1) and (4.3) through the conditional distribution, given y_{i1}, as in Equation (2.68). Then, the log likelihood or deviance is proportional to

$$N\left[1 + \log(|\mathbf{\Sigma}|) - \log\left(\frac{\xi}{1-\rho^2}\right)\right] = N[1 + (R-1)\log(\xi)] \quad (4.11)$$

where, as usual, N is the number of units observed, R the number of responses per unit, $\mathbf{\Sigma}$ the covariance matrix, and ξ the constant variance. Because $\mathbf{\Sigma}$ no longer appears explicitly in the likelihood function, we can fit the model even if it is nonstationary. This can be clearly seen from Equation (4.8) which can be estimated from fairly standard, if nonlinear, multiple regression techniques, even if $|\rho| > 1$. In other words, the problem in estimating Equation (4.2) under nonstationarity only arose from assuming the initial response to be stationary.

If the regression variables are factor levels, we may have, as a saturated model, the Potthoff and Roy (1964) model, which may be compared, as in the previous chapter. However, usually there will be no clearly defined saturated model, so that we must make comparisons to a null model with common mean across units and over time, constant variance and zero autocorrelation. Thus, we shall be trying to maximize the difference in deviance with respect to this null model, instead of the more usual minimization of the difference in deviance with respect to the saturated model.

Let us consider the data, from Hooper and Larin (1989), given in Table 4.1, on manufacturing compensation per hour in current U.S. dollars over a period of ten years from 1978 to 1987 for ten countries. If we plot the evolution of these wages, as in Figure 4.1, we see that the form of the change was different for the European countries than the others, in that the former were not (more or less) monotonely increasing. As well, Belgium, Germany, and Holland had higher wages than France, Italy, and the U.K. Of the four remaining countries, the North American ones had higher wages, although Japan was catching up rapidly within this period.

It will also be useful to look at the empirical covariance and correlation matrices, for the (unrealistic) null model with constant mean for all years and countries. These are

Table 4.1. Manufacturing compensation per hour in current U.S. dollars. (Hooper and Larin, 1989)

	1978	1979	1980	1981	1982
Belgium	10.14	11.84	13.15	11.31	9.49
Canada	7.25	7.69	8.47	9.32	10.20
France	6.43	7.69	8.94	8.02	7.85
Germany	9.65	11.29	12.33	10.53	10.28
Holland	9.98	11.41	12.06	9.91	9.78
Italy	6.09	7.12	8.00	7.39	7.30
Japan	5.54	5.49	5.61	6.18	5.70
Korea	0.80	1.06	1.01	1.06	1.13
U.K.	4.34	5.61	7.43	7.20	6.82
U.S.A.	8.27	9.02	9.84	10.84	11.64
	1983	1984	1985	1986	1987
Belgium	9.07	8.62	9.95	12.35	15.08
Canada	10.97	11.07	10.88	11.04	11.98
France	7.74	7.29	7.52	10.27	12.36
Germany	10.23	9.43	9.56	13.35	16.83
Holland	9.49	8.70	8.70	12.24	15.11
Italy	7.61	7.21	7.40	10.01	12.33
Japan	6.13	6.34	6.47	9.47	11.34
Korea	1.20	1.28	1.31	1.39	1.69
U.K.	6.39	5.95	6.19	7.50	9.07
U.S.A.	12.10	12.51	12.96	13.21	13.46

$$\begin{pmatrix}
10.52 & 9.85 & 8.89 & 8.19 & 8.03 & 7.74 & 7.70 & 7.68 & 6.20 & 4.81 \\
0.94 & 10.45 & 10.55 & 8.83 & 8.33 & 7.95 & 7.48 & 7.80 & 8.65 & 9.47 \\
0.80 & 0.96 & 11.66 & 9.33 & 8.58 & 8.16 & 7.36 & 7.99 & 10.66 & 13.25 \\
0.87 & 0.94 & 0.94 & 8.45 & 8.26 & 8.13 & 7.87 & 8.23 & 8.78 & 9.46 \\
0.84 & 0.88 & 0.86 & 0.97 & 8.60 & 8.65 & 8.56 & 8.67 & 8.41 & 8.38 \\
0.80 & 0.83 & 0.80 & 0.94 & 0.99 & 8.86 & 8.86 & 8.92 & 8.59 & 8.49 \\
0.79 & 0.77 & 0.72 & 0.90 & 0.97 & 0.99 & 9.08 & 9.06 & 7.83 & 6.91 \\
0.78 & 0.79 & 0.77 & 0.93 & 0.97 & 0.98 & 0.99 & 9.27 & 8.59 & 8.23 \\
0.52 & 0.72 & 0.84 & 0.82 & 0.78 & 0.78 & 0.70 & 0.76 & 13.71 & 18.30 \\
0.28 & 0.56 & 0.74 & 0.62 & 0.55 & 0.55 & 0.44 & 0.52 & 0.95 & 27.35
\end{pmatrix}$$

As would be expected, the correlation decreases with separation in time. Note how the variance increases for 1987.

At present, we have no explanatory variables, so let us just fit a simple autoregression, ignoring differences among countries. This is just a simple linear regression. Because the system is not in equilibrium at the first observation, we use the conditional model. The result is

$$\mathrm{E}[Y_{it}|y_{i,t-1}] = 0.3114 + 1.031y_{i,t-1}$$

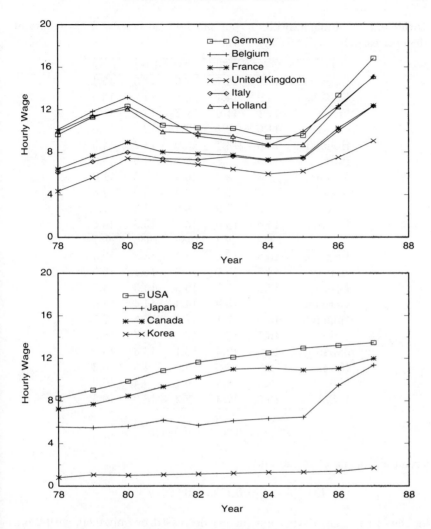

Fig. 4.1. Wage indices in six European and four other countries from Table 4.1.

In this conditional model, the observations for 1978 have not been used as responses. The variance, ξ, is estimated from the sums of squares for the remaining ninety observations. The autocorrelation is assumed identical for all countries, and is greater than one, indicating explosive growth. The deviance is decreased by 195.82 with 1 d.f. with respect to the null model, showing that the stochastic dependence is very important.

Suppose, now, that the autocorrelation is different for Europe and the other countries, reflecting the differences seen in the graphs. The fitted model is

Table 4.2.   Manufacturing output per hour in current U.S. dollars at 1980 prices. (Hooper and Larin, 1989)

	1978	1979	1980	1981	1982
Belgium	12.1	12.7	13.6	14.5	15.4
Canada	13.9	14.1	13.5	14.2	13.5
France	11.1	11.6	11.7	12.1	12.9
Germany	11.6	12.1	12.2	12.5	12.6
Holland	13.0	13.7	13.9	14.2	14.5
Italy	11.9	12.8	13.5	14.5	15.0
Japan	8.8	9.3	10.0	10.3	11.0
Korea	2.0	2.3	2.4	2.9	2.8
U.K.	6.9	7.0	6.9	7.3	7.7
U.S.A.	15.0	15.0	15.0	15.3	15.6
	1983	1984	1985	1986	1987
Belgium	16.8	17.6	18.1	18.8	19.4
Canada	14.5	15.7	16.2	16.2	16.6
France	13.2	13.5	13.8	14.3	14.7
Germany	13.4	13.9	14.4	14.4	14.6
Holland	15.5	17.1	17.7	17.6	17.7
Italy	16.1	17.1	17.6	17.7	18.3
Japan	11.6	12.4	13.1	13.3	13.9
Korea	2.9	3.6	4.2	4.6	5.3
U.K.	8.4	8.8	9.1	9.4	10.0
U.S.A.	16.5	17.4	18.2	18.9	19.5

$$E[Y_{it}|y_{i,t-1}] = 0.7629 + 0.9854y_{i,t-1}$$

for the European countries and

$$E[Y_{it}|y_{i,t-1}] = 0.1723 + 0.9299y_{i,t-1}$$

for the others. The deviance is further decreased by only 0.60, with 2 d.f. Although the change in deviance due to the difference between the two regions does not appear to be great, introducing it produces a model with the autocorrelations both less than one.

One thing to which wages may be related is the productivity. A measure of this is the output per hour, given in Table 4.2 and plotted in Figure 4.2. For these data, the evolutions are much more monotone in all countries, not showing the systematic fluctuation of the wage data for Europe. The basic differences of level among the regions are the same for both sets of data, although the U.K. has relatively lower productivity. Thus, we would not expect this variable to explain all of the changes in the European wages.

Let us first look at the relationship between wages and productivity without taking into account any autocorrelation. For all of the countries

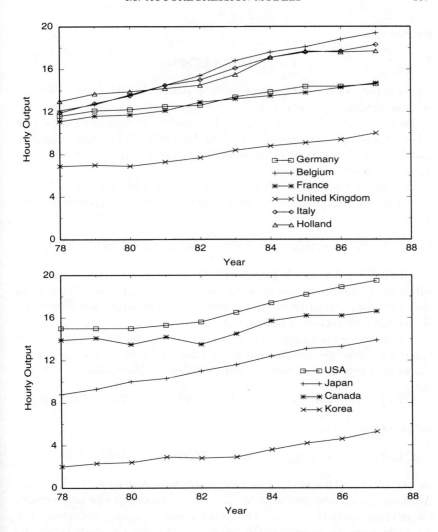

Fig. 4.2. Manufacturing output in six European and four other countries from Table 4.2.

together, the result is

$$\mu_{it} = 0.1668 + 0.6626 z_{it}$$

showing that wages appear to have been increasing on average only two-thirds as fast as productivity. The deviance is decreased by 100.50, with 1 d.f., as compared to the null model. When the differences between Europe and the others are taken into account, the fitted model becomes

$$\mu_{it} = 4.301 + 0.3819 z_{it}$$

for Europe and

$$\mu_{it} = -1.643 + 0.7829 z_{it}$$

for the others, with a further decrease in deviance of 19.00 on 2 d.f. Neither of these models fits nearly as well as the autocorrelation model above. As might be expected from the graphs, the wages of European workers were following less well their productivity than the others. If the North American and Asian countries are distinguished, the deviance only decreases by 1.03, not a significant improvement.

We can now return to our serial correlation model. First, we fit it for all of the countries together, using a nonlinear estimation procedure. This gives

$$E[Y_{it}|y_{i,t-1}] = 1.036(y_{i,t-1} + 9.250 + 1.167 z_{i,t-1}) - 9.250 - 1.167 z_{it}$$

The decrease in deviance is 118.42 on 1 d.f. with respect to the same model with only productivity and no autocorrelation and 23.10 with respect to that with autocorrelation but not productivity. The wages now seem to be decreasing with increased productivity.

Again, we can distinguish between the two regions. We shall do this in two steps. First, we leave the autocorrelation the same for the two but let the wage dependence on productivity be different. We obtain

$$E[Y_{it}|y_{i,t-1}] = 1.044(y_{i,t-1} + 2.527 + 1.992 z_{i,t-1}) - 2.527 - 1.992 z_{it}$$

for Europe and

$$E[Y_{it}|y_{i,t-1}] = 1.044(y_{i,t-1} + 4.685 + 0.396 z_{i,t-1}) - 4.685 - 0.396 z_{it}$$

for the others. This model has a deviance decreased by 8.33 with 2 d.f. Allowing the autocorrelation to be different for the two groups only reduces the deviance by 0.34 on 1 d.f. The model is nonstationary, indicating that we might consider successive differences in wages and productivity.

First, let us look at the graph of wage increases against productivity increases, as shown in Figure 4.3. We see that larger increases in productivity over a year are associated with smaller increases in wage. If we fix the autocorrelation, $\rho = 1$, we have a random walk. Let $\Delta y_{it} = y_{it} - y_{i,t-1}$ and $\Delta z_{it} = z_{it} - z_{i,t-1}$, the changes in wages and productivity. For all countries together, we obtain

$$E[\Delta Y_{it}] = 1.034 - 0.9680 \Delta z_{it}$$

This model has a deviance 14.52 larger, on 1 d.f., than the above model for the original observations, which had $\hat{\rho} = 1.036$. If we allow for differences between the two regions, we have

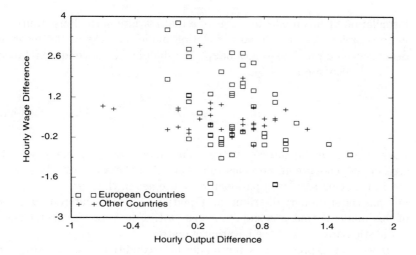

Fig. 4.3. Yearly differences in hourly wage against differences in productivity, from Tables 4.1 and 4.2.

$$E[\Delta Y_{it}] = 1.467 - 1.603\Delta z_{it}$$

for Europe and

$$E[\Delta Y_{it}] = 0.601 - 0.320\Delta z_{it}$$

for the others, with a decrease in deviance of 5.51 on 2 d.f. This is 17.68 larger than that given above with $\hat{\rho} = 1.044$. The important result to note is that increases in wages are inversely related to increases in productivity. The bigger is the increase in productivity, the smaller is the increase in wages. However, a random walk is not adequate to describe the process; ρ must be greater than one.

In general, the next step might be to try a higher order autoregression or to introduce lagged values of the explanatory variable(s), here (changes in) productivity in previous years. The problem is that we lose further observations for each unit. In long time series, this is not a major problem, but, in repeated measurements studies, it must be taken into account, especially if the distribution of the first response is not stationary.

We continue with the analysis of these data in the next section.

4.1.2 STATE DEPENDENCE

Let us now consider another, more complex, autoregression model. We keep the same stochastic dependence structure, the covariance matrix of Equation (4.2). But now, let the mean depend, in an ever-decreasing manner, on the previous states, as well as on the current state. Thus, the formulation of this model in terms of a multivariate normal distribution is more complex than the serial correlation model because of the recurring dependence on previous values of $\mathbf{z}_{it}$. We shall let the unconditional mean of the tth response be

$$\mu_{it} = \sum_{h=1}^{t} \rho^{t-h} \boldsymbol{\beta}^T \mathbf{z}_{ih} \qquad (4.12)$$

where ρ is again the autocorrelation. This allows for the progressively decreasing dependence on previous values of the explanatory variables, without introducing any extra parameters. Once again, the first response on a unit has the stationary distribution, Equation (4.3). One interesting aspect of this model is that ρ is simultaneously a true location parameter and a stochastic parameter of the covariance matrix.

If we now express this same model in a conditional form, using univariate multiple regression, things are much simpler. We just remove the lagged value of the explanatory variables from Equation (4.8) so that it becomes

$$\mathrm{E}[Y_{it}|y_{i,t-1}] = \rho y_{i,t-1} + \boldsymbol{\beta}_i^T \mathbf{z}_{it} \qquad (4.13)$$

Thus, paradoxically, this simplifying modification to the conditional model has the opposite effect to what might have been expected, giving the more complex multivariate model just described, but now with fixed initial conditions. The filtered difference in the response is being explained by the actual levels of the explanatory variables, instead of by the filtered differences in them, as in Equations (4.8) and (4.9). Thus, the regression coefficients, $\boldsymbol{\beta}_i$, have different values, and a different interpretation, in the two models.

In this state dependence model, the mean response at time t depends not only on the state at that time, defined by $\mathbf{z}_{it}$, but on the previous states. This is so because the previous response depended on the state at that time, and so on, and this part of the response is no longer being removed, as it was in Equation (4.8). Thus, as we saw in Equation (4.12), a change in the explanatory variables that affects the distribution of responses at time t will continue to affect it at time $t+1$, and so on, even although the change only occurred during the one period. Such was not the case in the first model, where previous states did not influence the present response. In Equations (4.1) and (4.8), there was only serial correlation; here, there is also state dependence over time.

When the initial response is taken to be fixed, this conditional model can be fitted much more easily than Equation (4.8), because it is a simple linear multiple regression. For this reason, it is again known as a conditional autoregression.

If the explanatory variables for a unit are not time dependent, then the models defined by Equations (4.1) and (4.13) are not distinguishable. On the other hand, the state dependence model is even more heavily influenced by the choice of initial conditions than that for serial correlation. This is so because the progressively decreasing stochastic dependence over time can only go back to the first observed value. Thus, it is much more critical either that the first observation be the real beginning of the process or that each unit be in the stationary state at the start of observation. Such conditions can most often only reasonably be attained under experimental conditions.

We now return to our wage example. Obviously, the first observations do not fulfil the conditions just stated. The year 1978 was not the first year wages were paid and there is little chance that they were in equilibrium at that time. Nevertheless, we shall fit the model.

As we have already seen, the state dependence model without time-varying explanatory variables is identical to the serial correlation model which we fitted above. When we add productivity, with the same relationship for all countries, we obtain

$$E[Y_{it}|y_{i,t-1}] = 0.9110y_{i,t-1} - 0.09375 + 0.1070z_{it}$$

When the wage is allowed to depend on the series of previous wages, its dependence on productivity is greatly reduced. Here, the deviance is decreased by 99.84 on 1 d.f. with respect to the model with only productivity and not the autocorrelation. This is 18.58 larger than the corresponding serial correlation model. With differences between the regions in dependence on productivity, we have

$$E[Y_{it}|y_{i,t-1}] = 0.9032y_{i,t-1} + 0.03439 + 0.1061z_{it}$$

for Europe and

$$E[Y_{it}|y_{i,t-1}] = 0.9032y_{i,t-1} - 0.1369 + 0.1094z_{it}$$

for the others, with deviance decreased by only 0.25 with 2 d.f. If the autocorrelation is also allowed to be different, we have

$$E[Y_{it}|y_{i,t-1}] = 0.9114y_{i,t-1} - 0.004037 + 0.1036z_{it}$$

for Europe and

$$E[Y_{it}|y_{i,t-1}] = 0.8479y_{i,t-1} - 0.2292 + 0.1507z_{it}$$

for the others, with deviance decreased by 0.07 on 1 d.f.. Thus, we find no difference between the regions for this model. These results are summarized in Table 4.3, along with those for serial correlation.

Table 4.3. Deviances for the serial correlation and state dependence models applied to the wage data of Table 4.1.

	Serial Correlation Deviance	State Dependence Deviance	d.f.
Null	322.9	322.9	88
Autocorrelation	127.1	127.1	87
Regional Autocorrelation	126.5	126.5	85
Productivity	104.0	122.5	86
Regional Productivity	95.6	122.3	84
Regional Autocorrelation	95.3	122.2	83

When state dependence is added to the model, the dependence of the wage level on productivity is greatly reduced as is the difference between the regions. However, the serial correlation model fits much better than the state dependence model. It appears that, when a disturbance in the wage level occurs, this only influences the wage level for the next year, not continuing to exert an effect into the future, but we should remember that the initial conditions of the state dependence model are not really appropriate for these data.

4.1.3   ANTE-DEPENDENCE

The two autoregression models so far considered were both second order stationary; the variance and the covariances remain constant over time. In standard time series analysis, this is a necessary assumption, because it is usually difficult to obtain a reasonable estimate of changing variance and covariances for a single time series. In that context, various gymnastics are performed, for example, by transformations of the data, in order to fulfil these conditions. In repeated measurements studies, this stationarity is usually neither a reasonable nor a necessary assumption. For example, the variability of the responses in a growth curve situation often increases with the growing response.

One possibility, which we saw in Section 3.4 above, is to use a random coefficients model. This can be manipulated to provide changing variances on the main diagonal of the covariance matrix. However, such a solution is not at all satisfactory, because it means arbitrarily choosing certain coefficients of the polynomial location model to have random distributions and it usually really tells us little about the mechanism structuring the stochastic dependence. It is most often much more informative to model directly the effects of interest as some form of generalized autoregression. To this end, Gabriel (1962) has introduced a nonstationary autoregression model, which he calls ante-dependence and which can be fitted when replications of a time series are available. Missing values may be present, as long as they take the form of dropouts or censoring; once a response is missing for a unit, there are no further values for that unit.

As a generalization of an AR(M), responses having a multivariate normal distribution are said to be M^{th} order ante-dependent if Y_{t+h+1} is independent of Y_t, for all t with $h \geq M$, conditional on all of the intermediate responses. This results in a covariance matrix which no longer has the usual Toeplitz form, but whose inverse has zero elements for all responses further apart than M, and no other constraints except symmetry. Thus, it will have a structure of zero elements the same as that for the usual autoregression, but with arbitrary symmetric nonzero elements. The inverse covariance matrix for first order ante-dependence will be similar to Equation (4.4), with three nonzero diagonals, but with no structure except symmetry placed on these nonzero elements.

This model, with $(M+1)(R-M/2)$ parameters in the covariance matrix, lies in between the full multivariate model of Section 3.1, which has $R \times (R+1)/2$ parameters, and an AR(M), which has $M+1$. When $M = R-1$, the full multivariate model is obtained. The additional parameters, as compared to the usual autoregression model, allow the correlations among successive responses to be different, so that the response times need no longer be equally spaced, as long as the spacing is the same for all units.

Let us consider further the first order ante-dependence model. This has $2R - 1$ stochastic parameters, instead of the 2 parameters of an AR(1). If the initial response is assumed to have a stationary distribution, the parameters can be obtained from the estimates of the unconstrained multivariate normal covariance matrix. The main and first minor diagonals give the maximum likelihood estimates. The rest of the matrix is filled in (Byrne and Arnold, 1983) using

$$\sigma_{t,t-h} = \frac{\prod_{m=t-h+1}^{t} \sigma_{m,m-1}}{\prod_{m=t-h+1}^{t-1} \sigma_{mm}} \qquad h > 1 \qquad (4.14)$$

where σ_{hm} are the elements of Σ. With this procedure, the location model, $\boldsymbol{\beta}_i^T \mathbf{z}_{it}$, has the usual maximum likelihood estimates. This is an extension of the serial correlation model of Section 4.1.1.

As a conditional location model, with the initial response taken as fixed, the model might be expressed as

$$E[Y_{it}|y_{i,t-1}] = \rho_t y_{i,t-1} + \boldsymbol{\beta}_i^T \mathbf{z}_{it} \qquad (4.15)$$

where the nonconstant variance of Y_{it} is ξ_t, yielding $2R - 2$ variance/covariance parameters, because y_{i1} is fixed and does not have a variance. If a

Table 4.4.   Deviances for the profile ante-dependence model applied to the wage data of Table 4.1.

	Different Variance Over Time		Constant Variance Over Time	
	Deviance	d.f.	Deviance	d.f.
Profile	23.7	63	42.0	71
Regional Profile	-46.6	54	-20.1	63
Regional Autocorrelation	-96.2	45	-59.4	53
Regional Variance	-480.1	36	-346.3	52

separate mean is fitted at each time period, possibly different for various treatment groups, j,

$$E[Y_{ijt}|y_{ij,t-1}] = \rho_t y_{ij,t-1} + \mu_{jt} \tag{4.16}$$

the likelihood function factors and the model may be estimated as $R - 1$ separate parts (Kenward, 1987). The series of means, μ_{jt}, is called the profile of the jth group. However, as they stand, these two models are extensions of the state dependence model of Section 4.1.2, and are not directly comparable to the serial correlation model given in the previous paragraph. However, Equation (4.15) can easily be modified to provide the extension of the conditional serial correlation model of Equation (4.8). Patel (1991) gives the relationship between the conditional and the usual multivariate parameters.

Two simpler models, with less than the $2R - 1$ parameters, are possible: one might make all R variances identical, leaving R parameters, or make all $R - 1$ autocovariances identical, giving $R + 1$ parameters. The first can easily be estimated from Equation (4.15).

Let us apply this ante-dependence model to our wage data in Table 4.1. As in previous sections, we use the conditional form for these data. With the profile approach of Equation (4.16), we can fit four different models. The most complex has a different mean, variance, and autocorrelation for each time and for each region, with a total of 54 parameters. Simplifications involve having the same variance for each region, but different in time, the same autocorrelation for each region, but different in time, the same mean for each region, but different in time. The resulting deviances are given in Table 4.4. All of the other models give a much poorer fit than the most complex, with regional differences in variance between the European and non-European countries. This is not surprising, because the variability among the non-European countries is much greater. This model can be considered to be the saturated model for first order stochastic dependence, because no further parameters can be added. The *conditional* parameters are

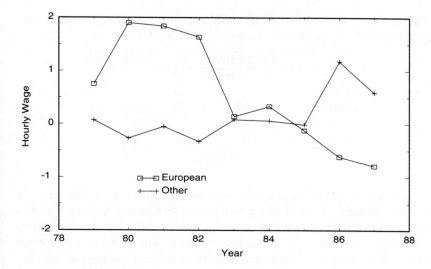

Fig. 4.4. Profiles of wage indices for the six European and four other countries with autocorrelation, from Table 4.1.

$$
\hat{\mu} = \begin{pmatrix} 0.75 & 1.90 & 1.84 & 1.63\ 0.14\ 0.33\ -0.12\ -0.62\ -0.79 \\ 0.07\ -0.27\ -0.05\ -0.33\ 0.08\ 0.06\ -0.01 & 1.18 & 0.59 \end{pmatrix}
$$

$$
\hat{\rho} = \begin{pmatrix} 1.08\ 0.92\ 0.70\ 0.77\ 0.97\ 0.90\ 1.06\ 1.41\ 1.30 \\ 1.05\ 1.12\ 1.11\ 1.09\ 1.05\ 1.02\ 1.01\ 0.96\ 1.03 \end{pmatrix}
$$

$$
\hat{\xi} = \begin{pmatrix} 0.02\ 0.10\ 0.04\ 0.23\ 0.06\ 0.01\ 0.19\ 0.37\ 0.03 \\ 0.06\ 0.02\ 0.00\ 0.19\ 0.02\ 0.01\ 0.05\ 1.48\ 0.41 \end{pmatrix}
$$

The profiles, $\hat{\mu}$, are plotted in Figure 4.4. They are, in fact, means for the filtered responses, i.e. mean wages minus $\hat{\rho}_{it}$ times the wage in the previous year. Differences in wages between successive years are generally decreasing and becoming negative in the European countries while they are increasing and becoming positive in the others.

If we compare our previous best model, in Table 4.3, the serial correlation model with regional differences in output, to this one, the difference in deviance of 575.7 with 48 d.f. shows that the earlier one was unacceptable.

One easily fitted simplification is to use Equation (4.15) with constant variance over time, but changing autocorrelation. The corresponding deviances are also given in Table 4.4. It is clear that a model with constant variance over time is also not acceptable. One possibility which might now be investigated is to increase the order of the ante-dependence, but, again, with attendant loss of observations.

Although it is not reasonable to fit Equation (4.14) to these data, because the wage level in 1978 obviously is nonstationary and depends on previous wage levels, let us do it to illustrate the procedure. Because the

Table 4.5. Deviances for the serial correlation and state dependence profile models applied to the wage data of Table 4.1.

	Productivity	d.f.	Profile	d.f.
Null	98.6	80		
Region	97.4	79		
Time	86.5	79	26.0	71
Time+Reg	75.8	77	-53.7	61
Multivariate	-95.5	51		

initial conditions are different, the deviances are not comparable with the previous models. We shall fit the null model with no explanatory variables and a common mean for all years and countries and, then, that for differences between Europe and the other countries. The deviance is decreased by 1.2. If we fit a model with productivity as the explanatory variable, but no differences between regions, the deviance is decreased by 12.1 with respect to the first model. When we allow the relationship between wage level and productivity to be different in the two regions, the deviance is smaller by 10.7 with respect to the previous model. These results are summarized in Table 4.5. This final model has covariance and correlation matrices

$$
\begin{pmatrix}
2.65 & 2.25 & 2.18 & 0.93 & 0.75 & 0.78 & 0.94 & 0.88 & -.07 & -.13 \\
0.83 & 2.77 & 2.69 & 1.15 & 0.92 & 0.97 & 1.16 & 1.09 & -.08 & -.17 \\
0.71 & 0.85 & 3.60 & 1.54 & 1.23 & 1.29 & 1.56 & 1.45 & -.11 & -.22 \\
0.53 & 0.63 & 0.74 & 1.20 & 0.95 & 1.00 & 1.21 & 1.12 & -.09 & -.17 \\
0.36 & 0.44 & 0.51 & 0.69 & 1.62 & 1.69 & 2.04 & 1.90 & -.15 & -.29 \\
0.34 & 0.41 & 0.48 & 0.65 & 0.94 & 2.00 & 2.41 & 2.25 & -.17 & -.34 \\
0.31 & 0.38 & 0.44 & 0.59 & 0.86 & 0.92 & 3.45 & 3.22 & -.25 & -.49 \\
0.30 & 0.36 & 0.43 & 0.58 & 0.84 & 0.89 & 0.97 & 3.19 & -.25 & -.49 \\
-.03 & -.04 & -.04 & -.06 & -.09 & -.09 & -.10 & -.10 & 1.75 & 3.46 \\
-.03 & -.03 & -.04 & -.05 & -.08 & -.08 & -.09 & -.09 & 0.87 & 8.96
\end{pmatrix}
$$

As compared to the full multivariate model, with dependence on productivity and differences between the two regions, the deviance is 171.3 larger, with a gain of 26 d.f.

In contrast to the serial correlation model of Section 4.1.1, here, we have nonstationarity for the variances. Both the variance and the auto-correlation are changing over time. For our data, the variance is fairly constant except for the large increase in 1987, already noted in the empirical covariance matrix for the null model above, while the correlations with the last two years are negative. This arises from the negative auto-correlation between the second and third last years. The poor fit of the

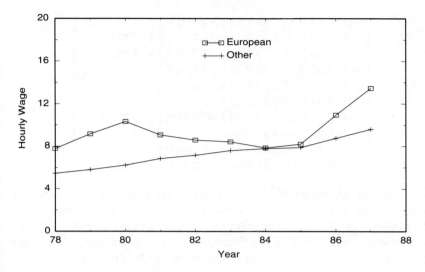

Fig. 4.5. Profiles of wage indices for the six European and four other countries, from Table 4.1.

location model is reflected in the stochastic structure of the covariance matrix. Note, also, that the variances in this model are much larger than the conditional variances given above, while the autocorrelations are smaller.

We can also look at the profile of changing wages for this model. Again, we fit a different mean for each year, no longer including the productivity variable. These profiles for the two regions are plotted in Figure 4.5. They simply reflect what we already observed in the plots of the raw data, but are very different from the conditional mean profiles of Figure 4.4, which involved the filtered data. The deviance is 79.6 smaller, with 10 d.f., for the model having two profiles than that with one profile for all countries. It is 129.5 smaller, with 16 more parameters, than our previous best ante-dependence model, which had the interaction between region and productivity. Although the profile model cannot be fitted for a general covariance matrix, because there are too many parameters and the matrix is not positive definite, we can compare our profile model to the full multivariate model with productivity, given above. The deviance is 41.8 more, with 10 fewer parameters. Note that, where the difference in number of parameters, instead of d.f., is given, the models are not nested. These results are also included in Table 4.5.

For this final model, whose profiles were plotted in Figure 4.5, the covariance and correlation matrices are

$$\begin{pmatrix}
6.32 & 6.74 & 6.87 & 6.48 & 6.64 & 6.89 & 6.91 & 7.04 & 7.13 & 7.76 \\
1.00 & 7.22 & 7.36 & 6.94 & 7.11 & 7.38 & 7.41 & 7.54 & 7.64 & 8.32 \\
0.99 & 0.99 & 7.64 & 7.21 & 7.38 & 7.66 & 7.69 & 7.83 & 7.93 & 8.64 \\
0.97 & 0.97 & 0.98 & 7.13 & 7.30 & 7.58 & 7.61 & 7.75 & 7.84 & 8.54 \\
0.94 & 0.95 & 0.95 & 0.98 & 7.82 & 8.12 & 8.15 & 8.30 & 8.40 & 9.15 \\
0.94 & 0.94 & 0.95 & 0.97 & 1.00 & 8.48 & 8.51 & 8.67 & 8.78 & 9.56 \\
0.94 & 0.94 & 0.95 & 0.97 & 1.00 & 1.00 & 8.57 & 8.73 & 8.83 & 9.62 \\
0.93 & 0.93 & 0.94 & 0.97 & 0.99 & 0.99 & 0.99 & 9.02 & 9.14 & 9.95 \\
0.89 & 0.89 & 0.90 & 0.92 & 0.94 & 0.94 & 0.94 & 0.95 & 10.24 & 11.16 \\
0.87 & 0.88 & 0.89 & 0.91 & 0.93 & 0.93 & 0.93 & 0.94 & 0.99 & 12.46
\end{pmatrix}$$

With this profile model, the variance among the countries is gradually increasing over time; the jump in the last year has disappeared. As suggested by Lindsey (1974b), one might want to construct a formal model to describe this change, analogous to a location model. The autocorrelation decreases very slowly with distance between the years. (The unit correlations are due to rounding.)

4.2   A general model

Let us now combine our models of the last section with those of the previous chapter to obtain one allowing for both heterogeneity and autocorrelation. Here, we shall assume, again, that the responses are equally spaced in time and that the process is stationary. The first assumption will be relaxed in the next section. Ante-dependence with heterogeneity will not be considered because such models will usually contain too many parameters to be useful.

We have multivariate normal responses for each of the N units,

$$\mathbf{Y} \sim \text{MVN}(\mathbf{M}, \mathbf{I}_N \otimes \boldsymbol{\Sigma}) \tag{4.17}$$

Suppose that the elements of $\mathbf{M}$, μ_{it}, have a suitable parametric specification such that

$$Y_{it}^* = Y_{it} - \mu_{it} \tag{4.18}$$

is a stationary, zero mean random process with the same structure for all units but independently for each. Examples of suitable location models include those used in the serial correlation and state dependence models given above. Then, the elements of $\boldsymbol{\Sigma}$ have the form

$$\sigma_{km} = \gamma(|t_k - t_m|) \tag{4.19}$$

where $\gamma(\cdot)$ is the autocovariance function of the process Y_{it}^*.

To specify a parametric form for $\gamma(\cdot)$, Diggle (1988) suggests that we must allow for

(1) some units giving consistently higher and some lower responses (the heterogeneity of Chapter 3),

(2) correlation between pairs of responses on the same unit decreasing with increased separation in time (the autocorrelation of this chapter), and

(3) variation among responses within a unit, even if remeasured instantaneously.

Then, we have

$$\gamma(u) = \delta + \xi + \psi^2 \quad u = 0$$
$$\gamma(u) = \delta + \xi\rho(u) \qquad u > 0 \tag{4.20}$$

where ψ^2 is the intra-unit variance, only estimable if there are repeated observations at the same time on a unit, δ is the additional component of variance across units, and ξ and $\rho(u)$ are the time series variance component and autocorrelation within a unit, the latter itself having some parametric form. Two commonly used forms for the autocorrelation function are given by

$$\rho(u) = \rho^u \tag{4.21}$$

with $|\rho| < 1$, so that it decreases as the distance in time, u, increases, and

$$\rho(u) = \exp(-\kappa u^\nu) \tag{4.22}$$

With $\rho = \exp(-\kappa)$, Equation (4.21) is just a special case of Equation (4.22), when $\nu = 1$. The latter has the advantage that it may be used in continuous time.

If the location model is allowed to be nonlinear, we have a general model which should be suitable in most situations where the repeated responses can be modelled with a normal distribution. We shall be concerned with various aspects of this model here and in the following sections of this chapter.

In this section, we shall look at the simplest case. The location model will be linear in the parameters and the observed responses will be equally spaced in time, with none missing. As an example, take the data of Grizzle and Allen (1969), reproduced in Table A2 of Appendix A, on the measurement of coronary sinus potassium in 36 nondescript dogs at two minute intervals after coronary occlusion under four treatment conditions, 1. control, 2. extrinsic cardiac denervation immediately prior to occlusion, 3. bilateral thoracic sympathectomy and stellectomy three weeks prior, and 4. extrinsic cardiac denervation three weeks prior. The average profiles for each treatment group, without taking into account stochastic dependence,

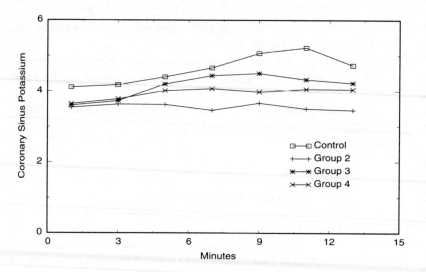

Fig. 4.6. Average response profiles of sinus potassium for the four treatments of 9 dogs each, from Table A2.

are given in Figure 4.6. Sinus potassium is slowly increasing after occlusion for all treatment groups except the last, with some evidence of nonlinearity.

First, we look at the empirical covariance matrix. With a different profile for each treatment, the covariance/correlation matrix is

$$
\begin{pmatrix}
0.201 & 0.153 & 0.151 & 0.173 & 0.178 & 0.174 & 0.165 \\
0.879 & 0.151 & 0.160 & 0.160 & 0.166 & 0.151 & 0.150 \\
0.559 & 0.685 & 0.362 & 0.320 & 0.237 & 0.173 & 0.188 \\
0.591 & 0.630 & 0.813 & 0.426 & 0.366 & 0.261 & 0.223 \\
0.592 & 0.636 & 0.585 & 0.834 & 0.452 & 0.357 & 0.309 \\
0.569 & 0.570 & 0.421 & 0.587 & 0.778 & 0.465 & 0.411 \\
0.541 & 0.567 & 0.458 & 0.503 & 0.677 & 0.885 & 0.463
\end{pmatrix}
$$

The correlation is slowly decreasing with increasing distance between responses, while the variance seems to be increasing over time. If we look at the raw data, we can see that the variability among dogs under a given treatment seems to be increasing over time, especially for the first two treatments.

Because we restrict ourselves to linear location models in this section, we use polynomials to represent the changing profile of response. As previously, we shall consider models based on an AR(1) serial correlation, using Equation (4.21), and compare them to the random effects and random coefficients (for linear time trend) models of the previous chapter, as well as combining the two.

Perhaps the best way to begin is by presenting the relative fit of the var-

Table 4.6.   Deviances for models fitted to the coronary data of Table A2.

	Indep.	Random Effect	Random Coefficient	AR(1)	Location Parameters
			Without Treatment Differences		
Mean	116.66	-38.18	-91.02	-160.61	1
Linear	105.79	-68.46	-107.10	-160.97	2
Quad	101.34	-81.77	-124.91	-172.40	3
Cubic	100.39	-84.68	-132.18	-177.25	4
6th Deg	100.20	-85.29	-133.08	-179.44	7
			With Treatment Differences		
			No Interaction		
Mean	38.69	-56.16	-107.43	-169.52	4
Linear	23.75	-86.44	-118.39	-177.91	5
Quad	17.57	-99.74	-137.86	-189.57	6
Cubic	16.25	-102.65	-142.00	-194.35	7
6th Deg	15.97	-103.27	-142.88	-196.47	10
			Interaction		
Linear	11.51	-113.28	-130.75	-183.55	8
Quad	2.53	-134.37	-158.18	-198.88	12
Cubic	-2.78	-147.50	-175.07	-217.07	16
6th Deg	-5.46	-154.33	-184.07	-233.10	28

ious models, as given in Table 4.6. Any model with stochastic dependence among responses on a unit, the dog, even one with constant mean, is better than the best model with independence. With one extra covariance parameter, the random coefficients model is consistently an improvement on the random effects model. However, the AR(1) model is always much better than any of the others, even although the random coefficients model always has one more parameter than either the AR(1) or the random effects models. An AR(2) does not improve the fit. It is interesting to note that a random effects model with a saturated location model does not fit as well as an AR(1) with a common mean for all responses, although the latter has 27 fewer parameters.

The more general models, with an AR(1) plus random effect, give virtually identical fits to the corresponding models with only an AR(1), so are not presented in the table. All of the stochastic variability among responses can be explained by the relationship over time of the AR(1) for each dog; there is no indication that some dogs give consistently higher or lower responses than others under the same conditions. In other words, the dogs used in this experiment were all very similar.

For the sixth degree polynomial with interaction, i.e. a different polynomial for each of the four treatments, which is equivalent to fitting a different mean for each time and treatment, the intra-dog variance, ξ, is estimated as 0.397, and the autocorrelation, ρ, as 0.801. This is reasonably consistent with the empirical covariance matrix given above. Skene and White (1992) propose that the dogs have a small number of distinct profiles over time and apply a latent class analysis to group them. Our analysis here seems to indicate that an AR(1) is superior.

Once we have established that the stochastic structure can be represented by an AR(1), we turn our attention to the location model. It is evident from Table 4.6 that an interaction between time and treatment is necessary. The average profiles are not parallel, but different under the four treatments. As Grizzle and Allen (1969) suggested, a cubic polynomial is required. The equations for the four treatments, with time centred on the mean of 7 minutes, are

$$
\begin{aligned}
\mu_{1t} &= 4.784 + .190t - .0102t^2 - .00388t^3 \\
&\quad\ (.189)\quad (.039)\quad\ (.0040)\quad\ \ (.00095) \\
\mu_{2t} &= 4.334 + .089t - .0176t^2 - .00102t^3 \\
\mu_{3t} &= 4.017 + .021t - .0048t^2 + .00033t^3 \\
\mu_{4t} &= 3.594 - .014t - .0026t^2 + .00019t^3
\end{aligned}
$$

The profiles for the last two treatments are very close to being straight lines. The standard errors for the parameters for the first treatment are also given in parentheses below the equation, the parameters for the other treatments having been calculated as differences from the first. The corresponding location models for independence among responses for a dog is

$$
\begin{aligned}
\mu_{1t} &= 4.758 + .189t - .0085t^2 - .00382t^3 \\
&\quad\ (.120)\quad (.053)\quad\ (.0057)\quad\ \ (.00177) \\
\mu_{2t} &= 4.369 + .087t - .0141t^2 - .00095t^3 \\
\mu_{3t} &= 4.025 + .016t - .0053t^2 + .00054t^3 \\
\mu_{4t} &= 3.585 - .009t - .0020t^2 + .00004t^3
\end{aligned}
$$

Although the parameter estimates are not very much different, those for the AR(1) model are generally more precisely estimated.

With its completely balanced, uniformly spaced design, this example is fairly typical of experimental growth studies. However, the units are not always so homogeneous, and often the random effects component is necessary. Nevertheless, Tiao and Tan (1966) demonstrate how important it is also to include autocorrelations, when necessary. In favourable cases, some scientific theory will also prescribe a mathematical form for the growth curve, so that the *ad hoc* polynomial approach will not be necessary.

4.3   The dynamic linear model

In general, we shall often wish to model responses which are unequally spaced in time, either because it was planned that way or because some responses are missing. When only heterogeneity across units is present, this creates no problem. However, time series methods will almost always be necessary and the problem becomes more complex. One possibility, when the series are nonstationary, is the ante-dependence model of Section 4.1.3. However, in the stationary situation, that model will have too many parameters and should not be used. The usual autoregression models, such as the AR(1) presented above, cannot directly be applied, because the autocorrelation parameter, ρ, measures the *constant* stochastic dependence among *equally spaced* observations. Thus, it cannot account for the different degrees of stochastic dependence among successive responses which are not the same distance apart.

What is required is a continuous time autoregression (Priestley, 1981, pp. 156-174; Harvey, 1989, pp. 479-501). An autocorrelation function for this model was already given in Equation (4.22) above. A continuous AR(1) process is defined by the linear differential equation

$$E[dY_{it}] = -\kappa_i y_{it} dt \qquad (4.23)$$

the mean of a normal distribution with variance ξdt. This gives an autocorrelation function

$$\rho_i(\Delta t) = e^{-\kappa_i |\Delta t|} \qquad (4.24)$$

as required. In the same way as for the discrete time AR(1), a location model, which, if linear, might be $\boldsymbol{\beta}_i^T \mathbf{z}_{it}$, will usually be included in the model to produce either a serial correlation or state dependence model. As in Section 4.2, heterogeneity should also be included in the covariance matrix.

In a general continuous time model such as this, there are a number of nonlinear parameters and estimation is usually difficult. Direct maximization of the likelihood function is possible (Jennrich and Schluchter, 1986; Jones and Ackerson, 1990). This approach is especially useful when the numbers of responses on a unit is not too large.

A more interesting approach is to consider the continuous autoregression as a special case of a dynamic linear model (see Section 2.8 and Jones and Ackerson, 1990). Estimation in this way gives the same results as by direct maximization of the likelihood function and is most suitable when the series of responses on a unit is fairly long. As well, it provides additional insight into the model.

The general observation or measurement equation (2.80) for responses following a normal distribution with an identity link can be written

$$E[Y_{it}] - \boldsymbol{\beta}_i^T \mathbf{z}_{it} = \boldsymbol{\lambda}_{it}^T \mathbf{v}_{it} \qquad (4.25)$$

with the state transition equation

$$E[\boldsymbol{\lambda}_{it}] = \mathbf{T}_{it}\boldsymbol{\lambda}_{i,t-1}$$

as in Equation (2.81).

Let us first look at the discrete time state space representation of a first order autoregression with a random effect. The measurement and state equations are

$$E[Y_{it}] - \boldsymbol{\beta}_i^T \mathbf{z}_{it} = \lambda_{1it} + \lambda_{2it} \qquad (4.26)$$

with conditional variance, ψ^2, fixed initial conditions, and

$$E[\lambda_{1it}] = \rho_i \lambda_{1i,t-1} \qquad (4.27)$$
$$E[\lambda_{2it}] = 0 \qquad (4.28)$$

with variances ξ and δ.

For a continuous AR(1), the measurement equation remains the same, while the state equation (Jones and Boadi-Boateng, 1991) for λ_{1it} is now assumed to be a continuous AR(1), so that

$$E[d\lambda_{1it}] = -\kappa_i \lambda_{1it} dt \qquad (4.29)$$

which is the mean of a normal distribution with variance ξdt. The state equation for the random effect is now

$$E[d\lambda_{2it}] = 0 \qquad (4.30)$$

with variance δdt. In this way, we have expressed our continuous time serial correlation model with a random effect for heterogeneity as a dynamic linear model.

The Kalman filter of Equations (2.84) to (2.86) is applied in order to obtain the estimates. Because only the mean and variance are required to define the normal distribution, these equations can be written in a simple closed form, and the procedure is relatively straightforward and efficient. The idea is to move forward in time from the first observation, estimating the expected value of each successive response before going on to the next. In this way, the total multivariate probability is built up as a product of conditional probabilities using Equation (2.69).

Consider a discrete time series. The one-step-ahead prediction or time update for $E[Y_{it}] - \boldsymbol{\beta}_i^T \mathbf{z}_{it}$ has mean

$$\hat{\boldsymbol{\lambda}}_{it|t-1} = \mathbf{T}_{it}\hat{\boldsymbol{\lambda}}_{i,t-1} \qquad (4.31)$$

and covariance matrix

$$\mathbf{A}_{it|t-1} = \mathbf{T}_{it}\mathbf{A}_{i,t-1}\mathbf{T}_{it}^T + \hat{\boldsymbol{\Xi}} \qquad (4.32)$$

where $\boldsymbol{\Xi}$ is a diagonal covariance matrix. In the random effects AR(1) above, this would contain the covariance elements ξ and δ. $\mathbf{A}_{i,t-1}$ is the prior covariance of the estimation error

$$\mathbf{A}_{it} = \mathrm{E}[(\boldsymbol{\lambda}_{it} - \hat{\boldsymbol{\lambda}}_{it})(\boldsymbol{\lambda}_{it} - \hat{\boldsymbol{\lambda}}_{it})^T] \tag{4.33}$$

The filtering or observation update, using the next response, y_{it}, has posterior mean

$$\hat{\boldsymbol{\lambda}}_{it} = \hat{\boldsymbol{\lambda}}_{it|t-1} + \mathbf{A}_{it|t-1}\mathbf{v}_{it}(y_{it} - \hat{\boldsymbol{\lambda}}_{it|t-1}^T\mathbf{v}_{it} - \boldsymbol{\beta}_i^T\mathbf{z}_{it})/c_{it} \tag{4.34}$$

and posterior covariance matrix

$$\mathbf{A}_{it} = \mathbf{A}_{it|t-1} - \mathbf{A}_{it|t-1}\mathbf{v}_{it}\mathbf{v}_{it}^T\mathbf{A}_{it|t-1}/c_{it} \tag{4.35}$$

where

$$c_{it} = \mathbf{v}_{it}^T\mathbf{A}_{it|t-1}\mathbf{v}_{it} + \psi^2 \tag{4.36}$$

As is usual for time series, the initial conditions must be chosen for $\boldsymbol{\lambda}_{i0}$ and for $\mathbf{A}_{i0}$.

To obtain the likelihood function, rewrite the observation equation (4.25) as

$$\mathrm{E}[Y_{it}] - \boldsymbol{\beta}_i^T\mathbf{z}_{it} = \hat{\boldsymbol{\lambda}}_{i,t-1}^T\mathbf{v}_{it} + (\boldsymbol{\lambda}_{it} - \hat{\boldsymbol{\lambda}}_{i,t-1})^T\mathbf{v}_{it} \tag{4.37}$$

from which the conditional distribution of $Y_{it}|\mathcal{F}_{i,t-1}$ has mean

$$\mathrm{E}[Y_{it}|\mathcal{F}_{i,t-1}] = \boldsymbol{\beta}_i^T\mathbf{z}_{it} + \hat{\boldsymbol{\lambda}}_{i,t-1}^T\mathbf{v}_{it} \tag{4.38}$$

with variance given by Equation (4.36). Then, the likelihood function for one unit is

$$\log(\mathrm{L}_i) = -\frac{1}{2}\sum_{t=1}^{R}\left[\log(2\pi c_{it}) + (y_{it} - \mathrm{E}[Y_{it}|\mathcal{F}_{i,t-1}])^2/c_{it}\right] \tag{4.39}$$

For a discrete serial correlation AR(1) with the first observation stationary, $\mathrm{E}[Y_{i1}|\mathcal{F}_0] = \boldsymbol{\beta}_i^T\mathbf{z}_{i1}$ and $c_{i1} = \xi/(1-\rho^2)$, while $\mathrm{E}[Y_{it}|\mathcal{F}_{i,t-1}] = \boldsymbol{\beta}_i^T\mathbf{z}_{it} + \rho y_{i,t-1}$ and $c_{it} = \xi$ for $t > 1$. These results are readily generalized to continuous time.

Let us now consider the application of this approach. The model is essentially the same as in Section 4.2, except that the times can be unequally spaced and different across the units. In the context of nonlinear growth curves, Heitjan (1991b) analyses data on the treatment of multiple sclerosis. In a randomized double-masked clinical trial, patients received either

1. two placebos (P), 2. real azathioprine (AZ) and a placebo, or 3. real doses of azathioprine and methylprednisolone (MP). Blood samples were drawn from the forty-eight patients one or more times prior to therapy, at initiation, and in weeks four, eight, and twelve plus every twelve weeks thereafter, and a measure of autoimmunity (AFCR) made. More details will be given below.

The data are given in Table A3 in Appendix A and plotted in Figure 4.7. The responses were measured at highly irregular periods in time. Except for the profiles of two or three patients with placebo, all three of the plots seem rather similar, although the placebo group does not approach as closely to the zero level of AFCR as do the two other groups. The responses are decreasing in time, except, perhaps, at the very end, and the average profiles appear to be nonlinear.

In this section, we shall fit a number of linear and quadratic polynomial models, with and without treatment effects, and with a number of covariance structures (random effects and/or AR(1)). For the polynomials, time has been centred at the mean of 501.5 days. We follow Heitjan in using a square root transformation on the AFCR responses. The deviances for these models are summarized in Table 4.7. For the moment, we are concerned only with the upper part of the table. We see that the random effects models fit better than the AR(1) models. The quadratic location model with treatment differences and interaction fits much better than the others. For this model, the AR(1) is not necessary. This indicates considerable variability among the patients, but little stochastic dependence among successive responses of a given patient. Of the polynomial curves, this model, with a deviance of 2876.0 and 11 parameters appears to be the best.

For our chosen model, the variance is $\hat{\xi} = 13.367$ and the additional variance component, and covariance, is $\hat{\delta} = 6.345$, giving a uniform intra-patient correlation of 0.32. The polynomial location models are

$$E[Y_{i1t}^{0.5}] = 14.53 - 0.0038t + 0.0000039t^2$$
$$E[Y_{i2t}^{0.5}] = 7.33 - 0.0096t + 0.000015t^2$$
$$E[Y_{i3t}^{0.5}] = 8.84 - 0.0095t + 0.000014t^2$$

The corresponding estimated profiles are plotted in Figure 4.8. The slight rise at the end of the time period is plausible in light of the plots in Figure 4.7. The double placebo group is much worse than the other two. This may be due, in part, to the two or three control patients having high profiles. The two groups receiving real medicine are fairly close, and show no significant difference. That the azathioprine and placebo group appears to be slightly better than the no placebo group may be due to one patient who had a higher profile in the latter group, as seen in Figure 4.7.

In the next section, we shall consider some other models for these data.

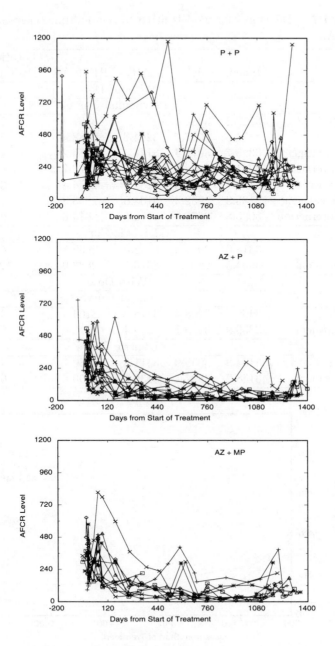

Fig. 4.7. Plots of AFCR level against time in days for the three treatments, from data in Appendix A3.

Table 4.7. Deviances for models fitted to the multiple sclerosis data of Table A3.

	Indep.	RE	AR	RE + AR	Location Parameters
			Without Dose		
			Polynomial Curve		
Null	3512.2	3434.1	–	3167.4	1
Linear	3374.6	3097.7	3284.7	2988.4	2
+ Treatment	3325.7	3074.4	3171.3	2986.2	4
+ Interaction	3201.8	3046.6	3159.4	2978.7	6
Quadratic	3306.1	2975.8	3251.1	2974.8	3
+ Treatment	3135.3	2952.3	3104.8	2943.8	5
+ Interaction	3080.8	2876.0	3066.6	2874.6	9
			Logistic Curve		
Null	3314.0	2984.6	3253.5	2969.8	4
Treatment	3094.5	2890.3	3076.7	2883.9	12
			With Dose		
			Linear Curve		
Null	3466.1	3208.4	3318.0	3164.4	3
Treatment	3300.9	3139.1	3225.4	3115.6	7
			Logistic Curve		
Null	3434.8	3055.9	3311.5	3030.7	4
Treatment	3158.2	2897.1	3135.1	2890.8	6

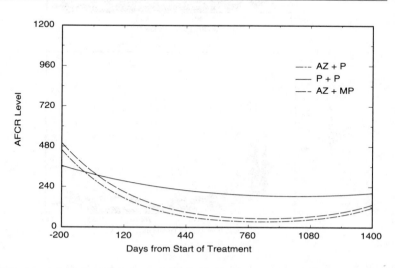

Fig. 4.8. Response profiles of AFCR for the three treatments, using a quadratic polynomial model, from Table A3.

4.4 Parametric growth curves

Up until now, in this chapter, we have only considered linear location models. These have the advantage of being relatively easy to fit. However, as we saw in Section 2.3.1, they are often only approximations to some more complex, but unknown functional form. One of the most common types of studies where nonlinear location models are usually necessary is in the analysis of growth curves. Sandland and McGilchrist (1979) provide a number of reasons why polynomials are unattractive for growth models:

(1) growth processes can undergo changes of phase which cannot be accommodated by polynomials,

(2) the stochastic structure of the model will be distorted if the polynomial is inappropriate, and

(3) polynomials cannot easily represent asymptotic behaviour of a growth curve.

They also emphasize that growth curve data are usually nonstationary and that polynomials do not represent biological mechanisms of growth. In the example to follow, the model will have up to 15 nonlinear parameters.

Nelder (1961 and 1962) has suggested a generalized form of the logistic growth curve, which Heitjan (1991a and b) has further generalized by adding autocorrelation and random effects. Suppose that the change in mean response obeys the following differential equation (Heitjan, 1991a and b):

$$\frac{d\mu_t}{dt} = \kappa_3 \mu_t [d(e^{\kappa_2}, \kappa_4) - d(\mu_t, \kappa_4)] \tag{4.40}$$

where

$$\begin{aligned} d(\mu, \kappa_4) &= \frac{\mu^{\kappa_4}-1}{\kappa_4} \quad \kappa_4 \neq 0 \\ &= \log(\mu) \quad \kappa_4 = 0 \end{aligned}$$

with initial condition $\mu_0 = \exp(\kappa_1)$ at t_0. The solution is

$$\begin{aligned} \mu_t &= e^{\kappa_2} \left[1 + \left(e^{(\kappa_2-\kappa_1)\kappa_4} - 1 \right) e^{-\kappa_3(t-t_0)e^{\kappa_2 \kappa_4}} \right]^{-\frac{1}{\kappa_4}} \quad \kappa_4 \neq 0 \\ &= e^{\kappa_2 + (\kappa_1-\kappa_2)e^{-\kappa_3(t-t_0)}} \quad \kappa_4 = 0 \end{aligned} \tag{4.41}$$

Then, $\kappa_1 = \log(\mu_0)$ is the initial condition and $\kappa_2 = \lim_{t\to\infty} \log(\mu_t)$, the asymptote. The parameters, κ_3 and κ_4 control the rate of growth. If $\kappa_3 < 0$ and $\kappa_2 > \kappa_1$ or $\kappa_3 > 0$ and $\kappa_2 < \kappa_1$, we have negative growth or decay. The parameter, κ_4, determines the type of the curve, varying from the Mitscherlich ($\kappa_4 = -1$) through the Gompertz ($\kappa_4 = 0$), and the logistic ($\kappa_4 = 1$) to the exponential ($\kappa_4 \to \infty$ and $d(e^{\kappa_2}, \kappa_4) \to$ constant).

Let us apply this generalized logistic location model to the multiple sclerosis data of Table A3. This will be a negative growth curve, because

Table 4.8. Location parameter estimates for the generalized logistic model fitted to the multiple sclerosis data of Table A3.

	P + P	AZ + P	AZ + MP
κ_{1j}	5.745	5.750	5.752
κ_{2j}	5.032	3.939	4.268
κ_{3j}	1.462	1.460	1.466
κ_{4j}	-1.563	-1.564	-1.561

the AFCR level is decreasing over time. We shall keep the same autocorrelation plus random effects covariance structure of the previous section. We fit models with identical generalized logistic curves for all three treatments and with completely different curves for each treatment. The covariance structure is kept the same for all treatments, as for the polynomial curve above. The deviances for the four possible covariance structures for each location model are given in the second part of Table 4.7 above. The treatment differences are significant. However, now, the autocorrelation cannot be set to zero. The variance is $\hat{\xi} = 4.33$, the additional variance component, and part of the covariance, is $\hat{\delta} = 6.69$, and the autocorrelation $\hat{\rho} = 0.82$. However, this model, with 4 more parameters, has a deviance larger by 7.9 than the quadratic model without autocorrelation, and, thus, is not as acceptable as that model. These parameters may be compared with those for the quadratic model with autocorrelation: 13.40, 21.82, and 0.51. The logistic model has higher autocorrelation, but the intra-patient correlation is about the same: 0.62 as compared to 0.61 for the quadratic model with autocorrelation and 0.32 for that without. The logistic and quadratic models without autocorrelation have almost identical estimates for ξ and δ.

The parameters for the three treatments are given in Table 4.8. The estimated parameter values in the curves for the two treatment groups are more similar than that for the placebo group. The parameter for the asymptote, κ_{2j} shows the main difference. The three location models are

$$g(\mathrm{E}[g^{-1}(Y_{i1t})]) = e^{5.03}\left[1 + \left(e^{1.16} - 1\right)e^{-1.46(t-t_0)e^{-7.87}}\right]^{0.64}$$

$$g(\mathrm{E}[g^{-1}(Y_{i2t})]) = e^{3.94}\left[1 + \left(e^{2.83} - 1\right)e^{-1.46(t-t_0)e^{-6.16}}\right]^{0.64}$$

$$g(\mathrm{E}[g^{-1}(Y_{i3t})]) = e^{4.27}\left[1 + \left(e^{2.32} - 1\right)e^{-1.47(t-t_0)e^{-6.87}}\right]^{0.64}$$

The function $g(\cdot)$, both the link function and the inverse of the transformation of the data, is here the square.

The curves are plotted in Figure 4.9. In contrast to the polynomial curves in Figure 4.8, these cannot rise at the end of the period, but instead reach an asymptote. This may be part of the explanation for the poorer

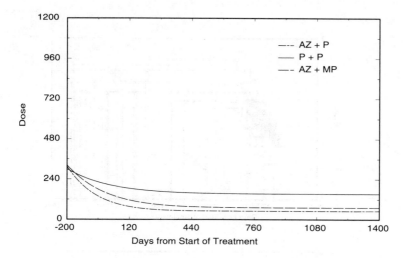

Fig. 4.9. Response profiles of AFCR for the three treatments, using a generalized logistic model, from Table A3.

fit of this model.

Often, in growth studies, as elsewhere, time-varying covariates are available. One way in which to modify Equation (4.41) to incorporate them is to have them influence the asymptote, κ_2 (Heitjan, 1991b). Suppose, for treatment, j, that this parameter is a function of time,

$$\kappa_{2jt} = \kappa_1 + \log\left(\frac{2}{1 + e^{\beta_j z_{jt}}}\right) \qquad (4.42)$$

Then, if $z_{jt} = 0$, the asymptote is constant at the initial condition. If $z_{jt} = z_j$ is constant in time, the mean grows or decays from its initial condition, e^{κ_1} to

$$\frac{2e^{\kappa_1}}{1 + e^{\beta_j z_j}}$$

a new equilibrium level. If $\beta_j = 0$, no growth or decay occurs for that treatment. If z_{jt} is a step function, the mean follows a piecewise generalized logistic curve. Then, if z_{jt} returns to zero, the mean goes back to the initial condition.

In our multiple sclerosis example, the dose of the medication for each patient was varied in time as a function of the patient's condition. As seen in Table A4 in Appendix A, the changes in dose occurred at irregular periods, which did not correspond to the times when the response, AFCR, was measured. The profiles of dose changes are plotted in Figure 4.10. The doses were all zero up until time zero, at which time a dose of one unit

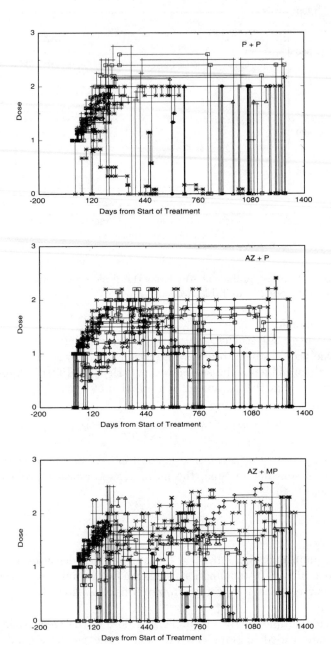

Fig. 4.10. Plots of dose level against time in days for the three treatments, from data in Appendix A4.

was given to all patients. This was rapidly increased to a maximum value of around two, but, subsequently, often drastically reduced and increased, following the condition of the patient. Patients in the complete placebo group were matched with patients in the other groups and their doses followed that of the paired patient, so that neither the patient nor the physician would know who was on placebo.

Dose appears to be following a stochastic process which may be interdependent with that for AFCR, an internal time-dependent covariate in the terminology of Kalbfleisch and Prentice (1980, pp. 122–127). However, here we follow Heitjan (1991b) in conditioning on the dose at a given time in modelling AFCR.

Our model has z_{ijt} as the strength of dose currently being administered, when the response is measured. Models with and without differences among groups were fitted using this model, with the four usual covariance structures. These give individual curves for each patient, instead of a mean curve for each group, because the dose profile varies among the patients. The curves have the general form of those in Figure 4.9, but with visible wobble, following the dose.

The deviances are given in the bottom section of Table 4.7. The model with differences in treatment has a consistently larger deviance than the corresponding model (quadratic or logistic) without the dose variable. However, the present model has fewer parameters than them. For example, it has a deviance 16.2 larger than the corresponding quadratic model, with only 3 fewer parameters, and 14.8 larger than the quadratic without autocorrelation, with 2 fewer parameters.

Heitjan (1991b) also fits a linear model in dose for comparison. Here, the response only depends on time through the dose being administered. The deviances are given in the second last section of Table 4.7. These are the poorest fitting models of all, although they have more parameters than many of the others.

These results may have several explanations.

- The generalized logistic model with dose may not be flexible enough, because it only allows the asymptote, κ_2, to vary among groups. The rate of decay may also be affected by the type of treatment.
- The dose may affect the rate of decay as well as or instead of the asymptote.
- Some other function of the dose, rather than the present strength, may be important. This might, for example, be a lagged value of the dose, given at some time in the past, or the total dose given up to the time the response is measured.
- Only the difference in treatment, and not the dose, may be important. As long as the patient is closely monitored, and the dose adjusted accordingly, the strength of the dose at a given time may not be

important.

Any of these possibilities would require special knowledge from the research workers involved, because the variability in the data would not permit us to choose among them.

The examination of this example shows that a number of different complex models can often be applied to the same data and give relatively comparable fits. Only theoretical considerations, from the branch of research involved, can, then, allow one to distinguish among them. However, each may provide its own illumination on the subject under study.

4.5   Crossover studies

When a time-varying covariate is an experimental variable, under the control of the research worker, we have a crossover or change-over design (Section 1.3). Because an excellent book on the subject (Jones and Kenward, 1989) already exists, I shall not go into this subject in great detail.

As we saw in Section 1.3, crossover studies involve applying two or more treatments to each unit at different times, and measuring the response after each. This response can depend on the time or period at which the treatment is applied and on the treatments which have already been applied, the sequence effect, as well as on the present treatment. In even more complex designs, a series of responses are measured over time after each treatment on each unit. We shall look at one such study here.

Ciminera and Wolfe (1953) give data on the comparison of NPH insulin mixtures. Two mixtures of insulin, the standard (A) and one containing 5% less protamine (B) were tested on rabbits. Two groups of eleven female rabbits were injected with the insulin at weekly intervals in the orders ABAB (group 1) and BABA (group 2). For each treatment, blood sugar level was measured at injection and at four post-injection times over six hours. The resulting responses are given in Table A5 of Appendix A and plotted separately for each treatment, but without distinguishing periods, in Figure 4.11. Blood sugar level goes down and then climbs back close to the pre-treatment level. The response for rabbit 3 in period 1 at 3 hours seems to be an outlier.

Many different analyses could be applied to these data. Here, we shall concentrate on the differences in profile of the responses under the two treatments. Note that, if the blood sugar level at time zero is included in the analysis as a response, the average differences in treatment will be attenuated, because the treatment will not yet have had an effect. If overall differences between treatments were of prime interest, it might be preferable instead to use it as a baseline covariate. However, here, for profiles, we use it as a response.

The model is similar to those we have been using. We allow for a random effect and for autocorrelation over the six hours. In a more complex model,

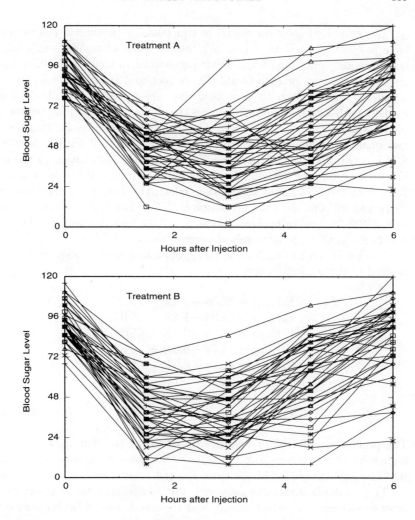

Fig. 4.11. Plots of blood sugar level against time in hours for the two treatments of rabbits, from data in Appendix A5.

a second level of autocorrelation could be introduced, across treatments over the several weeks, but I shall not do that here. If this were desired, Weissfeld and Kshirsagar (1992) show how to orthogonalize such data, in a way similar to that suggested in Section 4.1 for an AR(1) using the Cholesky decomposition.

The explanatory variables are the period, the treatment, and the group. Because there are only two treatments which alternate, the only sequence effect comes from which treatment is administered in the first period; this is covered by the group effect.

The saturated location model is equivalent to a quartic polynomial. Here, a quadratic will be fitted. We take as point of comparison a fairly simple model containing a quadratic polynomial in time, with differences in each parameter for each treatment, period, and group, but without interactions among them (in any case, they are not all identifiable). This model has 18 location parameters, plus the three covariance parameters. A model which has linear differences for period, no differences for treatment, and quadratic differences for group, fits almost as well with a deviance 8.72 larger on 6 d.f. On the other hand, complete elimination of the period effect raises the deviance by a further 32.38 with 6 d.f., and of group, 13.03 with 3 d.f. Without the random effect, the deviance is increased by 19.81 and without the autocorrelation by 6.31. The variance is estimated as $\hat{\xi} = 248.29$, the autocorrelation as $\hat{\rho} = 0.33$, and the random effect as $\hat{\delta} = 43.80$, so that the intra-rabbit correlation is 0.15.

The profile equations for the two groups and four periods, with time centred at three minutes, are

$$E[Y_{i11t}] = 38.56 - 3.23t + 4.57t^2$$
$$E[Y_{i21t}] = 32.05 - 1.48t + 5.64t^2$$
$$E[Y_{i12t}] = 41.26 - 2.84t + 4.57t^2$$
$$E[Y_{i22t}] = 34.75 - 0.09t + 5.64t^2$$
$$E[Y_{i13t}] = 41.10 - 0.97t + 4.57t^2$$
$$E[Y_{i23t}] = 34.55 + 0.77t + 5.64t^2$$
$$E[Y_{i14t}] = 55.69 - 0.70t + 4.57t^2$$
$$E[Y_{i24t}] = 49.18 + 1.05t + 5.64t^2$$

The profiles are plotted in Figure 4.12, where we see that they are much higher for the fourth period. This may indicate a cumulative effect and that the washout period of one week was not effective enough.

This example illustrates only one part of the analysis of these data. Interactions among period, group, and treatment should be investigated, as well as more complex stochastic structures. However, once period and sequence effects are allowed for, the modelling of crossover studies is similar to that for the other simple types of data already discussed in this chapter.

4.6   Spectral analysis

Up until now, in this chapter, we have studied how series of responses evolve over time. Sometimes, what is of interest is the frequency with which the series vary. However, to study and compare the periods of cyclic phenomena, fairly long series will be necessary. To do this, we use Fourier or spectral analysis. The readers not familiar with this approach may consult Bloomfield (1976) or Priestley (1981) for details.

The *spectral density function* can be written

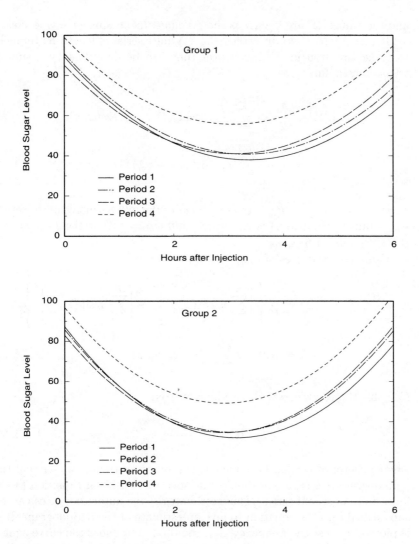

Fig. 4.12. Response profiles of blood sugar for the two groups in the four periods for the crossover study, from Table A5.

$$s(\omega) = \sum_{h=-\infty}^{\infty} \gamma(h)e^{-ih\omega}$$

$$= \gamma(0) + 2\sum_{h=1}^{\infty} \gamma(h)\cos(h\omega) \qquad (4.43)$$

where i is the root of minus one, $\gamma(h) = \text{cov}(Y_t, Y_{t-h})$ is the autocovariance, and the values, ω, are known as the *fundamental* or *Fourier frequencies*.

In any empirical study, we only have a finite series of length R. Suppose that the variation in the mean over time can be described by a sum of trigonometric functions

$$E[Y_t] = \frac{\beta_0}{2} + \sum_{m=1}^{[R/2]} [\beta_{1m} \cos(\omega_m t) + \beta_{2m} \sin(\omega_m t)]$$

$$= \frac{\beta_0}{2} + \sum_{m=1}^{[R/2]} v_m \cos(\omega_m t + \nu_m) \tag{4.44}$$

where $v_m = \sqrt{\beta_{1m}^2 + \beta_{2m}^2}$ is the *amplitude* of the mth harmonic of the series, $\nu_m = \tan^{-1}(-\beta_{2m}/\beta_{1m})$ is the *phase*, and $\omega_m = \frac{2\pi m}{R}$ is the *frequency* of the mth cycle with *period* $\frac{R}{m}$.

Using the estimates of $\hat{\beta}_{lm}$, we obtain

$$\frac{R}{4}(\hat{\beta}_{1m}^2 + \hat{\beta}_{2m}^2) = \frac{1}{R} \left\{ \left[\sum_{t=1}^{R} y_t \cos(\omega_m t) \right]^2 + \left[\sum_{t=1}^{R} y_t \sin(\omega_m t) \right]^2 \right\}$$

$$= \hat{\gamma}(0) + 2 \sum_{h=1}^{R-1} \hat{\gamma}(h) \cos(h\omega_m) \tag{4.45}$$

Call this the *periodogram*. It can be rewritten

$$I(\omega_m) = \frac{R v_m^2}{4\pi} \tag{4.46}$$

where a factor of $1/\pi$ has been included. From these results, we see that the periodogram is a representation of the spectral density of Equation (4.43) for a finite observed series. Thus, estimates of the autocovariances can be substituted into the equation to give an estimate of the periodogram. If it is plotted against the frequency, ω_m, the total area under the curve equals the variance of the series. This is why Equation (4.46) contains the extra factor. Thus, the periodogram is a decomposition of the variance of the series.

The periodogram, as it stands, however, is not a good estimate of the spectral density function. It usually tends to fluctuate wildly. To be of use, it must be smoothed, for example, with some filter or weighted average.

Because the periodogram involves variances, it has a χ^2 or gamma distribution and periodograms from different units can be compared using this distribution (Diggle, 1985; Lindsey, 1992, pp. 149–154).

Let us consider three series of 48 observations of the level of luteinizing protein in the blood of women (Diggle, 1989, p. 228). Blood samples were

Table 4.9.   Levels (units unknown) of luteinizing hormone in blood samples at 10 minute intervals. (Diggle, 1989)

0	5.5	2.4	4.3	240	4.8	2.3	4.5
10	4.5	2.4	4.6	250	5.5	2.0	4.6
20	5.1	2.4	4.7	260	5.1	2.0	5.8
30	5.5	2.2	4.1	270	5.2	2.9	5.0
40	5.7	2.1	4.1	280	5.0	2.9	5.1
50	5.1	1.5	5.2	290	4.0	2.7	4.5
60	4.3	2.3	5.0	300	3.7	2.7	4.2
70	4.8	2.3	4.4	310	4.8	2.3	6.0
80	5.6	2.5	4.2	320	5.9	2.6	5.6
90	5.9	2.0	5.1	330	5.5	2.4	5.4
100	6.0	1.9	5.1	340	4.9	1.8	5.0
110	5.1	1.7	4.7	350	4.4	1.7	4.4
120	5.2	2.2	4.4	360	4.7	1.5	4.6
130	4.4	1.8	3.9	370	4.2	1.4	5.7
140	5.5	3.2	5.4	380	5.5	2.1	5.2
150	5.4	3.2	5.9	390	4.9	3.3	5.0
160	4.1	2.7	4.2	400	4.8	3.5	4.4
170	4.4	2.2	4.1	410	4.5	3.5	5.7
180	4.7	2.2	4.1	420	4.9	3.1	5.7
190	4.6	1.9	3.6	430	4.9	2.6	4.8
200	6.0	1.9	3.1	440	4.5	2.1	3.4
210	5.6	1.8	4.8	450	4.2	3.4	5.5
220	5.1	2.7	5.1	460	4.9	3.0	5.5
230	4.7	3.0	5.1	470	5.9	2.9	5.6

Columns 1 & 5: Minutes
Columns 2 & 6: Cycle 1, late follicular phase
Columns 3 & 7: Cycle 2, early follicular phase
Columns 4 & 8: Cycle 2, late follicular phase

taken at ten minute intervals over a period of eight hours. One series was taken at the early follicular phase of the menstrual cycle and two at the successive cycles of the late follicular phase. All three series come from the same woman. The two at the same phase would be expected to be similar. The data are given in Table 4.9 and plotted in Figure 4.13. We can see a fairly regular cyclic effect. At the late follicular phase, the level of luteinizing protein is higher than at the early phase. This could be modelled with the procedures discussed in this chapter. However, what interests us here is the frequency of the cyclic phenomenon, and if it is the same at the two phases.

The estimated periodograms are plotted in Figure 4.14. These show higher values at the lower frequencies, but very little else because of the

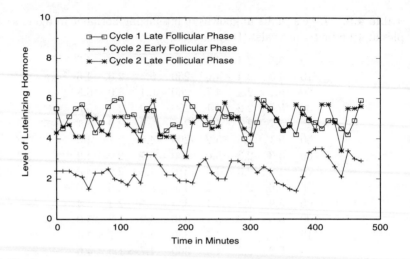

Fig. 4.13. Levels of luteinizing hormone in blood samples, from Table 4.9.

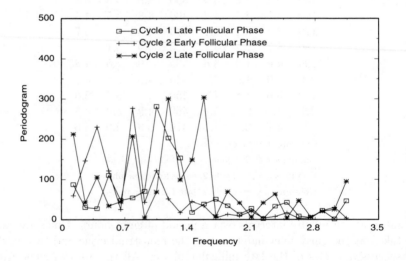

Fig. 4.14. Estimated periodograms of luteinizing hormone in blood samples, from Table 4.9.

irregularity of the curves. To obtain a useful estimate of the spectrum, we smooth them using weights of one third on each triplet of adjacent points. The resulting three estimated spectrograms are given in Figure 4.15, along with the approximate 0.15 likelihood intervals (regions in which the likelihood is at least 0.15 times that of the estimated curve). These regions are wide, and each one encompasses the other spectra, so that

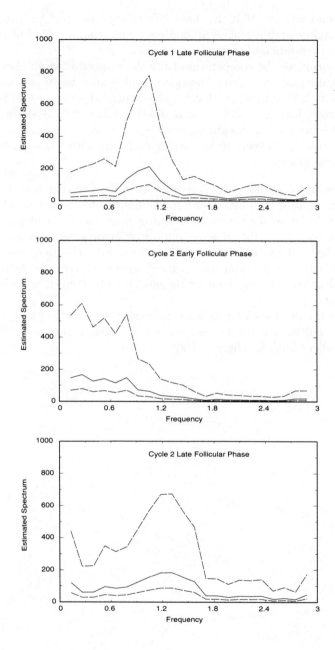

Fig. 4.15. Estimated spectra of luteinizing hormone in blood samples, from Table 4.9.

there is no evidence that the three spectra are different. All three spectra have maxima at frequencies around one, indicating a period of about 6.3 ($= 2\pi/1$) ten-minute intervals, or about an hour.

To formalize the comparison of the three spectra, we fit them with a gamma distribution. Using the raw periodograms, we obtain a deviance of 46.8 with 72 d.f., while that for the estimated spectra is 16.3 with 66 d.f. Again, there is no indication of difference among the spectra. In spite of the higher level of luteinizing protein in the blood in the late follicular phase, these levels seem to be varying cyclically with the same frequency in the two phases.

In concluding this part, it is important to note that models for longitudinal data should allow both for heterogeneity and for serial time effects. One, as in the sinus potassium in dogs and the multiple sclerosis data, or both, as in the crossover data, may be necessary, but both should be checked. The resulting intra-unit correlations and autocorrelations are directly interpretable and always much appreciated by the field worker. Equivalent remarks also hold for the models to be studied in the following two parts.

The standard work on variance components is Searle (1971a, Chapters 10–12). For normal theory time series type models, the reader will enjoy the excellent book by Diggle (1990).

Part III

Models for categorical data

5
Overdispersion

5.1 The multivariate multinomial distribution

If the response is an indicator of which of a number of events has occurred, we have *categorical* data. When no explanatory variables, including time, distinguish the responses on a unit, they will be aggregated as *counts*, a special case of categorical data. This will always occur if only one category is being observed; in some circumstances, such counts are called *rates*. In the simple case of counts, the data will often be collapsed into a contingency table. However, this does not mean that all contingency tables involve repeated measurements. We only have repeated measurements if repeated events are observed on the same units. Thus, I use *counts* to refer to repeated measurements of one or more categories of event on the same unit and *frequencies* for responses on different units aggregated into categories.

The log linear model for count and frequency data, and its special case, the logistic for binary data, has a unique place in statistical modelling. A conditional log linear model, with one variable as response and the others as explanatory variables, and a multivariate model, with all variables as interdependent responses, give identical results for the relationships of dependence among a given set of variables, i.e. independently of the manner in which conditioning is imposed. Here, this means that we can use standard log linear methods to construct multivariate models.

Log linear and logistic models for categorical data, one important special case of generalized linear models (Section 2.2.3), are now well known. Over twenty-five specialized books are available treating such models, some of which are given in the bibliography. Hence, details of their construction will not be given here; the reader may like to consult, for example, Agresti (1990). GLIM is used to analyse many of the models in this and the following chapters; most of the exact procedures used are given in Lindsey (1989, 1992).

In this section, we shall consider general multivariate models for discrete data, before going on to overdispersion in Section 5.2, and, finally, to more general heterogeneity models in the last sections of the chapter.

Table 5.1. Diagnostic classification regarding multiple sclerosis by two neurologists for two populations. (Landis and Koch, 1977a)

	Winnipeg Neurologist's Diagnostic			
	Class	1	2	3	4
	Winnipeg Patients			
	1	38	5	0	1
	2	33	11	3	0
	3	10	14	5	6
New Orleans	4	3	7	3	10
Neurologist's	New Orleans Patients			
Diagnostic	1	5	3	0	0
	2	3	11	4	0
	3	2	13	3	4
	4	1	2	4	14

5.1.1 COUNT DATA

First, let us look at an example of repeated counts. One situation in which such data arises is in the study of observer agreement or interviewer variability. Analysis by standard log linear models is possible if there is a small, fixed number of observers and few repeated counts. An example where these conditions do not hold will be given in Section 5.3.1 below. Here, we shall look at data provided by Landis and Koch (1977a) from Westlund and Kurland, reproduced in Table 5.1, on two neurologists who were asked to classify a series of patients in two locations into four categories concerning multiple sclerosis. These categories were 1. certain case, 2. probable case, 3. possible case (50/50), and 4. doubtful, unlikely, or definitely not.

Here, there are two levels of repeated responses. The unit of observation is the neurologist, who repeats the classification for each location, the two subunits. However, for each group, the neurologist must make an observation for each of the patients in order to obtain the counts in each category, giving the patients as subsubunits within location. A particularity of agreement studies is that observations across units are made on the same subunits, creating the two levels of repetitions. In this example, we shall ignore the second level of nesting; it will be the subject of the following sections of this chapter. Thus, the unit is the neurologist and an event is the diagnosis of a patient. Because patients are only distinguished by diagnosis, they are aggregated as counts.

We are interested in the profile of classifications as it may change between the two groups of patients for each neurologist. This is a relatively simple three-dimensional contingency table to which log linear models may be applied. A first question is if the relationship between the classifications of the two neurologists is the same for the two groups of patients. The model for this no interaction case is

$$\text{WINN} * \text{NO} + (\text{WINN} + \text{NO}) * \text{GROUP}$$

in the Wilkinson and Rogers (1973) notation of Section 2.3.2. Here, WINN and NO are four-level factor variables indicating the classification of each neurologist, while GROUP is a two-level variable for the group. The model has a deviance of 6.64 with 7 d.f., which is already very acceptable, indicating that this relationship is the same for both groups.

Two degrees of freedom are taken to be lost in this and the following models, because of two pairs of zeroes in the table. The isolated zero in the $(1, 4)$ position for New Orleans patients does not create a problem for our unsaturated models. However, such a reduction is controversial, because it makes the number of degrees of freedom random.

The next question is if each neurologist classifies a group in the same way. If they both agree, either that the two groups are similar or both different, the conclusion is clear. However, if one neurologist classifies the two groups differently and the other classifies them the same way, we have a problem. One possibility is that the two groups are actually the same and that one of the neurologists used different criteria of classification in the two cases. Another is that the groups are different and a neurologist has normalized the classification within each set to obtain similar distributions for both. However, if more than two observers were available, there would be little chance of such a problem arising.

For our data, it happens that we are in this second, problematic, situation. If we suppose that the Winnipeg neurologist used the same classification for the two groups,

$$\text{WINN} * \text{NO} + \text{NO} * \text{GROUP}$$

the deviance rises to 27.63 with 10 d.f., while, for the New Orleans neurologist,

$$\text{WINN} * \text{NO} + \text{WINN} * \text{GROUP} \tag{5.1}$$

it is 6.96 with 10 d.f. These models indicate that the New Orleans neurologist has the same classification profile for both groups of patients, while the Winnipeg neurologist does not. The second model has a parameter vector (WINN·GROUP) of $(0.00, 1.79, 2.03, 2.09)$ for the classification, by the Winnipeg neurologist, of the New Orleans patients with respect to the Winnipeg ones. The Winnipeg neurologist classifies significantly fewer New Orleans than Winnipeg patients into the first class.

A further refinement which we might want to apply to these data is to ask if there is a pattern in the disagreement between the neurologists in their classifications. In our present model, the probabilities of all possible combinations of disagreement are different. We can make them all the same by constructing a model of independence

$$NO + WINN + GROUP + WINN \cdot GROUP$$

between the two classifications for all disagreements, i.e. only for all off-diagonal elements. This gives a deviance of 38.89 with 13 d.f. indicating that there must be something systematic in their disagreement. Because we saw that the Winnipeg neurologist puts fewer patients from New Orleans into the first class, let us give a different probability to the disagreement combination, class one for the Winnipeg neurologist and class two, for the same patient, for the New Orleans neurologist. The deviance is reduced to 17.96 with 12 d.f., showing that the two neurologists have about equal probability of all disagreements except this one combination, which occurs much more often. Modifying the probability of no other disagreement combination significantly reduces the deviance further. Thus, the final model is

$$NO + WINN * GROUP + D \tag{5.2}$$

where D is a factor variable indicating the one distinct off-diagonal position, diagnostic one for the Winnipeg neurologist and diagnostic two, for the same patient, for the New Orleans neurologist.

One thing which we have had to ignore in this example is that relationships among the classifications for individual patients may vary between the two neurologists. For these data, the information is not available; when there are only two observers, such relationships may often be ignored. However, one possibility is to use the ordering inherent in the diagnostic (Agresti, 1988). The simplest possibility is to use a linear scale to represent the ordinal variable. Thus, in the models above, we can replace the factor variable for diagnostic by a linear scale in the interactions with group. A simple model is readily found to be

$$NO + WINN + LNO \cdot LWINN + GROUP + GROUP \cdot LWINN \tag{5.3}$$

where LWINN and LNO are the linear scale variables for the two neurologists. This model has a deviance of 30.29 with 22 d.f., as compared to the equivalent model of Equation (5.1) above, with 6.96 on 10 d.f. The interaction between the scales is estimated as 0.885 (s.e. 0.116), showing that they vary strongly together. For the Winnipeg neurologist, the interaction of diagnostic with group is 0.661 (s.e. 0.140), showing again that he classes New Orleans patients higher on the scale. However, if we again look only at disagreements, the model for equal probability of all disagreements except one,

Table 5.2.   Response to two drugs during four weeks after two different diagnoses. (Koch *et al.*, 1977)

		Week 1	N	N	N	N	A	A	A	A
		Week 2	N	N	A	A	N	N	A	A
		Week 4	N	A	N	A	N	A	N	A
Diagnosis	Treatment
Mild	Standard	16	13	9	3	14	4	15	6
Mild	New	31	0	6	0	22	2	9	0
Severe	Standard	2	2	8	9	9	15	27	28
Severe	New	7	2	5	2	31	5	32	6

N: Normal, A: Abnormal.

$$NO + WINN + GROUP + LWINN \cdot GROUP + D$$

equivalent to Equation (5.2) has deviance 25.64 with 14 d.f., which is not acceptable. The ordinal scale model can describe the overall symmetry among diagnostics parsimoniously, but it cannot pinpoint the disagreement between the neurologists. A more complete type of symmetry model will be applied to these data in the next section.

5.1.2   CATEGORICAL DATA

To perform multivariate analysis of categorical data, the individual events on each unit must be available, and distinguishable. This usually implies either that the events occur over time or that subunits have distinct characteristics. In the simple cases which we shall look at in this section, there are few events per unit and few possible different events. Then, all of the possible sequences of events or trajectories can easily be enumerated in a contingency table. These make up the multivariate response, with as many dimensions of the table as there are different events per unit. Units with the same sequence will not be distinguishable, except in so far as explanatory variables are available. The latter will constitute further dimensions of the contingency table.

Multivariate logistic models   An example will make the construction of multivariate models for categorical data clearer for the reader who is not too familiar with logistic and log linear models. Koch *et al.* (1977) give the hypothetical data on a longitudinal study, reproduced in Table 5.2, to compare a new drug to a standard one over a period of four weeks. The subjects are distinguished by the initial diagnosis, as well as by the treatments. Observations are made at weeks one, two, and four, when each subject is classified as normal or abnormal. This yields a trivariate binomial distribution with eight possible response sequence profiles. Thus, each unit has a count of from zero to three normal responses and a complementary number of abnormal ones. However, we also have the information on the order in which the events composing these counts occurred.

A multivariate logistic model for contingency tables can be fitted by standard log linear techniques. If we fit a multivariate model where the response sequence is independent of the diagnosis and treatment,

$$W1 * W2 * W4 + TREAT * DIAG$$

we obtain a deviance of 159.72 with 21 d.f. Adding the main effect of diagnosis

$$(W1 * W2 * W4) \cdot (1 + DIAG) + TREAT * DIAG$$

reduces this by 78.59 with a loss of 7 d.f., whereas the main effect of treatment

$$(W1 * W2 * W4) \cdot (1 + TREAT) + TREAT * DIAG$$

gives a reduction of 66.17 with the same d.f. A model with the main effects of both diagnosis and treatment

$$(W1 * W2 * W4) \cdot (1 + DIAG + TREAT) + TREAT * DIAG$$

has a deviance of 8.66 with 7 d.f., indicating an adequate fit or that the interaction between diagnosis and treatment is not necessary. The difference in response profile over time between the standard and new drugs is the same for the two initial diagnoses.

So far, in our model, we have a complete trivariate binomial distribution, with all interactions. It may be possible to simplify this further. The first thing to investigate is whether the three-way relationship among the observation times is necessary. If we eliminate it,

$$(W1 * W2 + W1 * W4 + W2 * W4) \cdot (1 + DIAG + TREAT) + TREAT * DIAG$$

we obtain a deviance of 14.25 with 12 d.f. This still gives us a model which is symmetric in the three successive responses. Next, we try eliminating all two-way relationships among the times,

$$(W1 + W2 + W4) \cdot (1 + DIAG + TREAT) + TREAT * DIAG$$

for a deviance of 19.50 with 19 d.f. Thus, for these data, the successive responses are independent of each other, conditional on the initial diagnosis, on the treatment, and on the time. One further simplification of the model is possible. As might be expected, the response at week one does not depend on the treatment. We end up with a model

$$W1 \cdot (1 + DIAG) + (W2 + W4) \cdot (1 + TREAT + DIAG) + DIAG * TREAT$$

having a deviance of 19.55 with 20 d.f.

The scientist with such data is usually primarily interested in the difference in response between the two drugs. But, first, let us look at the average response profile over time and the influence of different initial diagnoses. In the parameter values which follow, the first category is taken as basis: normal for response, mild for diagnosis, standard for drug. The parameter vector for the average response profile over time is $(-0.080, -0.354, -0.839)$, indicating that the proportion of abnormal responses is decreasing with time. The vector, over time, for difference in response profile of abnormalities with initial diagnosis is $(1.500, 1.299, 1.074)$ for the severe diagnosis. Those with a severe diagnosis consistently have more abnormalities, although the difference is decreasing with time. The parameter vector for average difference in treatment is $(-0.945, -2.00)$ for the second and fourth weeks, so that the new drug yields fewer abnormalities and this difference with respect to the standard is increasing over time. This is so for both diagnoses, because there is no interaction present in the model.

The model for these categorical data proved relatively simple because of the independence among successive responses. Only the margins of the trivariate binomial distribution were necessary. More complex relationships in time will be studied in the next chapter.

As an alternative, Rosner (1992b) proposes to divide the repeated binary responses into groups on a unit, usually in time, and fit separate correlations within each such set of responses, using a beta-binomial compound model (Section 5.2.1).

Multivariate log linear models When the repeated data have more than two categories, the cross-classified table can become very large. If there is not a large number of observations, the problem of small and zero frequencies will often be encountered and it may not be possible to fit the general multivariate models of this section. On the other hand, when there are few repetitions, it is possible to model certain forms of symmetry among the responses on the units.

Symmetry in the model is most clearly seen with only two responses per unit, although it is easily extended to more than two. Two common questions are if the overall or marginal distribution of responses is the same for each of the two responses on a unit

$$\pi_{h.} = \pi_{.h} \qquad (5.4)$$

and if differences between the two reponses are symmetric, with the same probability of change in each direction:

$$\pi_{hm} = \pi_{mh} \qquad (5.5)$$

Table 5.3. Eye tests of employees aged 30-39 in Royal Ordinance factories, 1943–46. (Stuart, 1955)

Right Eye Grade	Left Eye Grade				
	4	3	2	1	Total
	Female				
4	1520	266	124	66	1976
3	234	1512	432	78	2256
2	117	362	1772	205	2456
1	36	82	179	492	789
Total	1907	2222	2507	841	7477
	Male				
4	821	112	85	35	1053
3	116	494	145	27	782
2	72	151	583	87	893
1	43	34	106	331	514
Total	1052	791	919	480	3242

These are known, respectively, as the *marginal homogeneity* and *complete symmetry* models. In fact, complete symmetry encompasses marginal homogeneity, because the former implies that the margins are symmetric. A third model occurs when there are symmetric differences in the two responses, in so far as this is possible, given an observed asymmetric marginal distribution of each response. This is called *quasi-symmetry*. Thus, marginal homogeneity plus quasi-symmetry equals complete symmetry. Of these three, marginal homogeneity cannot be specified as a log linear model.

One common application of such models is to paired data, such as eyes or ears. Consider data from Stuart (1955), reproduced in Table 5.3, on the classification of vision for the two eyes of men and women employees of Royal Ordinance Factories in Britain in 1943–46. The questions, then, are if both eyes of a person generally have equally good vision and if left eye being poorer than right is equally as common as the reverse.

A positive answer to the first question would require that the two margins of the square table be fixed to have equal frequencies: marginal homogeneity. The deviance of this model is 15.67 with 6 d.f. However, this can be partitioned into 11.99 for the women and 3.68 for the men, each with 3 d.f. A glance at the data would already have shown that the marginal totals for the men are somewhat more symmetric than those for the women.

The second question leads to the complete symmetry model. We can fit this as a single factor variable having the form

$$\begin{pmatrix} - & 1 & 2 & 3 \\ 1 & - & 4 & 5 \\ 2 & 4 & - & 6 \\ 3 & 5 & 6 & - \end{pmatrix} \tag{5.6}$$

Note that the margins are not fitted in this model. Here, the deviance is 495.81 with 18 d.f. for a model with the same symmetry for the two sexes. It would not be surprising that this model is not acceptable, because it encompasses the marginal homogeneity which was just rejected for the women. However, the deviance is much too large just to be due to this factor. The model with different symmetry for each sex has a deviance of 24.02 with 12 d.f. Again, this can be decomposed, with 19.25 for the women and 4.77 for the men, each with 6 d.f. These values can be further decomposed using the quasi-symmetry model. For this last model, the deviance for the women is 7.26 and, for the men, 1.09, each with 3 d.f. Thus, the eyes of the men appear completely symmetric. In contrast, those of the women are asymmetric in that the right eye is, on average, better than the left, although, given this, there is no evidence of asymmetric differences in grades between pairs of eyes of a female subject.

Such symmetry models can also be applied to count data, such as that in Table 5.2 for inter-observer agreement (Darroch and McCloud, 1986). For those data, a quasi-symmetry model, the same for the two groups, but allowing for the difference in diagnostic of the two groups by the Winnipeg neurologist, is

$$NO + WINN + SYM + GROUP * WINN$$

where SYM is the symmetry factor variable which replaces the linear interaction in Equation (5.1). This model, with diagonal counts removed, has a deviance of 10.96 with 11 d.f. However, here it is not acceptable to use a linear scale for the ordinal variable in this interaction between the Winnipeg neurologist's diagnosis and group, as in Equation (5.3), because there is an increase in deviance of 7.69 with 2 d.f.

Marginal models The log linear model which we have been using is, in a certain sense, a conditional model; all dependencies are conditional on the appropriate marginal *frequencies*. For example, in the multivariate binomial distribution above, each binomial is conditional on the others. Another approach, suggested by McCullagh and Nelder (1989, pp. 219–235), is to model the multinomial distribution of binomial responses directly. Here, we model the marginal *probabilities*. Thus, a bivariate binomial is a four category multinomial, with three independent parameters. These will be the two marginal probabilities and the odds ratio, measuring dependence between the two binary responses.

Let π_{ikl} represent the four multinomial probabilities, which sum to one on subject i. The marginal probabilities are $\pi_{i.l} = \pi_{i1l} + \pi_{i2l}$ and $\pi_{ik.} = \pi_{ik1} + \pi_{ik2}$, the conditional probabilities $\pi_{ikl}^1 = \pi_{ikl}/\pi_{i.l}$ and $\pi_{ikl}^2 = \pi_{ikl}/\pi_{ik.}$, and the odds ratio $\pi_{i11}\pi_{i22}/(\pi_{i12}\pi_{i21})$. Then, the conditional bivariate binomial distribution has

$$\log\left(\frac{\pi_{i11}\pi_{i12}}{\pi_{i21}\pi_{i22}}\right) = \log\left(\frac{\pi_{i11}^1\pi_{i12}^1}{\pi_{i21}^1\pi_{i22}^1}\right)$$

$$= 2\log\left(\frac{\dot\pi_{i1.}}{\dot\pi_{i2.}}\right) \tag{5.7}$$

$$= \beta_{10} + \beta_{11}x_i$$

$$\log\left(\frac{\pi_{i11}\pi_{i21}}{\pi_{i12}\pi_{i22}}\right) = \log\left(\frac{\pi_{i11}^2\pi_{i21}^2}{\pi_{i12}^2\pi_{i22}^2}\right)$$

$$= 2\log\left(\frac{\dot\pi_{i.1}}{\dot\pi_{i.2}}\right) \tag{5.8}$$

$$= \beta_{20} + \beta_{21}x_i$$

$$\log\left(\frac{\pi_{i11}\pi_{i22}}{\pi_{i12}\pi_{i21}}\right) = \beta_{30} + \beta_{31}x_i \tag{5.9}$$

where $\dot\pi_{i.l}$ and $\dot\pi_{ik.}$ are the geometric means, while the marginal bivariate binomial distribution has

$$\log\left(\frac{\pi_{i11} + \pi_{i12}}{\pi_{i21} + \pi_{i22}}\right) = \log\left(\frac{\pi_{i1.}}{\pi_{i2.}}\right) \tag{5.10}$$

$$= \beta_{10} + \beta_{11}x_i$$

$$\log\left(\frac{\pi_{i11} + \pi_{i21}}{\pi_{i12} + \pi_{i22}}\right) = \log\left(\frac{\pi_{i.1}}{\pi_{i.2}}\right) \tag{5.11}$$

$$= \beta_{20} + \beta_{21}x_i$$

with the same Equation (5.9) for the odds ratio.

We see that, in the conditional model, the construction is symmetrical in the conditional binomial probabilities and the odds ratio; all probabilities are multiplied. Although the conditional probabilities must be adjusted to follow the linear regression across tables, indexed by i, this allows the marginal *frequencies*, upon which they are conditioned, to be held constant at their observed values. However, the corresponding marginal probabilities are not binomial.

On the other hand, in its construction, the marginal model is not symmetric; some probabilities are added and others multiplied. In this model, the marginal probabilities are binomial while the conditional ones are not. More important, the observed marginal frequencies are not held fixed, unless a saturated model is fitted, because those in individual tables, indexed by i, must be estimated so as to follow the linear regression model. Because

Table 5.4. Number of boys classified obese in 1977 and 1979 by age in 1977. (Woolson and Clarke, 1984)

1977	N	N	O	O
1979	N	O	N	O
5- 7	99	10	1	9
7- 9	165	16	17	27
9-11	163	18	14	24
11-13	126	11	17	27
13-15	105	9	8	21

N: Normal, O: Obese.

inference is not made conditional on the observed marginal frequencies, this limits the applicability of any empirical conclusions concerning dependence, drawn from this model, to tables with the same marginals.

The marginal approach may be useful when we wish to study each marginal response, taking into account the other(s), without conditioning on them. However, although the stochastic dependence among responses, measured in both cases by an odds ratio, may often be similar, when changes in this dependence are the centre of interest, it will usually be preferable to use the conditional model.

Let us look at data on the obesity of boys of different ages, measured at two points in time, given by Woolson and Clarke (1984) and reproduced in Table 5.4. First, we fit the conditional bivariate binomial distribution, used above, to the data. We shall only look at linear dependence on age. The independence model

$$O77 * O79 + AGEF$$

where O77 and O79 are binary indicator variables for the two years, AGEF is a five-category age variable, and AGEL will be the corresponding linear variable, has a deviance of 19.24 with 12 d.f. The model with linear dependence on age

$$O77 * O79 \cdot (1 + AGEL) + AGEF$$

has a deviance of 12.75 with 9 d.f. for a reduction of 6.49 with 3 d.f. However, neither the (odds ratio) stochastic dependence between the responses on the two years nor the marginal frequency for 1979 need vary with age. The resulting model

$$O77 * O79 + O77 \cdot AGEL + AGEF$$

has a deviance of 12.93 with 11 d.f. The parameters are given in Table 5.5.

The independence model for the marginal bivariate binomial is the same as above. The model with all three parameters depending linearly on age

Table 5.5. Parameter estimates for the conditional and marginal bivariate binomial distributions for the data of Table 5.4.

	Conditional	Marginal
O77	-3.480	-2.132
O77.AGEL	0.085	0.064
O79	-2.330	-1.425
O77.O79	2.969	2.977

has a deviance of 12.86 with 9 d.f., but the same two parameters can again be eliminated, yielding a deviance of 13.97 with 11 d.f. The parameters for this model are also given in Table 5.5. This marginal model fits slightly more poorly than the conditional model for these data. The interaction parameter estimates are very similar, although, as might be expected, those for the margins differ. In 1977, the proportion of obese boys was increasing with age. There is a strong relationship between obesity in the two years, but this relationship is the same for all ages.

For these data, the fits are very similar for the two models. This will not generally be the case. One of the two models may often fit much better than the other. When there is a strong reason not to want to condition on the marginal frequencies, the marginal bivariate binomial, or, more generally, multivariate multinomial, distribution will be useful. This will particularly occur when dependence of the marginal distributions on explanatory variables, and not the evolution of the relationship among responses, is of primary interest. However, the model has a number of inconveniences. Besides being relatively difficult to fit, because it is not a log linear model, it does not have the flexibility of application and interpretation of log linear models, as described at the beginning of this section.

5.2 Overdispersion in count data

If the events counted on a unit were independent, we would expect to be able to use a univariate multinomial distribution as the basis for our model. Two special cases are binary data, with the binomial distribution, and counts of one type of event, with the Poisson distribution. However, because events on a unit will not usually be independent, we must construct some model for the relationship among them.

As we saw in Section 2.2.2, the common distributions for count data, the binomial and Poisson, have strict relationships between the mean and the variance. Often, this is not sufficiently flexible to model possible variation in count data. Consider aggregated binomial counts on subunits and suppose that the probability is estimated as one half. As two extremes, this could mean either that each subunit has the same probability of the event or that, for some reason, one half of them have unit probability and the other half zero probability of the event occurring. Obviously, the variance is very different in these two cases.

Thus, when the data are aggregated so that only the counts are available, approaches analogous to those of Chapter 3 must be used. Often the underlying hypotheses of the model cannot be tested. Because a count is obtained from one unit, the relationship among the events producing that count may often be stronger than that for similar events on different units. The events will not follow a multinomial or Poisson distribution because of this stochastic dependence, but this will not be evident from the aggregated counts on one unit. It will only manifest itself as greater variability among counts on different units than would be expected for that distribution. This is called *overdispersion* or *extra-variability*. In an example which we shall look at below, counts of eggs hatching in tanks under supposedly the same experimental conditions show greater variability than would be expected under a binomial model, probably because of uncontrollable external factors.

Many forms of stochastic dependence among events on a unit, giving rise to heterogeneity among units, have been modelled, especially in the biological literature. These include *proneness*, *frailty*, or *apparent contagion* (units have constant but unequal probabilities of an event), *true contagion* (units initially have equal probabilities, but these change as a function of previous events, as through contagious contacts or learning), and *spell* (events occur in clusters) models. Only some of the more common models can be covered here. Many are indistinguishable for aggregated count data; we shall only really be able to model them in Chapters 6 and 8, when evolution over time is taken into account.

An additional complication is that missing explanatory variables can induce an effect identical to apparent contagion or proneness. If a missing variable can be assumed constant over all events on a unit, but differs among units, this will yield an apparent stochastic dependence among the events on each unit.

In the following two subsections, we look at some models for such overdispersion. We shall consider the special case of conjugate compound distributions (see Section 2.6.1) and then more general random effects models (see Section 2.6.2).

5.2.1   CONJUGATE COMPOUND MODELS

Binomial data   Let us first look at overdispersion in binomial count data. Consider a set of n_{ij} Bernoulli responses, y_{ijk}, on unit, i, under treatment, j, with constant probability, π_j, for all units under the same treatment. We only observe the total count, $y_{ij.} = \sum_k y_{ijk}$. Recall from Section 2.2.2 that the variance for binomial observations is defined in terms of the mean:

$$\begin{aligned} \mathrm{E}[Y_{ij.}] &= n_{ij}\pi_j \\ \mathrm{var}[Y_{ij.}] &= n_{ij}\pi_j(1 - \pi_j) \end{aligned} \tag{5.12}$$

If the variability among units is greater than can be accounted for by the available explanatory variables, we have overdispersion. This corresponds to reduced variability within a unit. But, remember that, due to aggregation, we have no direct information about relationships among events on the unit.

We can model this phenomenon in a variety of ways.

(1) Suppose that the variance is proportional to the variance function, instead of being equal to it:

$$\text{var}[Y_{ij.}] = \phi\tau_j^2$$
$$= n_{ij}\phi\pi_j(1 - \pi_j) \tag{5.13}$$

If the number of events on each unit is constant, $n_{ij} = n$, the constant correlation among them is

$$\rho = \frac{\phi - 1}{n - 1} \tag{5.14}$$

This scale parameter, ϕ, is often called the *heterogeneity factor*.

(2) As with normal distribution components of variance, let us now directly model the common correlation among the (unobserved) binary events on a unit for a given treatment (Prentice, 1986):

$$\rho_j = \frac{\Pr[Y_{ijk} = 1, Y_{ijl} = 1] - \pi_j^2}{\pi_j(1 - \pi_j)} \qquad k \neq l \tag{5.15}$$

Then, the resulting counts will have variance

$$\text{var}[Y_{ij.}] = n_{ij}\pi_j(1 - \pi_j)[1 + (n_{ij} - 1)\rho_j] \tag{5.16}$$

Because $-1/(n_{ij} - 1) < \rho_j < 1$, underdispersion, or negative correlation, is possible with this model, something which can occur in practice (Brooks, James, and Gray, 1991). If n_{ij} is constant, the model is identical to that of point (1).

Usually, the correlation is taken to be constant over treatments. However, in contrast to the case of the components of variance in normal distribution models, here, definition of the first two moments is not sufficient to determine fully a unique multivariate distribution.

(3) Suppose instead that we have a random effects model (Section 2.6.2) so that the probability varies over all units having the same treatment. $\Pr(Y_{ijk} = 1) = \lambda_{ij}$, with λ replacing π, because it is now random. We can interpret the model as being based on some unspecified compound distribution, with

$$\text{E}[\lambda_{ij}] = \pi_j \tag{5.17}$$

$$\text{var}[\lambda_{ij}] = \rho_j \pi_j (1 - \pi_j) \tag{5.18}$$

so that the correlation, ρ_j, is now restricted to be non-negative. That is, we have a binomial distribution, but where the binomial parameter varies over the units within each treatment according to some unspecified distribution. This is a quasi-likelihood situation (Williams, 1982a). The resulting marginal variance of $Y_{ij.}$ is the same as in Equation (5.16), but with the additional restriction on values of ρ_j that they be positive.

In point (2), we postulated a correlation among events on a unit. Here, we assume a changing probability across units under the same treatment, because the increased variability across units implies less variability within units (for $\rho_j > 0$).

(4) Suppose, as well, that this unspecified distribution for λ_{ij} is a beta distribution, with parameters, κ_j and υ_j, for randomly selected units. This is convenient for a number of reasons. The beta distribution is very flexible and can take many different forms. It can describe proportions, which are limited to values in the range zero to one. As well, it is the conjugate distribution (Section 2.6.1) to the binomial, so that we can obtain a closed form for the resulting compound distribution. And we have seen that its variance arises naturally from the correlation among responses.

The probability parameter, λ_{ij}, of the binomial distribution,

$$\Pr(y_{ij.}|\lambda_{ij}) = \binom{n_{ij}}{y_{ij.}} \lambda_{ij}^{y_{ij.}} (1 - \lambda_{ij})^{n_{ij} - y_{ij.}} \tag{5.19}$$

is taken to have a beta distribution, with density

$$p(\lambda_{ij}) = \frac{\lambda_{ij}^{\kappa_j - 1}(1 - \lambda_{ij})^{\upsilon_j - 1}}{B(\kappa_j, \upsilon_j)} \tag{5.20}$$

over the units i within treatment j, where $B(\cdot, \cdot)$ is the beta function. We have

$$\text{E}[\lambda_{ij}] = \pi_j$$
$$= \frac{\kappa_j}{\kappa_j + \upsilon_j} \tag{5.21}$$
$$\text{var}[\lambda_{ij}] = \rho_j \pi_j (1 - \pi_j)$$
$$= \frac{\rho_j \kappa_j \upsilon_j}{(\kappa_j + \upsilon_j)^2} \tag{5.22}$$

where

$$\rho_j = \frac{1}{\kappa_j + \upsilon_j + 1} \tag{5.23}$$

Then, integration over λ_{ij} yields the marginal distribution of $Y_{ij.}$,

$$\Pr(y_{ij.}) = \binom{n_{ij}}{y_{ij.}} \int \frac{\lambda_{ij}^{y_{ij.}}(1-\lambda_{ij})^{n_{ij}-y_{ij.}} \lambda_{ij}^{\kappa_j-1}(1-\lambda_{ij})^{\upsilon_j-1}}{B(\kappa_j, \upsilon_j)} d\lambda_{ij}$$

$$= \binom{n_{ij}}{y_{ij.}} \frac{B(\kappa_j + y_{ij.}, \upsilon_j + n_{ij} - y_{ij.})}{B(\kappa_j, \upsilon_j)} \tag{5.24}$$

Thus, the counts $Y_{ij.}$ have a compound beta-binomial distribution with the mean and variance as in point (2).

The unconstrained model is useful if the mean is allowed to be different for each treatment; the correlation can, then, also be different for each treatment. However, if a simpler location model, such as linear regression, is to be used, the model is less attractive, because any relationship among the response means places constraints on κ_j and υ_j, implying an unusual compounding distribution for the binomial probabilities. For example, with the logistic regression, $\log(\kappa_j/\upsilon_j)$ is constrained to vary with the linear predictor. In such cases, it is usually preferable to have a constant correlation over units. Then, the beta distribution parameters are $\kappa_j = (1 - \rho)\pi_j/\rho$ and $\upsilon_j = (1 - \rho)(1 - \pi_j)/\rho$, so that $\log(\kappa_j/\upsilon_j)$ does not depend on ρ (see Williams, 1975; Crowder, 1979).

(5) The binomial probability parameter, or its logit, may be considered to have some other specified distribution. Most often, the latter is taken to be normal, so that we have a compound normal-binomial distribution. However, here, this resulting compound distribution has no closed form and the likelihood function contains an integral, which must usually be evaluated by some numerical means, such as quadrature.

This rather *ad hoc* approach is most useful when we know that certain explanatory variables, which would explain the overdispersion, are not available. On the other hand, it may easily be extended to the case of nested counts, in analogy to the normal distribution random effects model. This approach is presented, in detail, in Section 5.2.2 below.

All of these models are population-averaged models. All may be used with any link function, including the logistic, probit, and complementary log log. In the following examples, the logistic link will be used.

As an example of overdispersion in binomial count data, consider data on the proportion of English sole (*Parophrys vetulus*) eggs hatching in sets of four tanks under a number of different conditions of salinity and temperature (Lindsey, 1975; see Table 5.6). This is a response surface experiment in which the optimum combination of these two factors for the fertility of the eggs was sought. Each salinity-temperature combination was replicated four times in different tanks. The problem is the uncontrollable differences among tanks at the same salinity and temperature. In spite of the similar responses among the four tanks, the resulting variation is greater than what would be expected for a binomial distribution.

We shall begin by fitting two location models to these data, a model with a different factor level for each combination of salinity and temperature and a quadratic response surface model. The former is the best possible location model which can be fitted with the explanatory variables available. The latter model provides a simpler description of the response surface which can be visualized and interpreted. It is especially useful when the goal of the experiment is to explore optimal conditions for the response.

For the standard binomial distribution, we obtain a deviance of 670.61 with 55 d.f. for the factor level model and 2314.8 with 66 d.f. for the response surface model. Neither model fits very well, although the response surface is much worse. The difference in deviance is 1644.2 with 11 d.f. The parameter values and standard errors are given in Table 5.7. Because the models fit so badly, the standard errors are greatly underestimated.

As we have seen, a simple solution is to estimate a scale parameter or heterogeneity factor in the variance from the mean deviance of the best binomial model, here the factor model. This is not a maximum likelihood estimate, but a moment estimate (see McCullagh and Nelder, 1989, pp. 124–128). For this model, the estimate is 12.19 (670.61/55). This obviously does not change the deviance or the parameters estimates. The standard errors will be increased by a factor which is the square root of this scale parameter. These standard errors are given in the third column in the table. Now, differences in deviance can no longer directly be compared. However, an *ad hoc* means of calibration is to take ratios of mean deviances as approximate F distributions. Thus, to compare the factor and response surface models, we have $(2314.8 - 670.6)/11/12.19 = 12.27$ with 11 and 55 d.f.

A more elaborate approach is to use the beta-binomial distribution, Equation (5.16) with ρ a constant intra-class correlation. Because one role of this parameter is to adjust for missing explanatory variables, only the estimate from the 'best' full model really measures intra-class correlation. For the factor level model, this is estimated as 0.0197 with a deviance of 45.22 on 54 d.f. We now fit all simpler models using this estimate obtained from the best model. Then, the difference in deviance for the response surface model is 129.07 with 11 d.f. In Table 5.7, we see that the parameter estimates have not changed very much from those for the binomial model. As well, the standard errors for the beta-binomial model are very similar to those that we obtained above using a heterogeneity factor.

Table 5.6. Effects of salinity and temperature on the proportion of eggs of English sole hatching. (Lindsey, 1975)

Salinity	Temperature	Hatch	Total	Hatch	Total
15	4	236	666	203	724
		183	764	212	723
15	8	600	656	697	747
		615	746	641	703
15	12	407	566	343	603
		365	560	302	394
25	4	203	717	177	782
		155	852	138	590
25	8	591	621	564	640
		714	754	532	570
25	12	475	622	465	645
		506	608	415	532
35	4	1	738	3	655
		10	742	3	743
35	8	526	616	419	467
		410	484	374	606
35	12	272	362	352	478
		392	590	382	459
10	10	303	681	329	710
		262	611	301	700
10	6	277	757	234	681
		263	647	287	801
40	10	387	450	389	553
		388	564	318	604
40	6	276	662	247	542
		248	527	149	591
20	10	351	391	559	650
		527	603	476	548
20	6	585	643	620	671
		437	497	667	771
30	10	447	491	462	530
		475	545	499	556
30	10	522	573	615	680
		539	581	517	561
30	6	563	666	600	704
		562	656	615	723

Table 5.7. Parameter estimates and standard errors for the response surface and factor level models for Table 5.6.

Factor Level Model					
Binomial Distribution			Beta-Binomial		
Estimate	s.e.[1]	s.e.[2]	Estimate	s.e.	Parameter
-0.232	0.0387	0.1353	-0.234	0.147	1
-0.664	0.0565	0.1972	-0.653	0.217	Salinity(2)
2.157	0.0749	0.2615	2.178	0.266	Salinity(3)
-0.983	0.0585	0.2044	-0.968	0.226	Salinity(4)
2.456	0.0634	0.2212	2.454	0.228	Salinity(5)
-4.893	0.2434	0.8499	-4.899	0.952	Salinity(6)
0.998	0.0602	0.2103	1.050	0.217	Salinity(7)
-0.310	0.0547	0.1910	-0.305	0.210	Temp(2)
6.485	0.2461	0.8593	6.539	0.958	Temp(3)
6.172	0.2460	0.8588	6.208	0.955	Temp(5)
-3.444	0.2569	0.8970	-3.499	1.000	S(2).T(3)
-4.579	0.2536	0.8855	-4.587	0.981	S(2).T(5)
0.520	0.1058	0.3694	0.499	0.387	S(3).T(2)
-2.701	0.2614	0.9128	-2.775	1.014	S(4).T(3)
-3.731	0.2545	0.8888	-3.774	0.986	S(4).T(5)
-0.169	0.0915	0.3196	-0.171	0.341	S(5).T(2)
-0.877	0.0832	0.2904	-0.923	0.303	S(7).T(2)

Response Surface Model					
Binomial Distribution			Beta-Binomial		
Estimate	s.e.[1]	s.e.[2]	Estimate	s.e.	Parameter
-13.03	0.1851	0.6464	-12.75	0.6811	1
2.403	0.0348	0.1214	2.340	0.1270	Temp.
0.403	0.0081	0.0285	0.395	0.0300	Salinity
-0.150	0.0021	0.0072	-0.145	0.0074	$Temp^2$
-0.010	0.0001	0.0005	-0.011	0.0005	$Salinity^2$
0.012	0.0005	0.0016	0.011	0.0017	Sal.Temp

[1]Standard errors for binomial distribution.

[2]Standard errors corrected by the variance estimate.

Because the response surface model fits much more poorly than the full factor level model and because we are interested in estimating the response *surface*, we proceed now to fit a nonlinear location model. Following Lindsey (1975), we estimate power transformations of the two explanatory variables. The deviance for the binomial model is reduced to 1280.6 with 64 d.f., half of that for the quadratic surface, but still almost twice the deviance of the factor level model. The estimated power transformations are 0.22 for the temperature and 0.16 for the salinity. This gives a location model

$$\log[\pi_i/(1 - \pi_i)] = -310.7 + 172.7x_{1i}^{0.22} + 215.8x_{2i}^{0.16} - 65.4x_{1i}^{0.44}$$
$$-77.6x_{2i}^{0.31} + 23.6x_{1i}^{0.22}x_{2i}^{0.16}$$

The resulting surface is plotted in Figure 5.1. We see that there is a flat plateau-like region where the proportion of eggs hatching is relatively constant. The standard errors of the linear parameters may be corrected in either of the two ways already used above. Again, both give essentially the same results.

Here, most of the n_i are of similar size. As we saw above, Equation (5.16) collapses to Equation (5.13) when the n_i are identical. In such cases as ours, we may conclude, with Cox (1983), that correcting the standard errors by a variance estimate from the mean deviance may be sufficient to account for binomial overdispersion. With much variation in the n_i, a more elaborate procedure, such as the beta-binomial distribution, may become necessary.

Poisson data   Models similar to those just described for binomial count data are also available for Poisson count data. The approach is identical to that given above. The conjugate distribution to the Poisson is the gamma distribution; their compounding yields a negative binomial distribution. The mean parameter, λ_{ij}, of the Poisson distribution,

$$\Pr(y_{ij}|\lambda_{ij}) = \frac{e^{-\lambda_{ij}}\lambda_{ij}^{y_{ij}}}{y_{ij}!} \tag{5.25}$$

is taken to have a gamma distribution with density

$$p(\lambda_{ij}) = \frac{v_j^{-\kappa_j}\lambda_{ij}^{\kappa_j-1}e^{-\frac{\lambda_{ij}}{v_j}}}{\Gamma(\kappa_j)} \tag{5.26}$$

over the units i in treatment j, where $\Gamma(\cdot)$ is the gamma function. We have

$$E[\lambda_{ij}] = \kappa_j v_j \tag{5.27}$$
$$\text{var}[\lambda_{ij}] = \kappa_j v_j^2 \tag{5.28}$$

Integration over λ_{ij} yields the marginal distribution of Y_{ij},

$$\Pr(y_{ij}) = \frac{v_j^{-\kappa_j}\int e^{-\lambda_{ij}}\lambda_{ij}^{y_{ij}}\lambda_{ij}^{\kappa_j-1}e^{-\frac{\lambda_{ij}}{v_j}}\,d\lambda_{ij}}{\Gamma(\kappa_j)y_{ij}!}$$
$$= \frac{\Gamma(y_{ij} + \kappa_j)}{y_{ij}!\Gamma(\kappa_j)}\left(\frac{1}{1 + v_j}\right)^{\kappa_j}\left(\frac{v_j}{1 + v_j}\right)^{y_{ij}} \tag{5.29}$$

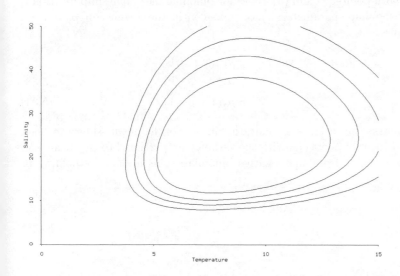

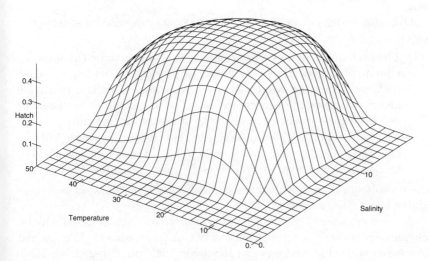

Fig. 5.1. Contours of the nonlinear binomial response surface for the fish egg data of Table 5.6 and the surface in three dimensions.

Then, the counts, Y_{ij}, have a negative binomial distribution so that

$$E[Y_{ij}] = \mu_j \tag{5.30}$$

$$= \kappa_j v_j \tag{5.31}$$

$$\text{var}[Y_{ij}] = \kappa_j v_j (1 + v_j) \tag{5.32}$$

As with points (2) and (3) above for binomial data, one simplification is to hold the shape parameter, κ_j, or equivalently, the correlation, $\rho_j = 1/\kappa_j$, constant:

$$\text{var}[Y_{ij}] = \kappa v_j(1 + v_j)$$
$$= \mu_j(1 + \mu_j/\kappa) \tag{5.33}$$
$$= \mu_j(1 + \rho\mu_j) \tag{5.34}$$

which makes the variance a quadratic function of the mean. However, here, the equivalent of point (1), with the variance proportional to the mean, can also be obtained from the negative binomial. This is done by holding v_j constant:

$$\text{var}[Y_{ij}] = \kappa_j v(1 + v)$$
$$= \mu_j(1 + v) \tag{5.35}$$
$$= \phi\tau_j^2$$

where $\phi = 1 + v$, because $\tau^2 = \mu$ for the Poisson distribution. As with binomial count data, these two simpler models are the ones most often used.

Over- (or under-) dispersion in Poisson counts can arise for at least two reasons.

(1) The probability of an event occurring to a unit may be the same for all units, but depend on the number of previous events happening to that unit. In other words, it varies over time. This is a contagion model, whereby the occurrence of a first event either increases or decreases the subsequent probability of events for that unit.

(2) The probability of an event may remain constant over time but not necessarily be the same for all units. This is the heterogeneous population model of frailty or proneness.

These two models are not distinguishable from a given data set only containing aggregated counts.

As an example of overdispersion in Poisson count data, consider data on the numbers of native species on the thirty Galapagos islands (Johnson and Raven, reproduced in Andrews and Herzberg, 1985, pp. 291–293; see Table 5.8). Here we have five variables, area, elevation, distance to the nearest island, distance to Santa Cruz, and area of the adjacent island, which may possibly be used to explain variations in numbers of native species. The six islands with missing observations will not be used in what follows, because the elevation will prove to be an important variable.

A first useful step is perhaps to examine the changes in deviance when each of these explanatory variables is added to the null model, which has a deviance of 582.26. These values are given in Table 5.9. We immediately

Table 5.8. Number of native species in the Galapagos Islands. (Andrews and Herzberg, 1985, pp. 291–293)

Number of Species	Area	Elevation	Nearest Island	Distance to Santa Cruz	Area of Adjacent Island
23	25.09	–	0.6	0.6	1.84
21	1.24	109	0.6	26.3	572.33
3	0.21	114	2.8	58.7	0.78
9	0.10	46	1.9	47.4	0.18
1	0.05	–	1.9	1.9	903.82
11	0.34	119	8.0	8.0	1.84
–	0.08	93	6.0	12.0	0.34
7	2.33	168	34.1	290.2	2.85
4	0.03	–	0.4	0.4	17.95
2	0.18	112	2.6	50.2	0.10
26	58.27	198	1.1	88.3	0.57
35	634.49	1494	4.3	95.3	4669.32
17	0.57	49	1.1	93.1	58.27
4	0.78	227	4.6	62.2	0.21
19	17.35	76	47.4	92.2	129.49
89	4669.32	1707	0.7	28.1	634.49
23	129.49	343	29.1	85.9	59.56
2	0.01	25	3.3	45.9	0.10
37	59.56	777	29.1	119.6	129.49
33	17.95	458	10.7	10.7	0.03
9	0.23	–	0.5	0.6	25.09
30	4.89	367	4.4	24.4	572.33
65	551.62	716	45.2	66.6	0.57
81	572.33	906	0.2	19.8	4.89
95	903.82	864	0.6	0.0	0.52
28	24.08	259	16.5	16.5	0.52
73	170.92	640	2.6	49.2	0.10
16	1.84	–	0.6	9.6	25.09
8	1.24	186	6.8	50.9	17.95
12	2.85	253	34.1	254.7	2.33

Missing observations are indicated by –.

see that two of the variables, area and elevation, may be of most use in describing the variation in numbers. On the other hand, the area of and the distance to the nearest island explain very little. If we inspect the plots (not given) of species numbers against area and against elevation, we may suspect that the relationships are nonlinear. If we add a quadratic term to the linear regression for these two variables, the deviance is reduced even more, as can also be seen from Table 5.9.

Because area and elevation stand out among the explanatory variables, it may be interesting to fit a response surface similar to that for binary

Table 5.9.   Differences in deviance for the Galapagos variables of Table 5.8.

Variable	Change in Deviance	
Area	146.69	
Area^2	360.60	213.91
Elevation	288.84	
Elevation^2	427.67	138.83
Nearest Island	5.90	
Santa Cruz	86.26	
Adjacent Island	3.68	

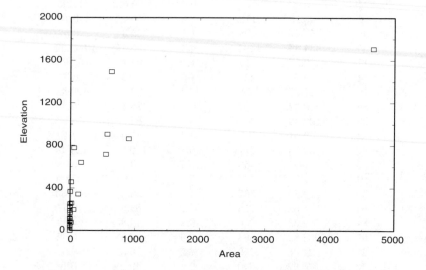

Fig. 5.2.   Distribution of the Galapagos Islands by area and elevation, from Table 5.8.

counts above. Note that, in contrast to that planned experiment, here, the observations are not well distributed over the space of the explanatory variables. In fact, as we see in Figure 5.2, there is only one island with an area greater than 1000; this influential observation will determine all results in that part of the explanatory variable space. However, we shall keep it in our models in order to explore what counts of species might be found if other intermediate-sized islands had existed. As well, there are no replications at the different points in the space. For these reasons, we cannot fit a factor level model as a base point of comparison, as we did above.

The response surface model will likely have overdispersion, because

Table 5.10. Parameter estimates and standard errors for the nonlinear response surface model from Table 5.8.

Poisson Distribution			Negative Binomial		
Estimate	s.e.[1]	s.e.[2]	Estimate	s.e.	Parameter
-18.96	4.091	6.423	-19.38	6.627	1
39.66	7.953	12.48	40.65	13.22	$\log(\text{Area})$
-0.04037	0.008627	0.01354	-0.04284	0.01541	Elevation
-18.19	3.817	5.992	-18.67	6.481	$\log^2(\text{Area})$
-2.943e-05	4.696e-06	7.372e-06	-2.930e-05	8.747e-06	Elevation^2
0.04283	0.008247	0.01295	0.04440	0.01485	E.log(Area)

[1]Standard errors for Poisson distribution.
[2]Standard errors corrected by the variance estimate.

other available variables, especially distance from Santa Cruz, can probably reduce the deviance further. The quadratic response surface in these two variables has a deviance of 95.88 with 18 d.f. If we estimate power transformations for these variables, we obtain a deviance of 44.36 with 16 d.f. The power transformation for area is estimated to be 0.066, which indicates a logarithmic transformation; that for elevation is 0.88, indicating that no transformation is probably necessary. The final equation is then

$$\log[\mu_i] = -18.96 + 39.66 \log(x_{1i}) - 0.0404 x_{2i} - 18.19 \log^2(x_{1i})$$
$$-0.0000029 x_{2i}^2 + 0.00428 \log(x_{1i}) x_{2i}$$

with a deviance of 46.69, where x_1 is area and x_2 elevation. The resulting surface is plotted in Figure 5.3. We see that, for this model, the number of different species increases with area but reaches a maximum for the islands of medium elevation. A relationship, such as this, between log counts and log area, is called an Arrhenius equation (Williamson, 1985).

Obviously, the model is not yet satisfactory. We could go in two possible directions: either we continue to add explanatory variables, because we have not used them all, or we allow for overdispersion as we did with the binomial data above. Because we are interested in overdispersion, let us first go in this latter direction, although this is not a generally recommended procedure. As we have seen, assuming that the variance is proportional to the mean gives a simple model to calculate, because all that it does is to change the standard errors of the estimates. Secondly, as stated above, the compounding distribution is the gamma, which yields a negative binomial distribution. The parameter estimates for both of these models are presented in Table 5.10. As for the binomial distribution, the parameter estimates do not change too much from the Poisson to the negative binomial distribution. As well, the model with variance proportional to the mean gives standard errors very similar to those for the negative binomial distribution. Both are about one and a half times the size of those for the

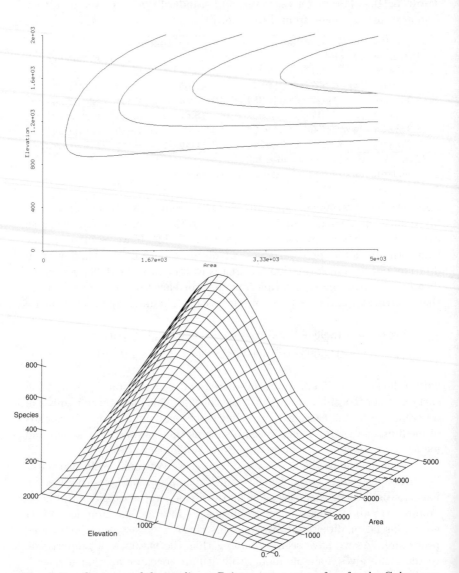

Fig. 5.3. Contours of the nonlinear Poisson response surface for the Galapagos Islands data of Table 5.8, and the surface in three dimensions.

Poisson distribution. The deviance for the negative binomial distribution is 18.92 with 15 d.f. and the constant shape parameter is estimated as 0.093. Note, however, that we have not used a 'best' model to estimate either the mean deviance or the shape parameter, because no full location model is available, so that this procedure is not really acceptable. The standard errors of the estimates are now somewhat *over*-estimated.

Thus, we should not stop here because we still have other variables not in the model. Let us go directly to a three variable nonlinear response surface by adding the distance from Santa Cruz, its square and all of the necessary interactions among the three transformed variables. We obtain a Poisson model with a deviance of 19.94 on 11 d.f., which is a considerable improvement on that with only two variables. The power transformations are estimated as -1.02 for area, 1.28 for elevation, and 1.60 for distance from Santa Cruz. That for area has changed considerably. These seem to suggest that we should now use the reciprocal of the area and the square for the distance to Santa Cruz, while not transforming the other variable. However, a model where only area is transformed is slightly better. This gives a model with a deviance of 21.43:

$$\log[\mu_i] = 2.59 + 0.43/x_{1i} + 0.033x_{2i} + 0.0045x_{3i}$$
$$-0.0010/x_{1i}^2 - 0.00000011x_{2i}^2 - 0.0000011x_{3i}^2$$
$$-0.0056x_{2i}/x_{1i} - 0.0044x_{3i}/x_{1i} - 0.0000012x_{2i}x_{3i}$$

We should now also fit the negative binomial, which would give a deviance of about 14 for either of the models just mentioned.

Four dimensional models are difficult to visualize graphically. One possibility is to make a series of graphs similar to Figure 5.3, for a number of different fixed values of the remaining variable, in this case, the distance from Santa Cruz (Lindsey, Alderdice, and Pienaar, 1970). For a series of three dimensional graphs, this is equivalent to contours in two dimensions, although one's imagination must be stretched a bit further to visualize the actual surface.

One should remember, as already pointed out, that all of these results depend very much on one island, the largest. However, this observation is not an outlier, given the goal which we set ourselves. We may hope that the models might indicate what sort of distribution of species could occur if intermediate-sized islands had existed in the region.

5.2.2 NORMAL COMPOUND MODELS

Although the beta and the gamma are the most commonly used compounding distributions for counts, being the conjugate distributions which yield analytically tractable distributions, they are by no means the only ones possible. The random effects model for normal distributions of Chapter 3 was also a compound distribution of two normal distributions, which resulted in a multivariate normal distribution, the normal distribution being its own conjugate. The same procedure may be applied here, although the result is not as simple, because the normal distribution is not conjugate.

Suppose that the Y_{ij} have either the binomial or the Poisson distribution. Then, a simple random effects model is

$$g(\mu_{ij}) = \lambda_j + \sum_{l=1}^{P} \beta_{il} x_{il} \qquad (5.36)$$

$$\lambda_j \sim N(0, \delta)$$

and the resulting compound distribution is

$$\Pr(y_{ij}) = \int \Pr(y_{ij}|\lambda_j) \phi(\frac{\lambda_j}{\sqrt{\delta}}) d\lambda_j \qquad (5.37)$$

where $\phi(\cdot)$ is the standard normal probability density function and the chosen distribution is represented by $\Pr(y_{ij}|\lambda_j)$ (Hinde, 1982; Brillinger and Preisler, 1983). Note that the variance of the count distribution, τ_j^2, and hence the intra-class correlation, $\rho_j = \delta/(\tau_j^2 + \delta)$, will be a function of μ_{ij}.

As in the previous section, this is a population-averaged model, with no explanatory variables distinguishing responses on a unit. However, here, it is not the mean of the count distribution which is given a normal distribution. Instead, it is the link-transformed mean which has the random distribution. Thus, for example, in logistic regression, the logit of the mean minus the linear predictor has a normal distribution. This is logical, because the probability parameter of the binomial distribution is limited to the range [0,1], whereas a logit can lie anywhere on the real line and is more suitable for a normal compounding distribution. On the other hand, we see yet another example where the normal (random effects) model, with its identity link, is deceptively simple.

In contrast to the compound distributions of the previous section, here it is not possible to obtain an explicit expression for the distribution from the integration. The result is a complex likelihood function which can only be evaluated by numerical integration. Because of this, it is equally easy to use any other distribution, instead of the normal, for compounding. The selection should be made on theoretical grounds, in terms of how the mean response is thought to vary in the population, although the normal distribution proves to be a robust all around choice.

Because the results of using compound normal distributions for overdispersion are similar to those obtained in the previous section, a simple example should be sufficient. Let us consider the data (Bissell, 1972) which Hinde (1982) originally used in introducing this model. There, the number of faults in rolls of fabric was related to the lengths of the rolls (Table 5.11). Thus, we use a simple linear Poisson regression for the counts. Hinde regresses the counts on the logarithm of the length, which, with the log link of the Poisson distribution, gives a linear relationship. Here, we shall use the raw lengths, because the fit is slightly better.

The data and the fitted curve are plotted in Figure 5.4. The location model is

Table 5.11. Number of faults in 32 rolls of fabric of various lengths. (Bissell, 1972)

Length	Faults	Length	Faults	Length	Faults
551	6	441	8	657	9
651	4	895	28	170	4
832	17	458	4	738	9
375	9	642	10	371	14
715	14	492	4	735	17
868	8	543	8	749	10
271	5	842	9	495	7
630	7	905	23	716	3
491	7	542	9	952	9
372	7	522	6	417	2
645	6	122	1		

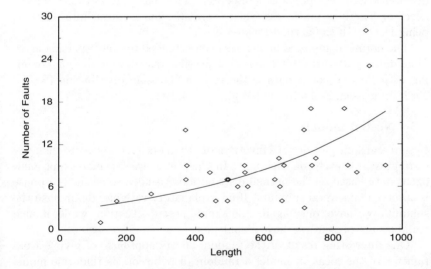

Fig. 5.4. Number of faults in rolls of fabric, from Table 5.11, with the simple linear Poisson regression curve.

$$\log(\mu_j) = 0.9718 + 0.001930x_j$$

and the deviance is 61.758 with 30 d.f., indicating overdispersion. The difference in deviance from the null model with zero slope is 39.17 with 1 d.f. From the plot, we can see that the variability in numbers of defaults seems to be increasing with the length of the roll, as would be expected for Poisson data. We also see that there is quite a bit of dispersion about the fitted line.

If we fit the negative binomial model of the previous section, we obtain a deviance of 28.957 with 29 d.f. and the constant shape parameter is estimated as 0.1208. The difference in deviance from the null model with the same shape parameter is 19.31. The location model is now

$$\log(\mu_j) = 1.003 + 0.001880x_j$$

Next, we fit the compound normal-Poisson model, which gives a deviance of 51.06 with 29 d.f. (which is not comparable, in value, to those given above) and a variance of 0.0984. The difference in deviance from the null model is 13.13 and the location model is

$$\log(\mu_j) = 1.031 + 0.001739x_j$$

For these data, all three models give very similar parameter estimates for the linear regression. This would be expected from our results for conjugate distributions in the previous subsection. The change in deviance for the effect of length is much less when overdispersion is taken into account, being least with the normal-Poisson model.

The normal compound model has been extended to overdispersion in ordinal data by Jansen (1990). Another possible distribution for compounding with Poisson count data is the inverse Gaussian distribution (Sichel, 1982; Jørgensen, Seshadri, and Whitmore, 1991).

5.3 Nested models

Nested and split plot designs involve more than one count or more than one distinguishable event on each unit. In Chapter 3, we saw models for analogous data based on the normal distribution. Usually, some such response is observed on several subunits. If no explanatory variables distinguish the subunits, we have, once again, the simple nested situation, while, if such variables are present, we have a split plot.

One interesting result of the random effects approach, of giving a parameter of the location model a random distribution, is that this model can be immediately extended to nested data and to data with changing covariates on each unit. Here, we have more than one count or distinguishable event for each unit so that, as in Equation (2.57), the compound distribution is multivariate

$$\Pr(y_{ij1} \ldots y_{ijR}) = \int \left[\prod_{k=1}^{R} \Pr(y_{ijk}|\lambda_j) \right] \phi(\frac{\lambda_j}{\sqrt{\delta}})d\lambda_j \qquad (5.38)$$

Because the integral contains a product of probabilities, care must be taken in the numerical integration to avoid underflow. This is now a subject-specific model, because we are distinguishing among responses on a unit.

Thus, the compound normal distributions of the previous section may

Table 5.12. Efficacy of two topical cream preparations in curing infection as tested in 8 clinics. (Beitler and Landis, 1985)

Drug		Control	
Favourable	Unfavourable	Favourable	Unfavourable
11	25	10	27
16	4	22	10
14	5	7	12
2	14	1	16
6	11	0	12
1	10	0	10
1	4	1	8
4	2	6	1

easily be extended to this case. The only change is that the mean of the normal compounding distribution rests constant for all subunits, that is for all responses on a unit, instead of being different for each observation.

5.3.1 COUNT DATA

If the design for the explanatory variables is balanced, nested and split plot data are usually tabulated in a multi-dimensional contingency table, where one or more dimensions will represent repeated responses on the same unit.

Let us look at an example for binomial counts which can be displayed in a three-way contingency table. We have a multicentre randomized clinical trial. The unit of observation is the clinic and the two treatments in the randomized block design are a drug, a cream for curing infection, and the control (Beitler and Landis, 1985; see Table 5.12). The binomial response is favourable and unfavourable. Interest centres on differences between the drug and the control, while the differences among clinics is a nuisance factor. Given the great variability in response among clinics, similar to that between neurologists in the problem of observer agreement of Section 5.1.1 above, there is reason to suspect that the responses are not independent within a given clinic.

A first reaction to such a table might be to analyse it in the standard way with a logistic or log linear model, looking at the significance of differences in response between control and drug treatments (Peto, 1987). In such an approach, the variability among clinics could be taken into account by the appropriate factor variable. Let us first try this. The deviance for a no-interaction logistic model with treatment and clinic as explanatory variables

$$\text{RESP} \cdot (1 + \text{TREAT} + \text{CLINIC}) + \text{TREAT} * \text{CLINIC}$$

is 9.75 with 7 d.f., indicating an acceptable fit; the response to treatment is the same in all clinics. This rises to 16.42 when difference in treatment is removed from the model,

$$RESP \cdot (1 + CLINIC) + TREAT * CLINIC$$

for a difference of 6.67 with 1 d.f., indicating that the effect of the drug is significantly different from the control. The estimated effect is -0.777 on the logit scale, with standard error, 0.306, showing that the controls had less probability of giving a favourable response. We might also add that the model with treatment differences, but without differences in clinics,

$$RESP \cdot (1 + TREAT) + TREAT * CLINIC$$

has a deviance of 90.96 with 14 d.f. This strongly indicates that the results from the eight clinics cannot be directly aggregated; we must keep clinic in our model to control for the variability among them.

In this first model, we have taken into account the exact differences among the clinics, but at the expense of a relatively large number of parameters: seven for the differences among the clinics. One important problem with this model is that the *number* of parameters is not fixed; it will increase with the number of observations, i.e. with each additional clinic included.

The idea of the random effects model is to allow for this heterogeneity among clinics by the compounding with the normal distribution. Then, the one parameter for the variance should have somewhat the same effect as the seven clinic contrasts above, giving the desired variability. The compound normal-binomial model gives an estimated treatment effect of -0.694 with standard error of 0.295. The variance is estimated as 1.117. The change in deviance when the treatment effect is eliminated is 5.6 with 1 d.f., which is still significant. Note that this model has 13 d.f. as compared to seven for the standard logistic model. When a large number of units is involved, the simplification can be very considerable as compared to a model where all differences among units are fitted.

We might wonder what would happen if we fit the overdispersion model of the previous section. With that model, we assume variability among all responses, that is both among clinics and between drug and control. Thus, the clinic is no longer taken as a homogenous unit. Rather, it is the treatment-clinic combination. This gives an estimated effect of -0.236 with standard error of 0.292 and variance of 1.371. The change in deviance for treatment effect is 0.5. We find very much less difference between the two treatments because most of it is absorbed by the variability allowed between them by the compound distribution. The conclusion from this incorrect approach would be that there is no significant difference indicated between drug and control.

Finally, we can compare these results to the (equally incorrect) beta-binomial model. Remember that, here, the probability parameter is assumed to have a beta distribution, instead of its logit having a normal distribution, as in the previous models. With this model, the estimated treatment effect is -0.525, with a standard error of 0.680, and the shape parameter is 0.364. The change in deviance for treatment effect is 0.6 and again would lead us to believe that there is no difference between control and drug. This is the danger of fitting an inappropriate model.

We may conclude that the standard logistic analysis and the compound normal-binomial random effects model give similar results for these data. The latter has a number of advantages:

- it is simpler in that many fewer parameters are required,
- the number of parameters does not increase with the number of units (here clinics),
- it allows inferences to be made over a population of units instead of just those observed, and
- it can allow for variability for which one has no explanatory variables or for which one wishes to allow without trying to explain, as is the case here with differences among clinics.

However, one should not ignore the possibility that certain distinctions among clinics, such as level of response under placebo, could be used as supplementary explanatory variables to explain the differences (Brand and Kragt, 1992).

Another approach to the problem of increasing numbers of parameters in such situations will be given in Section 5.4 below. It involves conditioning on these parameters to remove them from the likelihood function. Chamberlain (1980) gives a good overview of the different approaches. Miller and Landis (1991a) look at models for nested multinomial counts.

5.3.2   CATEGORICAL DATA

The same compound normal distributions may also be applied to categorical data, if the unaggregated events are available, usually as Bernoulli responses. They will give the same results as in Section 5.2.2, but it will be possible to go further and actually explore the nature of the stochastic dependence among events. Instead of pursuing this, let us look at a different model here.

For binary data, another distribution, besides the multivariate binomial, is of considerable interest for repeated measurements. As usual, let π_i be the common probability of all R binary events, Y_{ik}, on unit i and let ρ_i be the pairwise correlation among these events. Then, the joint distribution might be

$$\Pr(y_{i1}, \ldots, y_{iR}) = \frac{\prod_{k=0}^{y_{i.}-1}(\pi_i + k\nu_i) \prod_{k=0}^{R-y_{i.}-1}(1 - \pi_i + k\nu_i)}{\prod_{k=0}^{R-1}(1 + k\nu_i)} \quad (5.39)$$

where $\nu_i = \rho_i/(1 - \rho_i)$ and $y_{i.} = \sum_k y_{ik}$. The marginal distribution for one observation $(R = 1)$ is binomial

$$\Pr(y_{ik}) = \pi_i^{y_{ik}}(1 - \pi_i)^{1 - y_{ik}} \tag{5.40}$$

and the conditional distribution of any response, given all the others, is

$$\Pr(y_{ik}|y_{i1}, \ldots, y_{i,k-1}, y_{i,k+1}, \ldots, y_{iR}) =$$
$$\frac{[\pi_i + (y_{i.} - y_{ik})\nu_i]^{y_{ik}}[1 - \pi_i + (R - y_{i.} - 1 + y_{ik})\nu_i]^{1 - y_{ik}}}{1 + (R - 1)\nu_i} \tag{5.41}$$

which is also a binomial distribution, but now with probability $[\pi_i + (y_{i.} - y_{ik})\nu_i]/[1 + (R - 1)\nu_i]$. Thus, we have closed forms for the multivariate, marginal, and conditional distributions. The first is a beta-binomial distribution (Section 2.4.2) with $\pi_i = \kappa_i/(\kappa_i + \upsilon_i)$ and $\nu_i = 1/(\kappa_i + \upsilon_i)$, but without the combinatorial, $\binom{R}{y_{i.}}$, because we are distinguishing among different orders for the binary events. The latter two are binomial. When there are no explanatory variables distinguishing the observations on a unit, the total, $Y_{i.}$, also has a beta-binomial distribution (with the combinatorial) which we used in the Section 5.2.1 above.

When there are only two responses per unit, the bivariate distribution can be written

$$\Pr(Y_{i1} = 1, Y_{i2} = 1) = \frac{\pi_i(\pi_i + \nu_i)}{1 + \nu_i}$$
$$\Pr(Y_{i1} = 0, Y_{i2} = 0) = \frac{(1 - \pi_i)(1 - \pi_i + \nu_i)}{1 + \nu_i} \tag{5.42}$$
$$\Pr(Y_{i1} = 0, Y_{i2} = 1) = \Pr(Y_{i1} = 1, Y_{i2} = 0)$$
$$= \frac{\pi_i(1 - \pi_i)}{1 + \nu_i}$$

Rosner (1984, 1989; see also Neuhaus and Jewell, 1990a) uses this as a basis for constructing a model for paired binary data. Suppose that Equation (5.42) represents the probability for units with a baseline treatment or condition, such as a control group. Let us condition each of the pair of responses of a unit on the other in a logistic model

$$\log\left[\frac{\Pr(Y_{i1} = 1|Y_{i2} = y_{i2})}{\Pr(Y_{i1} = 0|Y_{i2} = y_{i2})}\right] = \beta_0 + \beta_1 y_{i2} + \boldsymbol{\beta}_2^T \mathbf{x}_{i1}$$
$$\log\left[\frac{\Pr(Y_{i2} = 1|Y_{i1} = y_{i1})}{\Pr(Y_{i2} = 0|Y_{i1} = y_{i1})}\right] = \beta_0 + \beta_1 y_{i1} + \boldsymbol{\beta}_2^T \mathbf{x}_{i2} \tag{5.43}$$

where $\mathbf{x}_{ik}$ can contain variables both within and between units. Now, the bivariate distribution, with arbitrary $\mathbf{x}_{ik}$, is

Table 5.13. Classification of hospitalized sibling pairs by schizophrenic diagnostic and by sex. (Cohen, 1976)

	Diagnosis			
Sex	00	01	10	11
MM	2	1	1	13
MF	4	1	3	5
FM	3	3	1	4
FF	15	8	1	6

0: not schizophrenic

1: schizophrenic

Elder sibling first

$$\Pr(y_{i1}, y_{i2} | \mathbf{x}_{ik}) = \frac{e^{\beta_0(y_{i1}+y_{i2})+\beta_1 y_{i1}y_{i2}+\beta_2^T(\mathbf{X}_{i1}y_{i1}+\mathbf{X}_{i2}y_{i2})}}{1 + e^{\beta_0+\beta_2^T\mathbf{X}_{i1}} + e^{\beta_0+\beta_2^T\mathbf{X}_{i2}} + e^{2\beta_0+\beta_1+\beta_2^T(\mathbf{X}_{i1}+\mathbf{X}_{i2})}} \quad (5.44)$$

For $\mathbf{x}_{ik} = 0$, this reduces to Equation (5.42) with $\beta_1 = \log[(\kappa + 1)(\upsilon + 1)/(\kappa\upsilon)]$. When $\beta_1 = 0$, it reduces to ordinary logistic regression.

This model may also be extended to data with more than two responses per unit. However, as it stands, the model is asymmetric in the values of the explanatory variables. It is only really useful in certain situations, where one category of explanatory variables, with zero code, is privileged. Prentice (1988) shows how to introduce a geometric mean of the responses to remove this restriction.

Cohen (1976) gives data, reproduced in Table 5.13, on hospitalized sibling pairs classified by whether they were diagnosed as being schizophrenic and by sex. The two explanatory variables are the sex of the person concerned and that of his or her sibling. Let us take male as the zero code.

The full model has parameter estimates $\hat{\beta}_0 = -0.0187$, $\hat{\beta}_1 = 1.899$, and $\hat{\beta}_2^T = (-0.0994, -1.307)$. Elimination of the sex of the person concerned increases the deviance by only 0.07, giving parameter estimates $\hat{\beta}_0 = -0.0187$, $\hat{\beta}_1 = 1.897$, and $\hat{\beta}_2 = -1.391$, while also removing the sex of the other sibling increases it further by 16.45, each with 1 d.f. We see that the diagnoses of the two siblings are closely positively related, while there is less chance of positive diagnosis if the other sibling is female. This is clear from Table 5.13, where pairs of males have high frequency of both being diagnosed schizophrenic and pairs of females do not. If female is taken as the zero code, results change very little, with the same model being acceptable. The estimates for this model are $\hat{\beta}_0 = -1.411$, $\hat{\beta}_1 = 1.901$, and $\hat{\beta}_2 = 1.389$. The conclusions are the same.

5.4 Item analysis

In educational testing, items in a test often have a binary, true/false, response. Each subject, the unit, replies to a series of questions making up the test. Responses will vary according to the ability of the subject and to the difficulty of each item. Rasch (1960; see also Hatzinger, 1989) introduced a binary data model whereby the probability of response y_{ik} of the subject i to item k is given by

$$\Pr(y_{ik}|\kappa_i) = \frac{e^{y_{ik}(\kappa_i - v_k)}}{1 + e^{\kappa_i - v_k}} \qquad (5.45)$$

This is called the *item characteristic curve* for item k. The data are represented by a $N \times R$ matrix of zeroes and ones.

Several additional assumptions are made:

(1) The probability of correctly answering items is strictly monotone increasing with ability.

(2) The items on a given test are a sample from a larger body of such items. All should measure one given latent trait, κ_i.

(3) Variation among item responses by a subject depend only on this latent trait: $\Pr(y_{ik}, y_{im}|\kappa_i) = \Pr(y_{ik}|\kappa_i)\Pr(y_{im}|\kappa_i)$.

(4) The marginal totals of the table, $y_{i.}$, contain all of the information for estimating the trait.

As with the clinical trials example in Section 5.3.1, the problem is that the number of subjects is variable. Thus, a standard log linear analysis of the frequencies of different combinations of responses is not really acceptable. Rasch proposed to use a conditional likelihood approach, because conditioning on the marginal totals, $y_{i.}$, eliminates κ_i from the likelihood function. The other approach is to assume some distribution for κ_i, something which we have already seen in previous sections.

Tjur (1982) has shown that the conditional model can be fitted as a log linear model. The margins for total score, with $R + 1$ possible values, and for each item are fitted in a $2^R \times (R + 1)$ table. In the Wilkinson and Rogers (1973) notation of Section 2.3.2, this is

$$R_1 + \cdots + R_R + \text{SCORE}$$

a model for quasi-independence. This table contains structural zeroes, because each combination of item responses only gives one score. Differences among groups can also be introduced into the model.

Let us consider data on responses to four questions from the arithmetic reasoning test on the Armed Services Vocational Aptitude Battery, with samples from white and black males and females (Mislevy, 1985, reproduced in Table 5.14). The simplest model is that the test results are the same for all four groups:

$$R_1 + R_2 + R_3 + R_4 + \text{SCORE} + \text{SEX} * \text{RACE}$$

Table 5.14. Frequencies of response to four questions on arithmetic from the Armed Services Vocational Aptitude Battery. (Mislevy, 1985)

	White		Black	
Response	M	F	M	F
1111	86	42	2	4
0111	1	7	3	0
1011	19	6	1	2
0011	2	2	3	3
1101	11	15	9	5
0101	3	5	5	5
1001	6	8	10	10
0001	5	8	5	8
1110	23	20	10	8
0110	6	11	4	6
1010	7	9	8	11
0010	12	14	15	7
1100	21	18	7	19
0100	16	20	16	14
1000	22	23	15	14
0000	23	20	27	29

This has a deviance of 188.00 with 53 d.f. If we add the dependence of the response on sex and race, without the interaction,

$$(R_1 + R_2 + R_3 + R_4 + \text{SCORE}) \cdot (1 + \text{SEX} + \text{RACE}) + \text{SEX} * \text{RACE}$$

the deviance is reduced to 51.08 with 39 d.f., which is an acceptable model. We may, next, ask if the distribution of the latent variable is the same in the different groups. This is equivalent to checking if the total score varies among groups, obtained by eliminating the interaction between groups and score. For sex, this is

$$(R_1 + R_2 + R_3 + R_4) \cdot (1 + \text{SEX} + \text{RACE}) + \text{SCORE} \cdot (1 + \text{RACE}) + \text{SEX} * \text{RACE}$$

with a deviance of 59.71 on 42 d.f. and for race

$$(R_1 + R_2 + R_3 + R_4) \cdot (1 + \text{SEX} + \text{RACE}) + \text{SCORE} \cdot (1 + \text{SEX}) + \text{SEX} * \text{RACE}$$

with a deviance of 85.61 also on 42 d.f. The measured trait varies with sex, and even more strongly with race. We can verify if the item characteristic curve is the same for all groups by removing the interaction between the item responses and groups

$$R_1 + R_2 + R_3 + R_4 + \text{SCORE} \cdot (1 + \text{SEX} + \text{RACE}) + \text{SEX} * \text{RACE}$$

Table 5.15.   Parameter estimates for the Rasch model applied to the data of Table 5.14.

Estimate	s.e.	Parameter
3.072	0.188	1
-0.008093	0.2205	Sex(2)
0.2748	0.2178	Race(2)
-0.02141	0.1602	Sex(2).Race(2)
0.9109	0.0903	R1
0.5869	0.0896	R2
0.1242	0.0906	R3
-0.2629	0.0932	R4
-0.8142	0.1815	Score(2)
-1.930	0.1716	Score(3)
-1.552	0.1574	Score(4)
0.000	aliased	Score(5)
0.03619	0.2440	Score(2).Sex(2)
0.2845	0.2493	Score(3).Sex(2)
-0.1868	0.2674	Score(4).Sex(2)
-0.6396	0.2837	Score(5).Sex(2)
-0.5082	0.2451	Score(2).Race(2)
-0.3367	0.2499	Score(3).Race(2)
-1.253	0.2780	Score(4).Race(2)
-3.328	0.4650	Score(5).Race(2)

We obtain a deviance of 55.50 with 45 d.f., showing little difference from the previously acceptable model. The items appear to have the same relative difficulty within all groups. Finally, we check if all items have the same difficulty,

$$\text{SCORE} \cdot (1 + \text{SEX} + \text{RACE}) + \text{SEX} * \text{RACE}$$

which has a deviance of 168.57 with 48 d.f. The parameter estimates for our acceptable final model are given in Table 5.15. The items are ordered in difficulty from easiest to most difficult. Blacks have high scores much more rarely than whites, and females less often than males.

This model has much wider application than simply to the analysis of the results of individual questions on a test. We have a multivariate binary response for a series of units, under a number of conditions, so that the responses on a unit are completely differentiated. This is similar to the situation in Section 5.1 above, except that the indexing of the units is a nuisance parameter, which we are not interested in, and the binary responses are not ordered (in time). In such situations, the model originally proposed by Rasch (1960) is especially interesting because it can easily be fitted by standard log linear model techniques. For example, Agresti (1992a) applies it to observer agreement data.

6
Longitudinal discrete data

6.1 Markov chains

Categorical and count data which are observed over time can be modelled in a number of ways. Suppose that, for categorical data, the multinomial probabilities of events vary over time (or space), for example, depending on the event(s) which occurred immediately previously, on the total number of previous such events which have already occurred, or on other time-varying covariates. Among all the possibilities, one of the simplest models is the *Markov chain*.

Longitudinal records of the events on the same units are required in order to be able to study such phenomena. If the events are actually recorded separately, and if their order is also available, so that the data are not aggregated as counts, models can be constructed directly for relationships among the successive responses. Often, units with the same trajectory can be aggregated, so that we have contingency tables. Another possibility is that we have longitudinal series of counts of the events on the units. This is most common when only one type of event is being observed, for example in mortality rates.

Conditional analysis of categorical longitudinal, or panel, data can yield Markov chains. These may be studied using logistic or log linear models; the theory and practice are standard and well known. However, the interpretation of Markov chain models can be problematic (Finch, 1982).

Suppose that the individual response, here an event, at a given time point depends only on the event at the immediately preceding point, the hypothesis of a *first order* Markov chain (Section 2.10). We have a square transition matrix of conditional probabilities, $\mathbf{T}_t$, of the events, which may vary in time. If the rows are the states at the previous time point and the columns are the present states, then the row probabilities sum to one. Pre-multiplying the vector, $\mathbf{n}_t$, of frequencies of units in the different states (the marginal frequencies) at a given time point, t, by the transpose of this matrix will give the vector for the next time period, $t + 1$:

$$\mathbf{n}_{t+1} = \mathbf{T}_t^T \mathbf{n}_t \tag{6.1}$$

Thus, the transition matrix represents the pattern of change.

Table 6.1.   June days with measurable precipitation (1) at Madison, Wisconsin, 1961–1971. (Klotz, 1973)

Year	
1961	10000 01101 01100 00010 01010 00000
1962	00110 00101 10000 01100 01000 00000
1963	00001 01110 00100 00010 00000 11000
1964	01000 00000 11011 01000 11000 00000
1965	10001 10000 00000 00001 01100 01000
1966	01100 11010 11001 00001 00000 11100
1967	00000 11011 11101 11010 00010 00110
1968	10000 00011 10011 00100 10111 11011
1969	11010 11000 11000 01100 00001 11010
1970	11000 00000 01000 11001 00000 10000
1971	10000 01000 10000 00111 01010 00000

Consider the successive June days in Madison, Wisconsin, with (coded 1) and without (coded 0) rain (Klotz, 1973) given in Table 6.1. Here, we might take June in Madison as the unit, with repetitions for eleven years, but other similar data might compare different cities. The estimated unconditional probability of precipitation is 0.32, while the probability of rain, given rain the previous day, is 0.41 and given dry the previous day, 0.28. Thus, the transition matrix, if assumed constant over the month and years, is

$$\mathbf{T} = \begin{pmatrix} 0.72 & 0.28 \\ 0.59 & 0.41 \end{pmatrix}$$

These probabilities can be obtained from a simple logistic model with a zero/one indicator of precipitation and a lagged variable for what happened the previous day. The difference in deviance between models with and without the lagged variable is 5.25 on 1 d.f., while adding differences among years further reduces it by 11.82 on 10 d.f. The latter confirms that the transition matrix seems to be constant over years. Introduction of lagged variables further back in time changes the deviance very little.

Because there are no time-varying covariates, the same results could be obtained by classifying the 330 observations in Table 6.1 into a $2 \times 2 \times 11$ contingency table of counts and applying a log linear or logistic model. Thus, although data aggregated into contingency tables will be used in most of the following examples, for economy of space, similar, if not identical, procedures can usually be applied to the individual event data.

In most situations, we shall wish to check for certain characteristics of a Markov chain, including: 1. the order of the chain, 2. stationarity, 3. reversibility (quasi-symmetry), and 4. equilibrium (marginal homogeneity). Special cases of Markov chains, such as random walks (diagonal symme-

Table 6.2.   Labour force participation of married women, 1967–1971.
(Heckman and Willis, 1977)

				1971	
1970	1969	1968	1967	Yes	No
Yes	Yes	Yes	Yes	426	38
No	Yes	Yes	Yes	16	47
Yes	No	Yes	Yes	11	2
No	No	Yes	Yes	12	28
Yes	Yes	No	Yes	21	7
No	Yes	No	Yes	0	9
Yes	No	No	Yes	8	3
No	No	No	Yes	5	43
Yes	Yes	Yes	No	73	11
No	Yes	Yes	No	7	17
Yes	No	Yes	No	9	3
No	No	Yes	No	5	24
Yes	Yes	No	No	54	16
No	Yes	No	No	6	28
Yes	No	No	No	36	24
No	No	No	No	35	559

try) and the mover–stayer model, may be of interest. With R repeated
measures, if the $R - 1$ order Markov chain, which includes all interactions,
is fitted, we have the complete multivariate model for discrete data, which
we looked at in Section 5.1 above.

With time ordering, heterogeneity may be modelled without necessarily
introducing the random effects described above. A factor variable for units
may be used, as in the clinic example of Section 5.3.1, or conditioning may
be applied, as in the Rasch model of Section 5.4.

6.1.1   ORDER

One factor determining the complexity of a Markov chain is its order, the
length of its dependence into the past. Such a chain can be modelled as
a logistic (if there are two states) or log linear model and the extent of
stochastic dependence on previous states determined.

Let us look at the history of labour force participation of married women
between 1967 and 1971 (Heckman and Willis, 1977), given in Table 6.2.
1583 women were observed over a period of five years. Because only 32
different trajectories are possible, many women will have the same career
and the results have been aggregated into a contingency table. Note that
these are frequencies and not counts, because each value in the table refers
to a number of different units (women), grouped together.

These are binary data, so that a logistic model can be used. The changes

Table 6.3. Deviances for increasing Markov order in Table 6.2.

Year	Deviance	d.f.
None	1123.5	15
1970	97.47	14
1969	49.80	12
1968	9.02	8
1967	0.00	0

in deviance when the dependence of 1971 participation is extended to previous years are presented in Table 6.3. We see that a third order chain is necessary, with dependence back to 1968.

In the models presented so far, we have only looked at the stochastic dependence of labour participation in 1971 on previous participation. However, implicitly, these models include the highest possible order stochastic dependence among the other years. It would be more consistent to have the same order of dependence for each year on the previous ones. To study this, we use a log linear, instead of a logistic model. For a third order chain, we fit

$$P68 * P69 * P70 * P71 + P67 * P68 * P69 * P70$$

for a second order chain,

$$P69 * P70 * P71 + P68 * P69 * P70 + P67 * P68 * P69$$

and so on. The deviance for the third, second, and first order chains are, respectively, 9.02 on 8 d.f., 62.67 on 16 d.f., and 210.23 on 22 d.f., showing that the third order chain is again required.

The models which we just fitted assume a different transition matrix, of given order, for each year. They are nonstationary models. We can also fit the equivalent stationary models by creating subtables, starting in each year and going back the required number of years. For example, with our third order model, we can create subtables for 1970, going back to 1967, and 1971, going back to 1968. Then, we suppose, for the moment, that the two tables are sufficiently similar so that we can combine them. However, we shall examine this stationarity assumption below in Section 6.1.2.

If we fit a second order model to this combined table, i.e. with no link between the most recent year and the earliest year,

$$T1 * T2 * T3 + T2 * T3 * T4$$

where T1 is the earliest year, we can determine if a stationary third order chain is necessary simultaneously for both 1970 and 1971. The resulting deviance is 46.95 on 4 d.f., showing that a second order chain is not sufficient. If it had been adequate, we could create, in the same way, three subtables for 1969, 1970, and 1971, each going back two years, and combine them, again provisionally assuming that they are sufficiently similar. If we, then, check for a stationary second order chain,

$$T1 * T2 + T2 * T3$$

we obtain a deviance of 140.53 with 2 d.f., showing, as would be expected, that a first order chain is not sufficient for these years. We shall look at a simplification of this model in Section 6.2.1 below.

Exactly the same procedures can be used if the Markov chain has more than two states.

6.1.2 STATIONARITY

A second factor which one might want to study in longitudinal categorical data is if the same pattern of change occurs between each set of periods. If the transition matrix is the same over each set, we have *stationarity* of that matrix, i.e. of the conditional distributions. We shall now look at how to check for the stationarity of a first order Markov chain, assuming the first order hypothesis to be true.

A first order chain is stationary if the relationship between all pairs of periods is the same, i.e. if the state at the end of the period only depends on that at the beginning of the period and not on the period itself. In our logistic or log linear model, we shall have three variables: the observed state at the beginning of any period (T1), that at the end of any period (T2), and the time periods themselves. To check for stationarity, we look for independence between the observed state at the end of the period and the time period; this relationship is omitted from the model, which leaves

$$T1 * (T2 + PERIODS)$$

Consider observations on the behaviour of male beavers over a fifteen hour period in the summer (Rugg and Buech, 1990), presented in Table 6.4. At any given time, an animal was classified into one of four states: feeding, swimming, in the lodge, or other. Assume that we are only interested in first order transitions. In any case, here we have no information on higher order transitions; such a table would be even more sparse than that which we have before us.

If we assume stationarity, i.e. that the fourteen subtables are identical, then, the common transition matrix is

$$\mathbf{T} = \begin{pmatrix} 0.904 & 0.075 & 0.009 & 0.012 \\ 0.100 & 0.866 & 0.012 & 0.021 \\ 0.000 & 0.059 & 0.933 & 0.008 \\ 0.047 & 0.094 & 0.034 & 0.825 \end{pmatrix}$$

Table 6.4. Hourly transition frequencies for the behaviour of male beavers in summer. (Rugg and Buech, 1990)

Feed	Swim	Lodge	Other	Feed	Swim	Lodge	Other
	1800-1900				1900-2000		
2	0	0	0	176	15	5	0
2	13	0	0	25	281	2	2
0	0	0	0	0	0	24	0
0	0	0	0	2	3	0	7
	2000-2100				2100-2200		
503	32	1	10	321	34	0	1
35	243	1	7	34	292	3	3
0	6	149	0	0	2	76	0
6	10	1	35	1	3	1	24
	2200-2300				2300-0000		
156	12	0	2	66	11	0	0
14	82	0	1	9	81	1	2
0	0	9	0	0	1	6	0
2	1	0	5	0	0	0	50
	0000-0100				0100-0200		
41	2	0	2	45	8	0	1
5	26	1	3	5	50	1	3
0	1	3	0	0	0	6	0
1	3	0	38	1	1	0	4
	0200-0300				0300-0400		
16	3	0	0	28	2	0	0
1	39	0	0	2	21	0	0
0	1	1	0	0	0	0	0
0	0	0	0	0	0	0	0
	0400-0500				0500-0600		
0	0	0	0	11	0	0	0
0	0	0	0	0	15	0	0
0	0	0	0	0	1	24	0
0	0	0	0	0	0	0	0
	0600-0700				0700-0900		
78	3	5	2	33	1	3	1
9	63	6	4	1	19	2	5
0	12	107	2	0	4	40	2
1	5	3	58	0	2	5	24

This is an average matrix for the whole observation period. We see that, most often, a beaver continues the same activity over two consecutive observation points. The most common transitions are between feeding and swimming and the rarest between the lodge and feeding.

If we now check for stationarity, we obtain a deviance of 236.72 with 116 d.f. (taking into account the zeroes in the table), indicating that the first order transition matrix is changing over time. Rugg and Buech (1990) suggest there are six distinct transition matrices for different periods of the day. The full table can be broken up in this way and stationarity examined for each section.

Let us now return to our data on women working. In Section 6.1.1, we assumed stationarity for the two third order and the three second order subtables. If we check the models

$$T1 * T2 * T3 * (T4 + PERIODS)$$

and

$$T1 * T2 * (T3 + PERIODS)$$

to see if the subtables can, in fact, be combined, we obtain deviances of 10.97 and 27.43, respectively, each with 8 d.f., giving little indication that the third order model is not stationary.

If stationarity is rejected for the transition matrix, one will usually want to go on to model the way in which it is changing over time. In most cases, this also can be done with standard log linear models; for one such example of a discontinuous change, see Lindsey (1992, pp. 29–33).

6.1.3 REVERSIBILITY AND EQUILIBRIUM

In a Markov chain, if the conditional probability of transition between events is the same in each direction, we have what is known as *reversibility*. In terms of the analysis of contingency tables using log linear models, this is the quasi-symmetry of Section 5.1.2. In the same way, if the margins are not changing over time, the process described by the Markov chain is in the *equilibrium state*; the marginal distribution is stationary. In terms of contingency tables, this was marginal homogeneity. If we combine the two, we have a completely symmetrical table. In Section 5.1.2, we used these models to look at heterogeneity in categorical data, applying them to examples for paired responses on eyes and for observer agreement. There, we did not have a time ordering among observations on a unit.

As we saw in Equation (5.5), a system showing both reversibility and equilibrium will have a completely symmetrical matrix in which the probabilities in opposing cells across the diagonal are equal:

$$\pi_{km} = \pi_{mk}$$

The corresponding log linear model is

$$\log(\mu_{km}) = \alpha_{km} \text{ with } \alpha_{km} = \alpha_{mk} \qquad (6.2)$$

Note that mean parameters for the margins are not fitted.

A weaker hypothesis is that of reversibility or quasi-symmetry only: the table would be symmetric if it were not for the distorting effect of the marginal totals. In other words, the system is not in equilibrium, because the total number in each state is changing over time. To fit this model, we add the mean parameters to the model:

$$\log(\mu_{km}) = \kappa_k + \upsilon_m + \alpha_{km} \qquad (6.3)$$

The model with only equilibrium, or marginal homogeneity, is closely related to the previous two. Suppose that the marginal totals are symmetric but that the body of the table is not. The distribution of units among the states is identical at different times, but the probability of shift between each pair of states is not the same in both directions. Marginal homogeneity plus quasi-symmetry equals symmetry.

Let us look at the changes in voting choice in three consecutive elections in Sweden (Fingleton, 1984, p. 151), given in Table 6.5. The deviance for a first order Markov chain is 207.33 on 36 d.f., showing that the vote in 1970 depended on that in 1964, even after conditioning on that for 1968. This indicates that a reversibility model for the pairs of transitions, 1964–1968 and 1968–1970 will not fit well, and, indeed, the deviance is 73.94 with 22 d.f. That for equilibrium over the three elections is 34.95 with 3 d.f., telling us that the proportions of voters choosing each party changed over the three elections.

It is also possible to look at reversibility and equilibrium for each pair of consecutive elections. For 1964–1968, the deviances are, respectively, 14.71 and 24.97, while, for 1968–1970, they are 2.49 and 65.32, all on 3 d.f. Only reversibility between the latter two elections seems acceptable. The aggregated frequencies are given in Table 6.6. Apparently, between these two elections, given the overall modification of the distribution of votes, changes between each pair of parties occurred about equally often in both directions. For a reversibility factor variable

	SD	C	P	Con
SD	–	1	2	3
C	1	–	4	5
P	2	4	–	6
Con	3	5	6	–

the parameter estimates are $(0.00, -0.44, -0.53, -0.45, -0.36, -0.35)$, indicating that the highest probability of transfer of votes is between the Social

Table 6.5. Changes in votes in the 1964, 1968, and 1970 Swedish elections. (Fingleton, 1984, p. 151)

1964	1968	1970			
		SD	C	P	Con
	SD	812	27	16	5
SD	C	5	20	6	0
	P	2	3	4	0
	Con	3	3	4	2
	SD	21	6	1	0
C	C	3	216	6	2
	P	0	3	7	0
	Con	0	9	0	4
	SD	15	2	8	0
P	C	1	37	8	0
	P	1	17	157	4
	Con	0	2	12	6
	SD	2	0	0	1
Con	C	0	13	1	4
	P	0	3	17	1
	Con	0	12	11	126

SD — Social Democrat   C — Centre

P — People's   Con — Conservative

Table 6.6. Changes in votes between the 1968 and 1970 Swedish elections, from Table 6.5.

1968	1970				
	SD	C	P	Con	Total
SD	850	35	25	6	916
C	9	286	21	6	322
P	3	26	185	5	219
Con	3	26	27	138	194
Total	865	373	258	155	1651

Democrat and Centre parties, and the lowest between Social Democrat and Conservative, the other pairs being fairly similar.

6.1.4 RANDOM WALKS

A special case of reversibility is the model for a random walk (Section 2.10) on a line. A general random walk allows successive moves in either direction with probabilities π_1 and π_2 respectively ($\pi_1 + \pi_2 < 1$ if the possibility of no move is allowed). If $\pi_1 \neq \pi_2$, there is said to be drift. This may be reformulated as a Markov chain where different points on the line are different states of the process. If the line is unbounded, the Markov chain will have an infinite number of states. We shall be concerned with random walks on a finite line segment with reflecting boundaries, so that the finite number of states corresponds to the categories of the contingency table.

If moves can only be between adjacent states, the transition matrix will have the values π_1 and π_2 on opposing first minor diagonals, either side of the main diagonal, and zero elsewhere. If $\pi_1 + \pi_2 < 1$, there will also be $1 - \pi_1 - \pi_2$ on the main diagonal. More usually, moves to any state are possible, with probabilities depending on distance. If symmetrically opposite moves of equal distance are equally probable, random walk models yield *diagonal symmetry* of the square table of transitions, a *symmetric minor diagonals* model. Thus, while reversibility only imposes the same probability in each direction between two states, a random walk has the additional constraint that all states the same distance apart, in either direction, have the same transition probability.

If the equal probability (in each direction) random walk model does not fit, we can relax the assumption and return to the original model, with drift, allowing steps with different probabilities in each direction, the *asymmetric minor diagonal* model. Another possibility is to combine the random walk or minor diagonals model with symmetry, i.e. equilibrium. When this model fits, it indicates that we would have a random walk except for the constraints of the differing marginal totals. Details for fitting such models are given in Lindsey (1989, pp. 97–110).

If we fit a symmetric random walk to the Swedish election data for 1968–1970 in Table 6.6, we obtain a deviance of 35.85 with 6 d.f., indicating that this restriction of the reversibility model is unacceptable. The model with drift is only slightly better, with deviance 29.63 on 4 d.f. As we already saw, the transfer of votes between the Social Democrat and Centre parties is much higher than between the Centre and People's parties or the People's and Conservative parties.

6.1.5   THE MOVER–STAYER MODEL

Another modification of a Markov chain has frequently been applied in mobility studies. A mobility table is a square two- or higher-dimensional table with the same categorical variable observed at two or more points in time, just as for other transition matrices for Markov chains. Common examples include migrant behaviour, social mobility among social classes or professions, and voting behaviour across elections, such as for the Swedish elections, used above.

As usual with a Markov chain, we shall wish to see if position at each point depends on that at the previous point. If it does not, this is the standard log linear model for independence. Often, in mobility tables, too many units do not change state between two time points for such independence

Table 6.7. Migration among four regions of the U.K. (Fingleton, 1984, p. 142)

1966	1971			
	CC	ULY	WM	GL
Central Clydesdale	118	12	7	23
Urban Lancashire & Yorkshire	14	2127	86	130
West Midlands	8	69	2548	107
Greater London	12	110	88	7712

to be acceptable; too many observations appear on the main diagonal. The simple solution is to eliminate these diagonal elements, creating a model for *quasi-independence*, i.e. an independence model for those off the diagonal. More theoretically, this approach assumes that the diagonal contains two types of units, the movers, who might have changed, but did not happen to in the observed time interval, and the stayers who never move. Hence, the name: the *mover–stayer model*. The movers, then, may form a Markov chain, usually of first order, but we have only directly observed those who actually moved in the given period; the others are inextricably mixed up with the stayers.

Consider data on migrations among four areas of the U.K. between 1966 and 1971 (Fingleton, 1984, p. 142), given in Table 6.7. The mover–stayer model has a deviance of 4.37 with 5 d.f., while the ordinary independence model has 19884 on 9 d.f. The estimates of the number of movers on the diagonal are $(1.6, 95.2, 60.3, 154.6)$. These are the estimated numbers of people in each region who are potential movers, but who did not happen to move in the observation period. This indicates that 12,193 out of 13,171 people in the table, or 92.6%, are not susceptible to migrate.

Because the mover–stayer model, as fitted, has independence between successive states for the movers, these results imply that the place of migration does not depend on the place of origin. For these data, which are ordered from north to south in the country, distance does not seem to be a factor determining place of migration.

6.2 Autoregression

6.2.1 BINARY AUTOREGRESSION

In a Markov chain model of order M, present response depends on all available information about the M previous states. In higher order models, this means that interactions among previous states are included. For example, a third order chain would have

$$T1 * T2 * T3 * T4$$

Table 6.8. Deviances for increasing autoregression order in Table 6.2.

Year	Deviance	d.f.
None	1123.5	15
1970	97.47	14
1969	50.09	13
1968	11.77	12
1967	9.46	11

This may be distinguished from a simple autoregression model, where present response, T4, depends only additively on previous states, so that, for example, the third order model would now be

$$T1.(1 + T2 + T3 + T4) + T2 * T3 * T4$$

If such a model is suitable, it allows a considerable simplification. If there are two states, a binary response at each time point on a unit, we call this binary autoregression. When there are time-varying covariates, logistic regression on the individual binary responses can be used. Otherwise, the units with identical paths can be aggregated to form a contingency table.

Let us reconsider the data on women working in Table 6.2. We saw above, in Section 6.1.1, that a third order Markov chain was necessary for 1971. The analogous series of reductions in deviance for successively more complex autoregressions are presented in Table 6.8, which may be compared to Table 6.3. A third order model is still necessary, but now with 12 d.f. instead of 8, with a change of only 2.75 in deviance for the gain of 4 d.f. The logistic model parameters are $(3.17, 0.59, 1.21)$, showing that dependence is highest on the previous year, 1970, but lower on 1969 than on 1968.

As with the Markov chain of Section 6.1.1, we have only taken the autoregression for the year 1971. Because we now know that the chain is third order stationary, we can reuse the combined table for the years 1970 and 1971 to fit the third order autoregression. This gives a deviance of 6.15 and 4 d.f., an acceptable fit, and a simpler model than the full Markov chain.

6.2.2 POISSON AUTOREGRESSION

The construction of autoregression models for Poisson count data is straight forward and requires no additional explanation. Consider the data in Appendix A6 on the numbers of epileptic seizures in four successive two week periods (Thall and Vail, 1990). Two different treatments, placebo and the antiepileptic drug, progabide, were applied to 59 patients, in addition to the usual treatment; age and an eight week baseline count of seizures were also available. Inspection of the data reveals a great deal of heterogeneity among the patients.

Table 6.9.   Coefficients of the Poisson autoregression for the epilepsy data in Appendix A6.

Coefficient	s.e.	Term
-0.4157	0.6199	1
-0.002961	0.007037	Base
-0.000037	0.000026	Base2
0.1105	0.03881	Age
-0.002130	0.000633	Age2
0.001116	0.000218	Base.Age
-0.8080	0.1022	Treat(2)
0.01371	0.002756	Treat(2).Base
0.03810	0.004224	LagCount
-0.000471	0.000058	Base.LagCount
0.02743	0.005709	Treat(2).LagCount

Preliminary analysis shows that linear and quadratic terms in age and in the baseline count, as well as their linear interaction and an interaction between treatment and baseline count, are necessary in a model which has a deviance of 810.76 on 228 d.f. Thus, this model, which does not take into account any dependence among responses on a unit, fits very poorly. If we introduce a first order autoregression, at each period (letting the count at two weeks depend on one-quarter of the baseline count, because it is over eight weeks instead of two), we obtain a reduction in deviance of 9.57 on 1 d.f. A second order autoregression yields little improvement. Adding interactions between the lagged count and both treatment and baseline further reduces the deviance by 68.31 on 2 d.f., to 733.88 with 225 d.f., still a poor model, however. This is a response surface model for age and baseline, different for the two treatments and changing dynamically with the lagged count. The coefficients are given in Table 6.9. The extra drug treatment reduces the number of seizures, but has less effect when the baseline is higher and when the count in the previous two weeks is higher.

If we take into account the heterogeneity, by fitting a factor variable for the patients, the deviance is reduced by 356.04, with 51 d.f., to 377.84 on 174 d.f. The model still does not fit sufficiently well, and none of the explanatory variables is any longer necessary in the model, except for the lagged count and the interactions with it. The autoregression coefficient is negative (-0.0136), while the interaction with baseline is positive (0.000111) and that with treatment negative (-0.0110).

These results suggest that any reasonable random effects type model would not adequately adjust for the heterogeneity in these data. The conclusion that the explanatory variables show little effect is very similar to that obtained by Zeger and Liang (1992) using generalized estimating equa-

Table 6.10. Number of successive conceptions in ten samples of semen from each of six bulls. (Kleinman, 1973)

Bull	1		2		3		4		5		6	
Sample	N	C	N	C	N	C	N	C	N	C	N	C
1	13	6	10	7	23	12	3	1	12	5	17	6
2	13	4	8	3	16	7	15	7	11	7	22	15
3	19	7	9	4	14	8	9	3	16	8	39	23
4	6	3	8	5	15	6	14	3	16	11	16	6
5	13	8	1	1	6	5	10	7	13	10	21	12
6	9	3	2	0	15	10	5	4	16	13	7	6
7	2	0	5	1	3	1	4	3	15	13	17	13
8	9	5	1	0	6	3	8	4	8	5	14	8
9	10	3	8	5	14	9	24	11	8	6	34	10
10	7	3	17	10	23	16	21	3	2	2	30	18

N: number of services   C: number of conceptions

tions (see Section 6.6, below), although, with that approach, there is no means of realizing that the 'model' in fact fits very poorly. Thall and Vail (1990) use log baseline count and log age, but such models fit more poorly than those with linear and quadratic terms, even without the interactions.

6.2.3 BINOMIAL AUTOREGRESSION

The way to construct autoregression models for multinomial count data is not as clear as when only one event is being counted. Even for the simple example of binomial count data, should the number of each type of event in the previous period be used or the proportion of events?

As an example, consider the number of successful conceptions in artificial insemination from six different bulls over ten samples (Kleinman, 1973), given in Table 6.10. A model where the present conception rate depends on the number of successes and failures in the previous period has a deviance of 95.93 with 51 d.f., while that depending on the proportion of successes in the previous period has 91.94 on 52 d.f. Both of these can be compared with a null model having a deviance of 99.72 with 53 d.f. The model using proportions is better, but neither model fits the data very well because of variability among the bulls. However, adding a factor variable to take this into account still does not yield an acceptable model (deviance of 70.91 with 47 d.f. for the proportion of successes model). Thus, in this study, there seems to be as much random variability among samples from the same bull as among bulls. This could be due to variability among the cows inseminated.

6.3　Birth, death, and learning models

6.3.1　LEARNING MODELS

Markov chain models present events as stochastically depending on what type of events happened at one or more immediately preceding times. Sometimes, the occurrence of an event may better be modelled as depending on the total number of previous events of different types. This accumulation of events constitutes a learning model. For example, it can indicate which types of event experiences most influence present behaviour. In the simplest case, where only one type of event is being recorded, and its probability depends on accumulated previous such events, we have a birth or contagion process.

Consider the Solomon-Wynne experiment on dogs, whereby they learn to avoid a shock (Kalbfleisch, 1985, pp. 83–88). A dog is in a compartment with a floor through which a shock can be applied. The lights are turned out and a barrier raised; ten seconds later, the shock occurs. Thus, the dog has ten seconds, after the lights go out, to jump the barrier and avoid the shock. Each of 30 dogs is subjected to 25 such trials. The results, in Table 6.11, are recorded as a shock trial ($y_{ik} = 0$), when the dog remains in the compartment, or an avoidance trial ($y_{ik} = 1$), when the it jumps out before the shock.

Our model supposes that a dog learns from previous trials. The probability of avoidance will depend on the number of previous shocks and on the number of previous avoidances. Let π_k be the probability of a shock at trial k ($k = 0, \ldots, 24$), given its reactions on previous trials, and x_{ik} be the number of avoidances before trial k. Then $k - x_{ik}$ is the number of previous shocks. We use the model

$$\pi_k = \kappa^{x_{ik}} v^{k - x_{ik}} \tag{6.4}$$

or

$$\log(\pi_k) = \alpha x_{ik} + \beta(k - x_{ik}) \tag{6.5}$$

where $\alpha = \log(\kappa)$ and $\beta = \log(v)$.

Because

$$\frac{\pi_k}{\pi_{k-1}} = \begin{cases} \kappa \text{ if } y_{i,k-1} = 1 \\ v \text{ if } y_{i,k-1} = 0 \end{cases} \tag{6.6}$$

we see that the probability of shock changes by a factor of κ if there was an avoidance at the previous trial or v if there was a shock.

The parameters are estimated as $\hat{\alpha} = -0.236$ (s.e. 0.0226) and $\hat{\beta} = -0.0793$ (s.e. 0.0119), so we see that $\hat{\kappa} = \exp(-0.236) = 0.790$ and $\hat{v} = \exp(-0.0793) = 0.924$, indicating that an avoidance trial appears more

Table 6.11. The Solomon–Wynne dog experiment with 25 trials for each dog on a line. (Kalbfleisch, 1985, pp. 83–88)

00101	01111	11111	11111	11111
00000	00100	00001	11111	11111
00000	11011	00110	10111	11111
01100	11110	10101	11111	11111
00000	00011	11111	11111	11111
00000	01111	00101	11111	11111
00000	10000	00111	11111	11111
00000	00110	01111	11111	11111
00000	10101	10100	01111	10110
00001	00110	10111	11111	11111
00000	00000	11111	10111	11111
00000	11111	00111	11111	11111
00011	01001	11111	11111	11111
00001	01101	11111	11111	11111
00010	11011	11111	11111	11111
00000	00111	11111	11111	11111
01010	00101	11101	11111	11111
00001	01011	11101	11111	11111
01000	01000	11111	11111	11111
00001	10101	10101	11111	11111
00011	11101	11111	11111	11111
00101	01111	11111	10011	11111
00000	00111	11111	11111	11111
00000	00011	10100	01101	11111
00000	01011	11010	11111	11111
00101	11011	01111	11111	11111
00001	01111	11111	11111	11111
00010	10111	01011	11111	11111
00001	10011	10101	01011	11111
00001	11111	01011	11111	11111

effective than a shock in reducing the probability of future shock. Because $\hat{\kappa} \cong \hat{\upsilon}^3$, a dog learns about as much by one avoidance as by three shocks. The deviance is 552.2 with 718 d.f., but, because these are binary data, it does not provide a measure of goodness of fit (Williams, 1982b).

It is interesting to note that this model, with a nonstandard log link, but a clear theoretical interpretation, fits better than the corresponding model with a canonical logistic link, which has a deviance of 605.73.

Table 6.12. Coefficients of the Poisson autoregression for the epilepsy data in Appendix A6.

Coefficient	s.e.	Term
0.2499	0.6009	1
0.000621	0.007036	Base
-0.000009	0.000025	$Base^2$
0.05626	0.03767	Age
-0.001153	0.000616	Age^2
0.000754	0.000220	Base.Age
-0.5664	0.1026	Treat(2)
0.007687	0.002647	Treat(2).Base
0.02424	0.001591	CumCount
-0.000212	0.000019	Base.CumCount
0.009856	0.001808	Treat(2).CumCount

6.3.2 BIRTH AND CONTAGION PROCESSES

When a series of counts of events is observed on each unit, the same procedures may be used as for single events. The accumulated number of events may be a better explanatory variable than the counts of events in the immediately preceding period used in autoregression models. This is characteristic of a birth or contagion process, where the number of new events depends on the cumulated number of previous births or infections, but this model is much more widely applicable.

If we reconsider our data on the numbers of epileptic seizures in two week periods of Section 6.2.2 and Appendix A6, we may replace the lagged count variable with an accumulated count of seizures, which, of course, does not take into account seizures before the observation period. This model, with the same interactions as in Table 6.9, fits much better than the autoregression given above, with a deviance of 553.79 as opposed to 733.88, both with 225 d.f. The results are presented in Table 6.12, which may be compared with those in Table 6.9. Again, the extra drug treatment reduces the number of seizures, although the effect is less marked than with the autoregression, and, as before, has less effect when the baseline is higher and when the accumulated previous count is higher.

If we use accumulated counts for the insemination data of Table 6.10, the model fits even less well than the autoregression of Section 6.2.3. Here, as might be expected, there is no cumulative effect over time.

6.3.3 CAPTURE–RECAPTURE MODELS

Learning and birth models are closely related. In both, the probability of a given event increases with accumulated previous events, through the learning process or by more procreative units becoming available. Birth, and death, or the corresponding immigration and emigration, can also be studied in a different way, by capturing, marking, and recapturing members of a population of animals. Thus, each animal has a repeated measure at each capture session, a binary variable indicating whether or not it was captured at that time.

Consider a population of animals where a first sample is marked. The Jolly-Seber model (Jolly, 1965; Seber, 1965; Cormack, 1985, 1989) can be described as follows. At subsequent samples, each marked animal recaptured is recorded. Any new animals are marked and recorded. The probability of capture at sample k is π_k, independently for each sample, i.e. with no trap learning. For example, if N is the constant total population size, the expected number captured only on the second of three samples would be $N(1 - \pi_1)\pi_2(1 - \pi_3)$. This model is sufficient to estimate population size in a closed population. With an open population, we have births or immigration and deaths or emigration. Suppose that π_k^s is the probability that an animal survives or stays from the kth to the $(k + 1)^{\text{th}}$ sample and π_k^d is the probability that an animal is not seen again after the kth sample. Let $\kappa_k \geq 1$ be the proportional increase in the unmarked population at the $(k + 1)^{\text{th}}$ sample due to births or immigration. (An animal cannot be born or immigrate marked!) By definition, the reciprocal of κ_k is the probability that an animal alive and unmarked at the $(k + 1)^{\text{th}}$ sample was already in the population at time k. The probability of death or emigration can be similarly reparametrized as $\nu_k = \pi_k^d/[\pi_k^s(1 - \pi_{k+1})\pi_{k+1}^d]$, the reciprocal of which is the probability that an animal, alive at time k, but not observed afterwards, was actually alive at time $k + 1$. Thus, κ_k and ν_k refer to births and deaths between the kth and $(k + 1)^{\text{th}}$ samples.

The expected numbers for a three sample study would be

$$
\begin{aligned}
n_{111} &= N\pi_1\pi_1^s\pi_2(1 - \pi_2^d) \\
n_{211} &= N(1 - \pi_1)\pi_1^s\kappa_1\pi_2(1 - \pi_2^d) \\
n_{121} &= N\pi_1\pi_1^s(1 - \pi_2)(1 - \pi_2^d) \\
n_{221} &= N(1 - \pi_1)\pi_1^s\kappa_1(1 - \pi_2)\kappa_2(1 - \pi_2^d) \qquad (6.7) \\
n_{112} &= N\pi_1\pi_1^s\pi_2\pi_2^d \\
n_{212} &= N(1 - \pi_1)\pi_1^s\kappa_1\pi_2\pi_2^d \\
n_{122} &= N\pi_1\nu_1\pi_1^s(1 - \pi_2)\pi_2^d
\end{aligned}
$$

where the index (of n), 1, indicates captured and 2 not observed (Cormack, 1989). Thus, we have a standard log linear model.

Consider a study of Dunnock (*Prunella modularis*), with five captures over a seven month winter period (Cheke, 1985; Cormack, 1985), given in Table 6.13. Here, we have a population where there may be migration, but no births. The closed population model has a deviance of 25.51 on 25 d.f., which is an acceptable fit. That which also allows for death and emigration has 24.34 with 22 d.f., while that with only immigration has

Table 6.13. Capture–recapture of Dunnocks over a seven month period. (Cheke, 1985, Cormack, 1985)

Capture	Frequency	Capture	Frequency
yyyyy	0	nyyyy	1
ynyyy	0	nnyyy	0
yynyy	0	nynyy	1
ynnyy	0	nnnyy	1
yyyny	0	nyyny	0
ynyny	0	nnyny	1
yynny	0	nynny	0
ynnny	0	nnnny	14
yyyyn	1	nyyyn	1
ynyyn	0	nnyyn	2
yynyn	1	nynyn	2
ynnyn	1	nnnyn	16
yyynn	0	nyynn	2
ynynn	0	nnynn	11
yynnn	2	nynnn	13
ynnnn	10	nnnnn	–

19.89 also on 22 d.f. On the other hand, when both are allowed for, the deviance is 12.97 on 19 d.f., a significantly better fit. A model with a constant capture probability over the three middle samples ($\pi_2 = \pi_3 = \pi_4$) increases the deviance by only 1.84 with a gain of 2 d.f. Trap dependence is introduced as an interaction between successive capture probabilities, making probability of capture at time k depend on that at $k - 1$. With a deviance of 10.18 on 17 d.f., it does not seem necessary for these data.

The sizes of the population at the five capture dates are estimated as $(39.5, 61.0, 53.4, 69.8, 47.4)$, with constant capture probability of 0.38. The estimated numbers of immigrants between samples are $(39.0, 17.8, 26.0, 29.9)$, with probabilities of survival, without death or emigration, of $(0.56, 0.58, 0.82, 0.25)$. All of these estimates suppose a constant capture effort at each time point.

6.4 Crossover trials

As we have seen in Section 4.5, crossover trials involve the application of more than one treatment to each individual, with the effects of each being measured. The order of the treatments is randomized for the different individuals. A sufficient washout period is usually allowed between treatments to reduce carry-over from one treatment to the next. However, because the same individual is measured several times, there may still be sequence and period effects. The first arises from a given treatment following another; the second from the place in time that a treatment is administered. As

Table 6.14. Effect of placebo, low, and high dose analgesic on pain relief. (Jones and Kenward, 1987)

	Response Pattern Sequence							
	000	001	010	011	100	101	110	111
ABC	0	2	2	9	0	0	1	1
ACB	2	0	0	9	1	0	0	4
BAC	0	1	1	3	1	8	0	1
BCA	0	1	1	0	1	0	8	1
CAB	3	0	0	2	0	7	1	1
CBA	1	0	0	1	5	3	4	0

Response — 0: no relief, 1: relief.

Treatment — A: placebo, B: low dose, C: high dose.

well, there may be some form of intra-unit correlation, as in random effects models. In a binary crossover trial, the response measured takes only two values. Jones and Kenward (1987) and Kenward and Jones (1987a) show how such data can be handled by the use of log linear models.

Consider the data on three treatments for pain given in Jones and Kenward (1987) and reproduced in Table 6.14. The group effect will distinguish the six different sequences of application of the treatments. Three period effects indicate if pain was relieved in that period. Three treatment effects indicate for what treatments the individual found relief. The three first order lag effects take into account the possibility that a treatment at one period has carry-over effects in the next period, while the three second order lags do the same for two periods later. Finally, three intra-individual correlations measure the possibility that an individual reacts the same way in two different periods. A number of these parameters are, in fact, redundant and do not need to be included.

We fit the model with these parameters, leaving out those which are redundant, but including some interactions which may be of importance. The parameter estimates are given in Table 6.15. A deviance of 28.11 with 16 d.f. indicates that the model does not fit too badly. The changes in deviances for successive withdrawal of various terms in the model are given in Table 6.16. Only the deviance for group.intra-individual correlations indicates that these parameters may be significantly different from zero. A look at the standard errors of the parameters shows that only one of the ten parameters, in fact, is a problem. The final model, given in Table 6.17, has only the treatment differences. It has a deviance of 71.05 with 44 d.f. The lack of fit of this model, as we have seen, is primarily due to one interaction between intra-individual correlation and group.

We may now compare the odds of relief using the three treatments. The odds of relief under treatment three, the high dose of analgesic, are about $11.81 = \exp(1.401 + 1.068)$ times larger than under the placebo and those

Table 6.15.   Parameter estimates for the full binary crossover model of
Table 6.14.

Estimate	s.e.	Parameter
-2.321	1.314	1
-2.936	2.587	Period2(2)
4.377	2.190	Period3(2)
0.902	1.299	Treatment1(2)
-1.560	1.234	Treatment2(2)
-0.496	1.193	Treatment3(2)
-0.509	1.177	Corr12(2)
0.991	1.426	Corr13(2)
1.245	0.881	Group(2)
3.477	1.772	Group(3)
3.761	1.947	Group(4)
3.371	1.910	Group(5)
4.341	1.891	Group(6)
-0.643	0.715	Lag2(2)
-1.941	0.825	Lag3(2)
0.013	0.931	Lag22(2)
-0.648	0.994	Lag23(2)
0.471	1.870	Period2(2).Treatment1(2)
-3.826	2.469	Period3(2).Treatment1(2)
6.009	2.187	Period2(2).Treatment2(2)
-2.255	2.190	Period3(2).Treatment2(2)
2.044	2.423	Period2(2).Treatment3(2)
0.549	1.299	Corr12(2).Group(2)
-4.353	1.715	Corr12(2).Group(3)
-1.256	1.821	Corr12(2).Group(4)
-1.177	1.541	Corr12(2).Group(5)
-1.465	1.778	Corr12(2).Group(6)
0.598	1.315	Corr13(2).Group(2)
-0.078	1.754	Corr13(2).Group(3)
-1.078	1.531	Corr13(2).Group(4)
0.370	1.766	Corr13(2).Group(5)
-2.088	1.694	Corr13(2).Group(6)

for the low dose about $7.099 = \exp(0.8920 + 1.068)$, so that high dose is
about $1.664 = \exp(1.401 - 0.8920)$ times as effective as low. These results
are fairly close to those of Jones and Kenward (1987) but our final model is
much simpler than theirs. For a latent variable approach to such crossover
data, see Kenward and Jones (1991).

Table 6.16.   Changes in deviance for elimination of parameters from the binary crossover model of Table 6.14.

Parameters	Change in Deviance	d.f.
Lag22+Lag23	0.55	2
Lag2+Lag3	5.85	2
Group.Corr	26.62	10
Corr	3.10	1
Group	1.14	5
Period.Treatment	4.88	6
Period	0.81	2

Table 6.17.   Parameter estimates for the final binary crossover model of Table 6.14.

Estimate	s.e.	Parameter
-0.490	0.3012	1
-1.068	0.2468	Treatment1(2)
0.892	0.2370	Treatment2(2)
1.401	0.2702	Treatment3(2)

6.5   Point processes

A point process (Section 2.10) is the record of the occurrences of an event over time. Because an event either occurs or not, it is a special case of a two-state Markov chain. In the context of repeated measurements, it will be the series of occurrences of the event on different units. The usual function of interest is the rate, intensity, or risk of occurrence of the event. This will be compared across units, usually under different conditions. It can be studied in several equivalent ways: by the frequency of the event in given intervals of time, by a binary variable indicating whether or not an event has occurred in each unit interval of observation, or by the length of intervals between successive events. In this section, we look at the first of these, and in Chapter 8 the latter two.

The simplest point process is the Poisson process, because the intensity does not depend on the time since the previous event. It can be easily handled by all approaches. When the intensity function is more complex, most other types of point processes can usually be dealt with more easily by modelling the inter-event times or the binary indicator variable, as in Chapter 8. However, many other approaches, besides looking at intensities, can also be applied to point processes. We have already seen some, for example, with Tables 6.1 and 6.11 above.

Table 6.18. Number of deaths by horse kicks in the Prussian army from 1875 to 1894 for 14 corps. (Andrews and Herzberg, 1985, p. 18)

Corps	
G	0 2 2 1 0 0 1 1 0 3 0 2 1 0 0 1 0 1 0 1
I	0 0 0 2 0 3 0 2 0 0 0 1 1 1 0 2 0 3 1 0
II	0 0 0 2 0 2 0 0 1 1 0 0 2 1 1 0 0 2 0 0
III	0 0 0 1 1 1 2 0 2 0 0 0 1 0 1 2 1 0 0 0
IV	0 1 0 1 1 1 1 0 0 0 0 1 0 0 0 0 1 1 0 0
V	0 0 0 0 2 1 0 0 1 0 0 1 0 1 1 1 1 1 1 0
VI	0 0 1 0 2 0 0 1 2 0 1 1 3 1 1 1 0 3 0 0
VII	1 0 1 0 0 0 1 0 1 1 0 0 2 0 0 2 1 0 2 0
VIII	1 0 0 0 1 0 0 1 0 0 0 0 1 0 0 0 1 1 0 1
IX	0 0 0 0 0 2 1 1 1 0 2 1 1 0 1 2 0 1 0 0
X	0 0 1 1 0 1 0 2 0 2 0 0 0 0 2 1 3 0 1 1
XI	0 0 0 0 2 4 0 1 3 0 1 1 1 1 2 1 3 1 3 1
XIV	1 1 2 1 1 3 0 4 0 1 0 3 2 1 0 2 1 1 0 0
XV	0 1 0 0 0 0 0 1 0 1 1 0 0 0 2 2 0 0 0 0

6.5.1 POISSON PROCESSES

In an ordinary Poisson process, the intensity is constant for a given unit. If it changes, with covariates or simply at some point(s) in time, the process is nonhomogeneous. Such processes can easily be modelled as log linear models if the numbers of events in fixed intervals of time are recorded.

Consider perhaps the most classical example of a Poisson process, the number of deaths by horse kicks in the Prussian army from 1875 to 1894 (Andrews and Herzberg, 1985, p. 18), reproduced in Table 6.18. Corps G, I, VI, and XI are noted as having a different organization than the others.

A model with the same homogeneous Poisson process for all corps has a deviance of 323.23 with 279 d.f., while that for a different intensity for each corps reduces this by 26.14 on 13 d.f. However, a glance at the parameters (not shown) reveals that the four corps with a different organization have similar, higher rates of deaths than the most of the others, although the corps XI and XIV distinguish themselves by being the highest. Reducing the model to only two different intensities, for the two types of organization, raises the deviance by 18.47 with an increase of 12 d.f. Thus, the one d.f. for different organization accounts for a change in deviance of 7.67.

We may also allow the process to vary with the year, a nonhomogeneous Poisson process, the same within each of the two groups of corps (i.e. no interaction). This reduces the deviance by a further 38.50 with 19 d.f. However, a look at the parameters (again not shown) shows that the rate was lower in the first three years, as well as a scattering of other years. If we only allow nonhomogeneity as a different rate before and after this time, the deviance is raised by 28.68 on 18 d.f., so that the break between

1877 and 1878, with one d.f., accounts for 9.82 of the change in deviance. A model with a linear trend instead of this break fits much more poorly. Introduction of a lagged variable for the number of deaths in the previous year changes the deviance very little.

If we look at the frequencies of different numbers of horse kicks for the four categories, before and after 1877 combined with the two corps organizations, we find that a Poisson distribution describes each of them well. Thus, we end up with two nonhomogeneous Poisson processes, one for each type of corps.

The intensity is estimated as 0.47 deaths/year for the four corps with a different organization for the three years up to 1877. This is multiplied by a factor of 0.66 for the other corps (for all years) and by 2.13 after 1877 (for both corps).

If we look again at the rainfall data of Table 6.1, counting the number of rainy days in each consecutive set of five days, we find that the same homogeneous Poisson process fits well for all years. However, with such grouped data, we lose the relationship among successive days which we had discovered with our Markov model.

6.5.2   DYNAMIC GENERALIZED LINEAR MODELS

Dynamic generalized linear models are of particular use when one needs to predict the evolution of several series of responses. However, they also provide a general means of estimating parameters for trend and seasonal effects. The modelling procedure was described in Section 2.8 and the special case for the normal distribution was employed in Section 4.3.

Here, we are specifically interested in models applicable to counts and categorical data. For the former, the Poisson and negative binomial or Polya processes are often suitable, and, for the latter, the binomial process. The conjugate distribution for the Poisson is the gamma, yielding a negative binomial likelihood, while that for the negative binomial is the beta distribution, giving a hypergeometric distribution. The conjugate for the binomial is also the beta distribution, yielding a beta-binomial distribution. Here, we shall use an example of count data, so as to compare those two possible distributions for that type of data.

To estimate the parameters of the model, a Kalman filter type procedure is applied chronologically to the observations. Let us look, in more detail, at the steps for the gamma-Poisson process. The model is the same as given by Equations (5.25) and (5.26), but now the parameter estimates are updated dynamically. Thus, starting with some initial values, for one series, we have the prediction equations

$$\kappa_{t|t-1} = \zeta \kappa_{t-1} \tag{6.8}$$

$$\frac{1}{v_{t|t-1}} = \frac{\zeta}{v_{t-1}} \tag{6.9}$$

and the updating equations

$$\kappa_t = \kappa_{t|t-1} + y_t \tag{6.10}$$

$$\frac{1}{v_t} = \frac{1}{v_{t|t-1}} + 1 \tag{6.11}$$

where ζ is a discount factor between zero and one. If explanatory variables are present, Equation (6.11) becomes

$$\frac{1}{v_t} = \frac{1}{v_{t|t-1}} + e^{\beta^T \mathbf{x}_t} \tag{6.12}$$

where a log link has been used. Then, the deviance, derived from Equation (5.29), is given by

$$-2 \sum_t \left\{ \log[\Gamma(y_t + \kappa_{t|t-1})] - \log[\Gamma(\kappa_{t|t-1})] - \kappa_{t|t-1} \log[v_{t|t-1}] \right.$$

$$\left. -(\kappa_{t|t-1} + y_t) \log \left[\frac{1}{v_{t|t-1}} + 1 \right] \right\} \tag{6.13}$$

which must be minimized by some numerical procedure in order to estimate the regression parameters, and perhaps the discount.

In the case of repeated measurements, we have several series. The most complex model involves applying the above procedure separately to each series, so that all parameters are different. For 'parallel' series, v_t, and the regression coefficients, are the same for all series, while κ_t is allowed to be different. The latter are updated, in Equation (6.10), using the respective values of y_t for each series. To fit a common model to all series, the mean response may be used in Equation (6.10). Similar procedures may be used for the models for other types of data mentioned above.

Consider, now, the reported numbers of U.K. deaths from bronchitis, emphysema, and asthma (Diggle, 1990, p. 238) each month from 1974 to 1979, distinguished by sex, presented in Table 6.19. We shall fit models with a linear time trend to see if the number of deaths is changing over the years. As well, we require a seasonal component, because the number of deaths varies regularly over the year. We use seasonal harmonics for the twelve month period. Thus, three models will be fitted: 1. separately to the data for each sex, 2. with the same trend and seasonal, but a different level for each sex, and 3. with all components the same for each sex. The resulting deviances for the gamma-Poisson (negative binomial) and beta-negative binomial (hypergeometric) are displayed in Table 6.20. We arbitrarily take a fixed discount of 0.7, although this could also have been estimated.

Although there is no clear saturated model, we take our most complex model as a baseline, giving it zero deviance, so that it can easily be compared with the others. We immediately see that all of the gamma-Poisson

Table 6.19. Monthly numbers of deaths from bronchitis, emphysema, and asthma in the U.K., 1974-1979. (Diggle, 1990, p. 238)

Males									
2134	1863	1877	1877	1492	1249	1280	1131	1209	1492
1621	1846	2103	2137	2153	1833	1403	1288	1186	1133
1053	1347	1545	2066	2020	2750	2283	1479	1189	1160
1113	970	999	1208	1467	2059	2240	1634	1722	1801
1246	1162	1087	1013	959	1179	1229	1655	2019	2284
1942	1423	1340	1187	1098	1004	970	1140	1110	1812
2263	1820	1846	1531	1215	1075	1056	975	940	1081
1294	1341								

Females									
901	689	827	677	522	406	441	393	387	582
578	666	830	752	785	664	467	438	421	412
343	440	531	771	767	1141	896	532	447	420
376	330	357	445	546	764	862	660	663	643
502	392	411	348	387	385	411	638	796	853
737	546	530	446	431	362	387	430	425	679
821	785	727	612	478	429	405	379	393	411
487	574								

Table 6.20. Deviances for various models for monthly deaths of Table 6.19.

Effect	Separate Sexes		Different Levels		Sexes Together	
	Deviance	Par.	Deviance	Par.	Deviance	Par.
Gamma-Poisson						
Trend	7510.16	6	7510.18	5	29237.84	3
Seasonal	645.91	26	672.97	14	22400.61	13
Both	639.72	28	671.86	15	22399.55	14
Beta-Negative Binomial						
Trend	280.30	8	281.36	6	479.81	4
Seasonal	0.27	28	4.62	16	447.72	14
Both	0.00	30	4.62	17	447.72	15

models are unacceptable as compared to the beta-negative binomial ones. A trend is not necessary and the seasonal components can be the same for deaths of both sexes. However, the level must be different for the two sexes. This model has a deviance of only 4.62 greater than the most complex one, but with 14 fewer parameters. The only further simplification is to reduce the number of harmonics. It is only possible to eliminate the two highest ones, with a further in increase in deviance of 3.76. The fitted values of

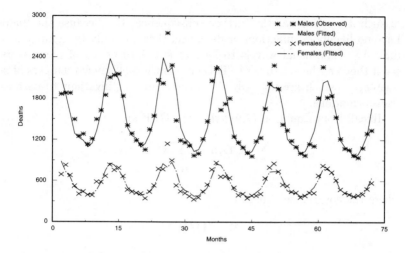

Fig. 6.1. Monthly deaths from bronchitis, emphysema, and asthma, from Table 6.19, with fitted DGLM.

the final model are plotted in Figure 6.1, along with the observed numbers of deaths. The fitted line follows the observed deaths fairly closely, with the exception of three high numbers of male deaths and one of females. Male deaths are consistently higher than female, but with the same seasonal variation. There is no indication of a change in the number of deaths over the years.

Procedures for dynamic generalized linear models are not yet well developed, other than in the normal case. There, as we have seen, they are very useful for fitting random effects and autoregression when the observation times are unequally spaced. In the more general case, the intractability of the integrals means that time-consuming numerical methods must be used or approximations made. Here, only the first two moments of the conjugate distribution were employed. However, the power of the procedure makes it one of the most promising avenues of research in repeated measures.

6.6 Generalized estimating equations

Another way to introduce correlation into a discrete data model is to use the generalized estimating equations (GEE) of Section 2.9. Unfortunately, in general, such a procedure destroys the interpretability of the model, as a representation of some physical mechanism which could have generated the data. This is because the estimating equations, which will be solved, in general, have no corresponding likelihood function, or probability distribution. As usual, a special exception is the normal distribution, where this approach yields the models of Chapters 3 and 4. This characteristic of such a method has the further consequence that because the 'models' have no likelihood function, or deviance, they can only be compared with difficulty. The usual way is to look at standard errors of parameters to see if they can be set to zero. This entails the well-known dangers of such inferences, which are not only asymptotic but not invariant to parameter transformations.

Recall from Equation (2.91) that the GEE are

$$\sum \mathbf{X}_i \mathbf{D}_i \mathbf{d}_i = \sum \mathbf{C}_i \boldsymbol{\Sigma}^{-1} \mathbf{d}_i$$
$$= 0$$

where the covariance matrix is

$$\boldsymbol{\Sigma}_i = \mathbf{U}_i^{\frac{1}{2}} \mathbf{R}_i \mathbf{U}_i^{\frac{1}{2}}$$

with correlation matrix, $\mathbf{R}_i$ and $\mathbf{U} = \mathrm{diag}(\tau_i^2 \phi / w_i)$, the variance. Remember also that the variance function, τ_i^2, is a function of μ. For $\mathbf{R} = \mathbf{I}$, we have the usual independence model, a generalized linear model.

For intra-unit stochastic dependence, or pairwise correlation, we require a matrix of the form of Equation (1.1), so that the $\mathbf{R}_i$ will have one unknown parameter, δ or $\rho = \delta/(\delta + \psi^2)$, to estimate. Zeger and Liang (1986) call $\mathbf{R}_i$ the working correlation matrix and show that the approach is robust to various choices of the parameters. Thus, they estimate a value for the parameter before estimating the regression parameters, instead of estimating all parameters simultaneously.

For binary data, Prentice (1988) sets up a second set of generalized estimating equations for the parameters of the correlation matrix. In our case, this reduces to

$$\sum_{i=1}^{N} \boldsymbol{\Sigma}_{Vi}^{-1} \mathbf{d}_{Vi} = 0 \qquad (6.14)$$

where

$$\mathbf{d}_{Vi} = \left[\frac{(y_{ik} - \pi_{ik})(y_{ih} - \pi_{ih})}{\sqrt{\pi_{ik}(1 - \pi_{ik})\pi_{ih}(1 - \pi_{ih})}} - \rho \right] \qquad (6.15)$$

is the $R \times (R - 1)$ vector of differences between sample correlations and their constant expected value or mean correlation, ρ, with variance $\boldsymbol{\Sigma}_{Vi}$. Here, a vector of partial derivatives is not necessary, because the pairwise correlation is constant for all responses on a unit.

As an alternative to the correlation coefficient, Lipsitz, Laird, and Harrington (1991) suggest the odds ratio. Now, the correlation matrix has the form

$$\mathbf{R}_i = \left[\frac{\pi_{ikh} - \pi_{ik}\pi_{ih}}{\sqrt{\pi_{ik}(1 - \pi_{ik})\pi_{ih}(1 - \pi_{ih})}} \right] \qquad (6.16)$$

where $\pi_{ikh} = \Pr(Y_{ik} = 1, Y_{ih} = 1)$, and, in Equation (6.14),

$$\mathbf{d}_{Vi} = y_{ikh} - \rho \qquad (6.17)$$

where $y_{ikh} = 1$ if $y_{ik} = y_{ih} = 1$ and zero otherwise. Again, $\boldsymbol{\Sigma}_{Vi}$ is an appropriate covariance matrix of the correlation coefficients. Because it involves a fourth order probability, the authors suggest using $\boldsymbol{\Sigma}_{Vi} = \mathrm{diag}[\pi_{ikh}(1 - \pi_{ikh})]$ instead.

Such an approach can also be used for correlation over time. In Section 4.1.1, we saw how to fit a serial correlation model by time filtering both response and explanatory variables. This was the same as multiplying the variables by the Cholesky decomposition of the inverse of the time series covariance matrix. In fact, this procedure can be applied to any type of data. Unfortunately, as for other forms of GEE than the normal distribution, the estimating equations so obtained again do not correspond to a likelihood function, and thus to no probability model (see Scallon, 1985; Zeger and Liang, 1986).

For $\mathbf{R}$ nondiagonal, let $\mathbf{R}^{-1} = \mathbf{A}^T\mathbf{A}$, where $\mathbf{A}$ is the Cholesky decomposition. Because all matrices except $\mathbf{R}$ are diagonal, this is the same as multiplying all variables by $\mathbf{AU}^{-\frac{1}{2}}$.

Consider an AR(1) model for $\mathbf{R}$ with Cholesky decomposition adapted from Equation (4.5):

$$\mathbf{A} = \begin{pmatrix} \sqrt{1 - \rho^2} & 0 & \cdots & 0 \\ -\rho & 1 & \cdots & 0 \\ \vdots & \vdots & \ddots & \vdots \\ 0 & 0 & \cdots & 1 \end{pmatrix} \qquad (6.18)$$

This is equivalent to applying an autoregressive filter to the data. Note that the covariances only depend on the mean through the variance; the correlation is constant. The procedure extends readily to an AR(M). The Yule-Walker estimates of Equation (4.7) may be used for the ρ_t when $M > 1$ (see Scallon, 1985; Zeger, 1988a).

One natural extension of the usual overdispersion model for Poisson data to the case with autocorrelation would be

$$\boldsymbol{\Sigma} = \mathbf{U} + \mathbf{URU} \qquad (6.19)$$

which, however, does not yield a simple Cholesky decomposition, so that we might again use

Table 6.21. Parameter estimates, with standard errors, of various dependence 'models' for the data of Appendix A6.

	Indep.	Overdisp.	Uniform	AR(1)	General
Intercept	0.563	0.563	0.563	0.455	0.623
	(0.136)	(0.276)	(0.449)	(0.433)	(0.365)
Age	0.023	0.023	0.023	0.023	0.023
	(0.001)	(0.001)	(0.002)	(0.002)	(0.001)
Baseline	0.023	0.023	0.023	0.026	0.019
	(0.004)	(0.008)	(0.013)	(0.013)	(0.011)
Treatment	-0.153	-0.153	-0.153	-0.165	-0.209
	(0.048)	(0.097)	(0.158)	(0.152)	(0.129)

$$\boldsymbol{\Sigma} = \mathbf{U}^{\frac{1}{2}} \mathbf{R} \mathbf{U}^{\frac{1}{2}} \tag{6.20}$$

where $\mathbf{U} = \mathrm{diag}(\mu_i + \phi\mu_i^2)$.

Because of the limitations of this approach, both for physical interpretation and for inference, we shall only consider one example. It is most useful to look at data which we have already modelled in other ways, so let us use the data on epileptic fits from Appendix A6. In Section 6.2.2, we fitted a response surface type autoregression model, and, in Section 6.3.2, a birth process, using the accumulated numbers of fits. As we saw there, a model which does not take into account stochastic dependence among the responses on an individual fits very badly, with only the three explanatory variables available, baseline count, age, and treatment.

Here, we shall not try to fit elaborate models such as those, but compare the results for several stochastic dependence structures. We include the three variables mentioned, but not their squares or interactions, and look at the standard errors of the estimates, our only means of inference. We can start first with the independence model, and the standard correction for overdispersion of Section 5.2. These parameter estimates are given in Table 6.21. The independence model has a deviance of 958.46 with 232 d.f., which yields an estimate of 4.13 for the dispersion (or variance) parameter. With this correction, the treatment effect seems no longer to be significantly different from zero, although age and baseline count still are.

For Poisson data, the variance is equal to the mean. We shall take $\mathbf{U}$ in its simplest form, as a diagonal matrix of the estimated means, so that

$$\boldsymbol{\Sigma} = \phi\mathbf{U}^{\frac{1}{2}} \mathbf{R} \mathbf{U}^{\frac{1}{2}} \tag{6.21}$$

with $\mathbf{U} = \mathrm{diag}(\mu_i)$ and the correlation matrix, $\mathbf{R}$, having a form to be specified. The standard errors will be estimated from the covariance matrix of the regression parameters, given by

$$\mathbf{C}\boldsymbol{\Sigma}^{-1}\mathbf{C}^T \tag{6.22}$$

(Prentice, 1988), instead of that recommended by Zeger and Liang (1986), because the latter does not reduce to that of the independence case, when $\mathbf{R}$ is the identity matrix and $\phi = 1$. This should allow a minimal amount of comparability. Moment estimators, based on the standardized residuals, but updated at each iteration, will be used for the variance and correlation parameters. Given the approximate nature of this approach, such estimators prove to be sufficiently accurate.

Three different forms of stochastic dependence will be used here in the GEE: an unstructured or general correlation, a uniform correlation, and an AR(1). The general covariance matrix, without the variance function, $\text{diag}(\phi_i)\mathbf{R}$, is estimated as

$$\begin{pmatrix} 4.387 & 1.488 & 2.417 & 1.119 \\ 1.488 & 4.213 & 3.796 & 1.473 \\ 2.417 & 3.796 & 10.790 & 2.953 \\ 1.119 & 1.473 & 2.953 & 2.228 \end{pmatrix}$$

Dependence seems to be decreasing with distance in time between the responses. From the last column of Table 6.21, with this GEE stochastic dependence structure, only age is significantly different from zero, although the point estimate of treatment effect is somewhat greater than in the independence model. In contrast to the other 'models' used here, this one allows the variance at different time points to vary independently of the mean, so that it is virtually a normal theory model.

For the uniform correlation GEE, the variance, ϕ, is estimated as 5.00 and the uniform correlation as 0.397. The regression parameter estimates are almost identical to those for the independence and overdispersion models. This is not surprising, because the latter also assumes a uniform correlation. However, the standard errors are much larger, even than those allowing for overdispersion, so that, again, only age seems to be significantly different from zero. Finally, for the AR(1) GEE, the variance estimate is 5.06 and that of the autocorrelation 0.497. The parameter estimates are fairly similar to those of the independence and overdispersion models and the uniform GEE, but the standard errors are similar to the latter.

These results confirm those of Section 6.2.2, that these data are very heterogeneous, and that, when this heterogeneity is taken into account, the other effects become nonsignificant. Zeger and Liang (1992) provide a slightly different analysis of these data, using an overdispersion parameter, with similar conclusions. However, as already mentioned, no adequate means of comparing our results to each other is available, so that we cannot judge which 'model' is superior for these data. It is not clear that the GEE approach has offsetting advantages over the true model building procedures of the previous sections, which would compensate for its lack of interpretability, and of comparability.

Among the large number of recent books on categorical and count data, very few treat the problem of overdispersion. An excellent exception is Collett (1991, Chapter 6), although it only concerns binary data. Agresti (1990, Chapter 10) looks at matched pairs data. In the same way, it is difficult to find an adequate treatment of longitudinal discrete data. Fingleton (1984, Chapter 6) gives a good account of many of the models connected with Markov chains; Collett (1991, Chapters 7 and 8) covers cohort studies, clinical trials, and binary time series; and Agresti (1990, Chapter 11) looks at certain models for repeated categorical responses.

Part IV

Models for duration data

7
Frailty

7.1 Characteristics of duration data

A duration is the waiting time to some event. When we have a series of successive events, with the accompanying durations, this is called an *event history*. Thus, such studies have at least two aspects, a continuous variable measuring the time that has past so far and a discrete variable indicating if an event has occurred or not. A third (set of) variable(s), called a mark, may indicate what kind of event it is, or other relevant information, yielding a marked point process. Such data can be studied in two equivalent ways: by the durations or by the frequency of events in given intervals. In the latter case, when the intervals are allowed to become very small, we obtain the intensity of the event which is strictly equivalent to the distribution of the durations, as we saw in Section 2.4.3. Modelling the intensity directly is often the most flexible approach, although the risk functions for certain common distributions are relatively complex.

In order to record a duration, units must be repeatedly, or continuously, observed over time, so that all duration data are inherently repeated measurements, measuring whether or not an event has occurred. Thus, survival and reliability data, where the first event is absorbing, can be considered to be repeated measurements. However, because there is already a vast literature on this subject, it will not be treated, in general, here.

In *survival studies*, observation on a unit begins when the characteristic, usually a disease or defect, of interest is first diagnosed and ends with an absorbing event, often death. *Incidence* studies are very similar, except that the duration is measured from the origin of the unit; it is a rate of occurrence, obtained by longitudinal follow up. In contrast, *prevalence* is the frequency in a population at a given time point. Naively, in a population of fixed size, prevalence = incidence × duration. Survival and incidence are modelled by an intensity, while prevalence is a probability. Population-averaged models for event data can be used to model prevalence, while conditional longitudinal models, such as Markov chains, model incidence. Subject-specific ones model individual risk.

In this section, some general remarks will be made on the specific problems of modelling duration data, before going on, in Section 7.2, to look at some simple preliminary models, and finally closing with models for heterogeneity in Section 7.3.

7.1.1 DISTRIBUTIONS

The construction of models for duration data immediately introduces a number of complications which were not encountered in the previous chapters. There, the models were essentially based on two different distributions, the normal in Part II and the multinomial, with its special cases, the binomial and Poisson with fixed total, in Part III. Here, things are not so simple, because many different distributions, and their corresponding intensity functions, exist to model the duration of time until an event. We looked at some of these in Section 2.4.3. Only some of the commoner, and simpler, ones can be covered in this text.

A first characteristic of observations of durations is that the response is a time which must be non-negative, and usually positive. This, then, implies that the distribution of responses will usually be skewed with a long tail on the right. Unless all durations are very long, the models based on the normal distribution of Part II can be excluded. A model should, optimistically, be selected in the knowledge of the mechanism generating the data. Some of these mechanisms were described in Section 2.4.3. Unfortunately, such a choice of a suitable distribution is rarely easy. Moreover, the data usually do not allow a useful direct evaluation of the goodness of fit, although the methods of Lindsey (1974a and b) can be used. More *ad hoc* graphical means, such as the study of residuals and empirical survivor curves, may be employed. In addition, as we shall see in several examples below, changing the distribution may drastically alter the conclusions to be drawn about the effect of explanatory variables on the duration or intensity.

Not only are many different distributions possible to describe durations, but no general family of multivariate distributions is available. Those mentioned in Section 2.5 are primarily for modelling heterogeneity and not for more general stochastic dependence relationships. Thus, it will perhaps be appropriate to present first some simpler models, those which assume no dependence among the durations on a unit. When the periods succeed each other in time, these will form renewal processes.

7.1.2 INTENSITIES AND COUNTING PROCESSES

Construction of the model directly through the duration distribution is only possible when the conditions on the unit remain constant over each period between events. If there are time-varying covariates which change within the period, then, this cannot be allowed for in the probability distribution. Instead, the intensity of the process must be modelled; of course, this can also be done in other simpler contexts as well. As we shall see, this will bring us back to extend the models of Chapter 6 for point processes. Because the intensity is the reciprocal of an 'instantaneous mean duration', this is similar to classical location models, with a reciprocal link function, but now in a dynamic context.

Models for intensities are most easily constructed in terms of a counting process (Section 2.10). As its name suggests, this is a random variable over time, N_t, which counts the number of events which have occurred up to t. As we saw in Equation (2.96), the corresponding intensity of the process is $\omega(t|\mathcal{F}_{t-};\boldsymbol{\beta})$, such that

$$\omega(t|\mathcal{F}_{t-};\boldsymbol{\beta})dt = \Pr[dN_t = 1|\mathcal{F}_{t-}]$$

where $\boldsymbol{\beta}$ is a vector of unknown parameters and $\mathcal{F}_{t-}$ is the filtration, or history, up to, but not including t. Then, the kernel of the log likelihood function for observation over the interval $(0,T]$ is

$$\log[L(\boldsymbol{\beta})] = \int_0^T \log[\omega(t|\mathcal{F}_{t-};\boldsymbol{\beta})]dN_t - \int_0^T \omega(t|\mathcal{F}_{t-};\boldsymbol{\beta})I(t)dt \quad (7.1)$$

where $I(t)$ is an indicator function, with value one if the process is under observation at time t and zero otherwise.

Now, in any empirical situation, even a continuous-time process will only be observed at discrete time intervals, once an hour, once a day, once a week. Suppose that these are sufficiently small so that at most one event is observed to occur in any interval, although there will be a finite nonzero theoretical probability of more than one, unless it is an absorbing event or a transition to another state. With R intervals of observation, not all necessarily the same size, Equation (7.1) becomes, by numerical approximation,

$$\log[L(\boldsymbol{\beta})] \cong \sum_{t=1}^R \log[\omega(t|\mathcal{F}_{t-};\boldsymbol{\beta})]\Delta N_t - \sum_{t=1}^R \omega(t|\mathcal{F}_{t-};\boldsymbol{\beta})I(t)\Delta_t \quad (7.2)$$

where Δ_t is the width of the tth observation interval and ΔN_t is the change in the count during that interval, with possible values zero and one. The implicit assumption, here, is that the intensity remains constant over each observation interval, Δ_t (or that an average is being used), but, in any case, we have no information about possible change within the interval. Equation (7.2) is now just the kernel of the log likelihood for the Poisson distribution of ΔN_t, with mean, $\omega(t|\mathcal{F}_{t-};\boldsymbol{\beta})\Delta_t$. Conditional on the filtration, it is the likelihood for a Poisson process. Any observable counting process is locally a Poisson process, conditional on the filtration. But, a Poisson likelihood is also equivalent to a multinomial likelihood for the joint distribution of the events over the complete observation period, conditional on the total observed number of events, $N_T = \sum \Delta N_t$.

The structure which we shall place on this multivariate likelihood will determine what stochastic process we are modelling. If it is multiplicative, we shall have a log linear model. For example, if the logarithm of the elapsed time since the previous event is included in the Poisson regression, we obtain a Weibull process, whereas, if the intensity is allowed to jump at each event time, we have the well-known Cox proportional hazards model.

The development just outlined assumes observation in continuous time. Sometimes, for some reason, a process may be taken to occur in discrete time, so that only one event can, even theoretically, occur to a unit in each time interval. This is most appropriate in survival situations, where only one (absorbing) event can occur, but has also been used for event history data, perhaps as an approximation. Then, the preceding analysis still holds, but the Poisson likelihood is replaced by a binomial one, with a binomial denominator of unity. However, in most situations, the results by the two methods are usually virtually indistinguishable.

7.1.3 CENSORING

By definition, responses which are measured as a duration occur over time. However, any research study must occupy a limited period, so that some durations will, almost invariably, be incomplete or *censored*. If observation must stop because the unit drops out of the study, there may be differential censoring in the population, which can bias the results. All appropriate measures should be taken to minimize this type of censorship. On the other hand, the general way in which observation is to stop for all units will be under the control of the research worker. Thus, a first question involves the nature of the stopping rule.

In the collection of survival data, two types of censoring are usually distinguished:

(1) all recording may be stopped after a fixed interval of time, called Type I or time censoring;

(2) recording may be continued until complete information is available on a fixed number of units, i.e. until a prespecified number of events has occurred, called Type II or failure censoring.

As examples of censoring in event histories, Aalen and Husebye (1991) suggest several possibilities:

(1) censoring time is fixed in advance, as in Type I above;

(2) censoring times are random variables, independent of the stochastic process;

(3) recording of each process stops when it has completed a fixed number of durations, as in Type II above;

(4) recording of each process stops either when it has completed a fixed number of durations or after a fixed time, whichever comes first;

(5) recording of each process stops at the first event after a fixed period.

The last three differ from the first two in that they depend on the event history, itself. However, because these three depend only on what has already been observed, the filtration, not depending on any as yet incompletely observed data, all five can be treated in the same way. This can be demonstrated using the concept of stopping time (Section 2.10) from martingale theory.

Suppose, now, that the last, incomplete, duration in Type I censoring of an event history is ignored, and not recorded. However, to know that the last duration recorded is the last complete one, one must look beyond it up to the fixed end point. This differs from the above examples, in that the cutting point depends on information after that point. It is not a martingale stopping time. One can also see that this will bias downwards the average duration time, because long intervals have a higher probability of being incomplete at the end.

The start of observation must also be at a 'stopping time'. Interestingly, this implies that incomplete durations at the beginning can be ignored. This will involve some loss of information, but no bias.

From Equation (2.35), the probability distribution of the durations is the product of the intensity times the survival function. This term will be included in the likelihood function for all complete durations. For a censored duration, we only know that there was no new event up to that point; this is given by the survival function. Thus, censored data will contribute to the likelihood function a term which is only the survival function. Then, for independent observations of durations, y_i, the likelihood function will be

$$L(\boldsymbol{\beta}; \mathbf{y}, \mathbf{x}) = \prod_i \omega(y_i; \boldsymbol{\beta}, \mathbf{x})^{I(y_i < t_i)} S(y_i; \boldsymbol{\beta}, \mathbf{x}) \qquad (7.3)$$

where t_i is the censoring time for the ith observation and $I(\cdot)$ is the indicator function. The estimation procedure is simpler when modelling the intensity directly, as in Equation (7.2) and the following discussion, because the discrete variable indicating occurrence of an event simply always remains at zero if there is censoring before an event. However, Equation (7.3) can be rewritten in the form of Equation (7.1) by using the relationship between the survival function and the integrated intensity function.

7.2 Models without dependence

7.2.1 RENEWAL PROCESSES

A renewal process (Section 2.10) is a point process for which successive durations are independent. Such models are reasonably realistic in certain situations, such as in industry, where a repaired machine might be

Table 7.1. Number of operating hours between successive failures of air-conditioning equipment in 13 aircraft. (Cox and Lewis, 1966, p. 6)

					Aircraft Number							
1	2	3	4	5	6	7	8	9	10	11	12	13
194	413	90	74	55	23	97	50	359	50	130	487	102
15	14	10	57	320	261	51	44	9	254	493	18	209
41	58	60	48	65	87	11	102	12	5		100	14
29	37	186	29	104	7	4	72	270	283		7	57
33	100	61	502	220	120	141	22	603	35		98	54
181	65	49	12	239	14	18	39	3	12		5	32
	9	14	70	47	62	142	3	104			85	67
	169	24	21	246	47	68	15	2			91	59
	447	56	29	176	225	77	197	438			43	134
	184	20	386	182	71	80	188				230	152
	36	79	59	33	246	1	79				3	27
	201	84	27	15	21	16	88				130	14
	118	44	153	104	42	106	46					230
	34	59	26	35	20	206	5					66
	31	29	326		5	82	5					61
	18	118			12	54	36					34
	18	25			120	31	22					
	67	156			11	216	139					
	57	310			3	46	210					
	62	76			14	111	97					
	7	26			71	39	30					
	22	44			11	63	23					
	34	23			14	18	13					
		62			11	191	14					
		130			16	18						
		208			90	163						
		70			1	24						
		101			16							
		208			52							
					95							

considered as good as new, so that it is starting its life afresh.

Let us look at the classical data set on machine repair, that for successive failures of air-conditioning equipment in 13 Boeing 720 aircraft, reproduced in Table 7.1 taken from Cox and Lewis (1966, p. 6, originally from Proschan). We shall look for differences in failure rates among the 13 planes. The counting processes of cumulated numbers of failures over time are plotted, for the 13 aircraft, in Figure 7.1. We see how some aircraft are accumulating breakdowns much faster than others.

Two models which might be considered for these data are the gamma and inverse Gaussian distributions (Section 2.4). For each, we can fit a model with a constant location or a constant scale parameter, as well as one

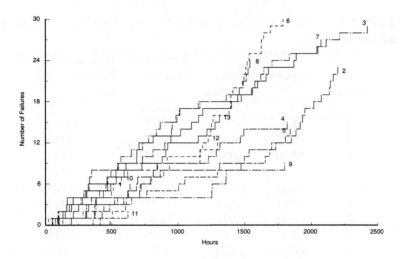

Fig. 7.1. Counting processes of failures for the 13 aircraft from Table 7.1.

Table 7.2. Changes in deviance, all with 12 d.f., for several models for the aircraft data of Table 7.1

	Gamma	Inverse Gaussian	Reciprocal Gamma	Reciprocal Inverse Gaussian
Location	22.3	6.5	83.0	28.9
Scale	73.2	64.9	103.6	16.9

where both parameters change with the aircraft. The changes in deviance for the two distributions are given in Table 7.2. For the inverse Gaussian distribution, no difference in mean duration seems to be present, while, for the gamma distribution, there is some indication of a difference. On the other hand, both models appear to require different scale parameters for the aircraft. The amount of variability among failure times varies among the planes. The parameters are presented in Table 7.3. Recall that a scale parameter of one for the gamma distribution yields an exponential distribution which corresponds to a Poisson process. If the gamma model were judged better than the inverse Gaussian, a Poisson process could be suitable for a number of the aircraft.

Two further models can easily be modelled in a similar way. These are the reciprocal gamma and reciprocal inverse Gaussian distributions, fitted by taking the reciprocal durations as response. The deviances and parameter values for these models are also presented in Tables 7.2 and 7.3. Here, we find differences in mean duration for both models, but there is now

Table 7.3. Scale parameter estimates for several models for the aircraft data of Table 7.1.

Aircraft	Gamma	Inverse Gaussian	Reciprocal Gamma	Reciprocal Inverse Gaussian
		Location Parameter*		
1	82.17	6752.	36.03	1298.0
2	95.72	9166.	30.73	943.9
3	83.52	6976.	42.45	1801.6
4	121.30	14717.	40.24	1619.3
5	131.51	17298.	61.75	3813.0
6	59.61	3554.	11.18	125.0
7	76.82	5901.	14.50	209.0
8	64.13	4112.	18.26	333.2
9	200.04	40000.	8.59	74.0
10	106.51	11344.	17.68	312.3
11	311.53	97087.	205.73	42331.0
12	108.11	11692.	14.78	218.0
13	82.00	6725.	42.89	1839.6
		Scale Parameter		
1	1.08	0.019	0.90	55.4
2	1.26	0.023	1.12	67.9
3	0.68	0.012	0.72	42.5
4	1.37	0.018	0.99	86.8
5	0.72	0.009	0.90	75.1
6	1.51	0.075	1.95	50.1
7	1.04	0.058	2.42	64.7
8	1.13	0.041	1.49	47.9
9	3.15	0.125	3.93	215.3
10	2.11	0.057	2.20	106.6
11	0.83	0.003	0.83	211.5
12	1.86	0.064	2.48	101.8
13	0.67	0.012	0.72	41.7

*For comparability, the first two columns are reciprocal location parameters.

little difference in scales for the reciprocal inverse Gaussian distribution. In fact, all four of our models are imbedded in a generalized inverse Gaussian distribution. Jørgensen (1982, pp. 116–153; see also Jørgensen, Seshadri, and Whitmore, 1991; Jørgensen, 1992) shows that the special case of that more general model which fits these data acceptably is the reciprocal inverse Gaussian distribution.

All models indicate differences among the aircraft. We may note, particularly, that aircraft 11 has consistently the largest location parameter for

all models which we have fitted. On the other hand, aircraft 9 is indicated
as having much greater variability in failure rates than the others by all of
the models, while its location parameter varies wildly among the models.
If we were not interested in modelling specific differences among aircraft,
a frailty model (Section 7.3) would be more appropriate for these data.

The most unsettling aspect of this example is that different models
can lead to quite different conclusions about the mean durations and the
variability of duration across units. Great care must be taken in selecting
a model and checking its assumptions with the data. This has led many
statisticians to rely primarily on the 'semi-parametric' Cox (1972a) proportional hazards model, a highly parametrized model, allowing the intensity
to change between all pairs of events.

7.2.2 SIMPLE EVENT HISTORIES

As an example where significant censoring is present, let us look at event
histories which are neither renewal processes nor survival times. In the
study of the disease progression of floating gallstones, biliary pain may
be experienced, and, if so, it may be followed by cholecystectomy. If the
former does not occur, i.e. is censored, the latter will also be censored. A
series of observations on 113 patients under two treatments, placebo and
the drug, chenodiol, from Wei and Lachin (1984), are presented in Table
7.4.

The generalized extreme value distribution

$$f(y; \mu, \kappa, \nu) = \kappa \mu^{-\kappa} e^{\kappa y^\nu} e^{-\mu^{-\kappa} e^{\kappa y^\nu}} \tag{7.4}$$

contains the Weibull ($\nu \to 0$) and extreme value ($\nu = 1$) distributions
(Section 2.4.3) as special cases. We fit this model with a different mean for
time to pain and subsequent time to cholecystectomy, with and without
a difference in mean for treatment. Because ν is estimated to be very
close to zero in all cases, the Weibull distribution is indicated. The power
parameter of this distribution is estimated as $\hat{\kappa} = 0.894$, indicating that an
exponential distribution might be adequate. And, indeed, the difference
in deviance is only 1.53 on 1 d.f. The change in deviance for treatment is
6.6 with 1 d.f. with either of these distributions, with parameter estimate,
-0.39, indicating a longer period between events when treatment is applied.
Introducing an interaction between treatment and type of event changes the
deviance very little, showing that the drug extends the time to both events
about the same amount. The Weibull intensities are plotted for the two
types of events under the two treatments in Figure 7.2. The corresponding
exponential intensities would be parallel horizontal straight lines.

In this example, we have assumed the same form of distribution for time
to pain and to cholecystectomy. Only the mean duration of each is allowed
to vary. The two times for an individual are not linked together in any

Table 7.4. Floating gallstone progression through time in days to experience of biliary pain and cholecystectomy. (Wei and Lachin, 1984)

Placebo				Treatment			
741*	741*	35	118	735*	735*	742*	742*
234	234	175	493	29	29	360*	360*
374	733*	481	733*	748*	748*	750	750*
184	491	738*	738*	671	671	360*	360*
735*	735*	744*	744*	147	147	360*	360*
740*	740*	380	761*	749	749	726*	726*
183	740*	106	735*	310*	310*	727*	727*
721*	721*	107	107	735*	735*	725*	725*
69	743*	49	49	757*	757*	725*	725*
61	62	727	727*	63	260	288	810*
742*	742*	733*	733*	101	744*	728*	728*
742*	742*	237	237	612	763*	730*	730*
700*	700*	237	730*	272	726*	360*	360*
27	59	363	727*	714*	714*	758*	758*
34	729*	35	733*	282	734*	600*	600*
28	497			615	615*	743*	743*
43	93			35	749*	743*	743*
92	357			728*	728*	733*	755*
98	742*			600*	600*	188	762*
163	163			612	730*	600*	600*
609	713*			735*	735*	613*	613*
736*	736*			32	32	341	341
736*	736*			600*	600*	96	770*
817*	817*			750*	750*	360*	360*
178	727			617	793*	743*	743*
806*	806*			829*	829*	721*	721*
790*	790*			360*	360*	726*	726*
280	737*			96	720*	363	582
728*	728*			355	355	324	324
908*	908*			733*	733*	518	518
728*	728*			189	360*	628	628
730*	730*			735*	735*	717*	717*
721*	721*			360*	360*		

First column: Time to Pain

Second column: Time to Cholecystectomy

Stars indicate censoring

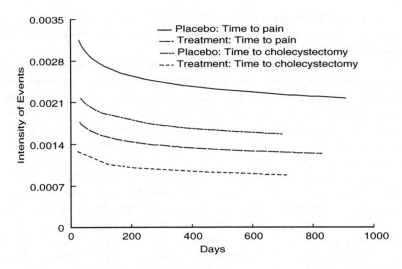

Fig. 7.2. Weibull intensity functions for time to pain and to cholecystectomy, from Table 7.4.

way and all individuals are taken to have the same susceptibility to each of them. These simplifying assumptions will be relaxed in various ways in the rest of this chapter and in the next. For example, here, a semi-Markov model might be more appropriate, because the first event brings about a change of state.

7.3 Compound models

7.3.1 RISK AND HETEROGENEITY

Comparison of event histories on different units may show heterogeneity among them. Care must be taken because this heterogeneity can distort the estimation of the intensity function, so that it does not represent the intensity of events for any possible unit. This is particularly true for survival data where the event is absorbing. Vaupel and Yashin (1985) and Blossfeld and Hamerle (1992) plot a number of vivid examples. Much recent research has been done in this area, especially in the econometrics literature (see, for example, Heckman and Borjas, 1980; Lancaster, 1990).

A heterogeneous population is one where each unit may have a different intensity function for the events: a different frailty or proneness. For survival data, as time passes, the more frail units will tend to have the event early and, thus, disappear, so that the overall intensity will descend towards that of the least frail subgroup, which has lower risk of failure. This result is well known in reliability theory, where a system, with all components having increasing failure rates, can eventually have a decreasing failure rate. In fact, a similar heterogeneity problem is inherent in any long term longitudinal repeated measurements study, no matter what type of response is being recorded. Frail units may disappear early, distorting trends later in the study period.

For event history data, the more frail units will have more events in a given observation period. In heterogeneous populations, although the individual intensity functions may have a very simple form, the resulting overall intensity can be more complex. The latter is the estimate of the intensity for the overall (heterogeneous) group, which, however, may be what is of interest. But, as for models for normally-distributed and categorical data, covariate parameters, such as those for treatment effects, will be estimated too precisely if the heterogeneity is not taken into account.

Whitmore and Lee (1991) suggest that intensity functions are often determined by a combination of two stochastic processes. In engineering, one process is environmental: the random levels of stress created by operating characteristics such as electrical voltage, ambient temperature, or vibration. The second process is specific to the manufacturing process which originally created the unit and imparts random imperfections to the physical structure of the unit. In the simplest case, a group of units operating under the same environmental conditions and having the same physical make–up will have a common independent constant intensity (i.e. an exponential distribution), which is a function of the stress and of the number of imperfections during manufacture. These intensities can, then, be assumed to vary randomly across groups of units according to some distribution.

They further argue that often the random intensity is determined by a stopping or first hitting time in some process. For example, if this is a Wiener process, as is often the case in chemical and physical reactions, the distribution of first hitting times will be inverse Gaussian (Sections 2.4.1 and 2.10). In this simple case, this results in an exponential distribution compounded with an inverse Gaussian distribution.

In medical applications, the randomly varying risk is determined by factors known as frailty, while in the study of accidents and crime victims, it is called proneness. As Hougaard (1984) notes for survival data, in the more complex case where the risk is not constant, the effect of age is greater on individual unit mortality than on population mortality. A large decrease in individual risk may give only a small decrease in population risk; continued decreases in unit mortality may give accelerated decreases in population mortality; if decreases in unit mortality stop, population risk may increase. When the compounding distribution is gamma, the relative frailty distribution among surviving units is independent of age, while, for the inverse Gaussian, the population becomes more homogeneous over time.

When several intensities are involved, as in semi-Markov processes with transitions among several states, it may be desirable to have a different frailty, or random effect, for each intensity.

Thus, heterogeneity in duration data is a complex issue, which must be handled with care. As in previous chapters, compound distributions can be used to account for heterogeneity of duration times among units, their differential frailty or proneness. Here, we shall give a few examples to illustrate this approach.

7.3.2   A MULTIVARIATE LOG LOGISTIC MODEL

In this first example, we shall model the duration times and introduce heterogeneity into the probability distribution. The logistic distribution can be generalized to the multivariate case (Scallan, 1987) for random effects in repeated measurements of durations. Consider the distribution function

$$F(\mathbf{y}) = \left[1 + \sum_{k=1}^{R} \frac{e^{(-\phi_k y_k - \sum_{l=1}^{P} x_l \beta_l)}}{\gamma} \right]^{-\gamma} \qquad \gamma, \phi > 0, -\infty \le y \le \infty \tag{7.5}$$

for R repeated measures and $P-1$ explanatory variables. The ϕ_k are scale parameters, homogeneous over units, but not necessarily over responses and the β_l are linear location parameters.

The moments of the standardized variable, $Y_k^* = \phi_k Y_k + \sum_l x_l \beta_l, k = 1, \dots, R$ are

$$E[Y_k^*] = \Psi(\gamma) - \Psi(1) - \log(\gamma) \tag{7.6}$$
$$\mathrm{var}[Y_k^*] = \Psi^1(\gamma) + \Psi^1(1) \tag{7.7}$$
$$\mathrm{cov}[Y_k^*, Y_m^*] = \Psi^1(\gamma) \tag{7.8}$$

where $\Psi^{n-1}(u) = \frac{\partial^n}{\partial u^n} \log[\Gamma(u)]$ is the polygamma function.

The correlation, $\mathrm{corr}[Y_k^*, Y_m^*] = \Psi^1(\gamma)/(\Psi^1(\gamma) + \Psi^1(1))$, between any two repeated measures on a unit is a function of the parameter, γ, and goes from 1 to 0 as $\gamma \to \infty$. Because the responses, $\mathbf{y}$, can take any real value, for duration data, it is more appropriate to take logarithms of the data, yielding a log logistic model.

The likelihood function for the R observations on one individual is

$$\frac{\partial F(\mathbf{y})}{\partial \mathbf{y}} = \tag{7.9}$$

$$\gamma(\gamma + 1) \cdots (\gamma + r - 1) \frac{\prod \exp(-\phi_k y_k - \sum_l x_l \beta_l - \log(\gamma))}{[1 + \sum \exp(-\phi_k y_k - \sum_l x_l \beta_l - \log(\gamma))]^{\gamma + R}}$$

For fixed ϕ and γ, this is a log linear model, so that it is relatively easy to estimate the model for various values of these parameters.

Consider a very simple example where 11 subjects are each given five probe words and their reaction times measured. The data, given in Table 7.5, are from Timm (1980). For this multivariate log logistic model, the

Table 7.5.   Reaction times (sec.) of eleven subjects to five probe words.
(Timm, 1980)

Probe Word				
1	2	3	4	5
51	36	50	35	42
27	20	26	17	27
37	22	41	37	30
42	36	32	34	27
27	18	33	14	29
43	32	43	35	40
41	22	36	25	38
38	21	31	20	16
36	23	27	25	28
26	31	31	32	36
29	20	25	26	25

correlation among the responses of a subject is estimated to be 0.69 when
the scale parameter is held constant for all probe words. A deviance plot
(not given) indicates that a zero correlation is very implausible. When
the scale parameter is allowed to vary, the deviance is only decreased by
5.93 on 4.d.f. Thus, there is a relatively high correlation among reaction
times on a subject, but with little difference in reaction time across probe
words. Some subjects, such as the first, have relatively long reaction times,
whereas others, such as the last, have consistently short times. This should
be taken into account if more complex models were to be constructed using
additional data.

7.3.3   MULTIVARIATE PROPORTIONAL HAZARDS MODELS

Next, we shall model, directly, the intensities, introducing heterogeneity
into that function, Suppose that the intensity for the ith unit is given by

$$\omega_i(t|\mathcal{F}_{t-};\boldsymbol{\beta}) = \lambda_i \omega_{i0}(t|\mathcal{F}_{t-};\boldsymbol{\beta}) \qquad (7.10)$$

where $\omega_{i0}(\cdot)$ is the part of the intensity involving the observed covariates
and λ_i is a random effects parameter, called the frailty, having a gamma
distribution with unit mean and variance δ. Responses on a unit will
be independent when $\delta = 0$ and increasingly stochastically dependent as
$\delta \to \infty$. In this way, we obtain a multivariate proportional hazards model
(Clayton, 1988; Clayton and Cuzick, 1985), which could also be derived
as a copula (Section 2.5.1), and which is, in fact, closely related to the
multivariate log logistic distribution of the previous section. Here, we shall
use the multiplicative intensity (Aalen, 1978b) extension of the Cox (1972a)
proportional hazards model for $\omega_{i0}(\cdot)$, which can be fitted as a log linear

Table 7.6.   Survival times (days) of closely and poorly matched skin grafts on the same person. (Holt and Prentice, 1974)

Close Match	Poor Match	% Surface Burned
37	29	30
19	13	20
57*	15	25
93	26	45
16	11	20
23	18	18
20	26	35
18	23	25
63	43	50
29	15	30
60*	42	30

Stars indicate censoring

model. Then, the compound distribution is negative binomial (Lawless, 1987b), as in Section 5.2.1.

For the probe word reaction times of Table 7.5, three models can be fitted: 1. a different intensity function for each word, 2. proportional functions for the words, and 3. identical intensities for all words. The differences in deviance are, respectively, 74.0 on 19 d.f. and 25.8 on 4 d.f., here indicating that the intensities are different for each word, in contrast to our previous results. For this model, we obtain an estimate, $\hat{\delta} = 0.80$, again indicating strong correlation among the responses by a subject. Although the deviance plot (not given) is fairly flat, especially around $\delta = 1$, a value near zero is clearly unlikely ($\delta = 0.25$ gives a difference in deviance of 3.8).

As a second example, consider the survival times in days of closely and poorly matched skin grafts on the same person, given in Table 7.6 (Holt and Prentice, 1974). For these data, we can again fit the three models, with changes in deviance of 5.42 on 3 d.f. and 3.40 on 1 d.f. There is but little indication of a difference between the two levels of matches in the risk function for losing the graft. The estimate δ is very close to zero, indicating less stochastic dependence than for the probe words, but this is not surprising, because here there are only two observations per unit, and some censoring. Again, the deviance plot (not given) is very flat, indicating that zero dependence cannot be excluded. In other words, from these data, there appears to be little variability in graft retention among the subjects.

We can now add the effect of percentage of body burned as an explanatory variable. It was of little use to do it before, because it was virtually confounded with the random effect. This covariate reduces the deviance of the model with no difference for type of match by 7.13 with 1 d.f. Inter-

estingly, when match is now reintroduced, it further reduces the deviance
by 5.54. The parameter estimates, on the log scale, are 1.18 for poor, as
compared to close match, and −0.093 for burn. The interaction between
the two variables is negligible. Poor matches have a higher risk of rejection,
but the risk decreases with increasing burn area.

7.3.4 NORMAL COMPOUND MODELS

As for discrete data, it is relatively simple to allow for frailty by introducing random parameters which have a normal distribution and performing
numerical integration. As in the previous sections, this can be done either
with the probability distribution or with the intensity function. Because the
latter is similar to what we saw in Section 5.2.2, we shall treat compounding
with the probability distribution here.

Because the Weibull distribution is not a member of the exponential
family, and hence does not have a simple conjugate distribution with which
it could be compounded, let us use it here. The general form of such
a compound distribution was given in Equation (2.57) and that specific
to the normal compounding distribution in Equation (5.38). Then, the
Weibull distribution of Equation (2.41) is applied to this to obtain the
required result.

For the reaction times to probe words, in Table 7.5 above, the model
with random effect has a deviance reduced by 11.82 (1 d.f.) with respect to
the ordinary (independence) Weibull model, when differences among the
words are taken into account. The estimate of the variance of the normal
compounding distribution is $\hat{\delta} = 0.264$. For this compound model, the
deviance is increased by 9.16 with 4 d.f. when the differences among words
are not taken into account. This is similar to, although less marked than,
the result which we obtained for the multivariate proportional hazards
model of Section 7.3.3 above. This is not surprising, because we have here
another multivariate proportional hazards model, but where the baseline
intensity is parametric instead of allowing discontinuous changes at each
event time.

Let us now look at a second example, with censoring. This concerns
the times to tumour occurrence, in weeks, of one rat exposed to a putative
carcinogen and two control rats in each of 50 litters, given in Table 7.7
(from Clayton, 1991). Because each set of three rats comes from the same
litter, we might expect more similar results within each litter than across
litters.

Here, the difference in deviance for the compound, as compared to the
independence, model, with treatment effect, is only 1.24 on 1 d.f., showing
that there is little variability among the litters. Under independence, the
deviance for the treatment effect is 7.99, again with 1 d.f., with a lower
intensity of tumour occurrence, −0.90 on a log scale, for the control rats.

Table 7.7. Times to tumour occurrence (weeks) of exposed and control rats from 50 litters. (Clayton, 1991)

Exposed	Controls		Exposed	Controls	
101*	49	104*	89*	104*	104*
104*	102*	104*	78*	104*	104*
104*	104*	104*	104*	81	64
77*	97*	79*	86	55	94*
89*	104*	104*	34	104*	54
88	96	104*	76*	87*	74*
104	94*	77	103	73	84
96	104*	104*	102	104*	80*
82*	77*	104*	80	104*	73*
70	104*	77*	45	79*	104*
89	91*	90*	94	104*	104*
91*	70*	92*	104*	104*	104*
39	45*	50	104*	101	94*
103	69*	91*	76*	84	78
93*	104*	103*	80	81	76*
85*	72*	104*	72	95*	104*
104*	63*	104*	73	104*	66
104*	104*	74*	92	104*	102
81*	104*	69*	104*	98*	73*
67	104*	68	55*	104*	104*
104*	104*	104*	49*	83*	77*
104*	104*	104*	89	104*	104*
104*	83*	40	88*	79*	99*
87*	104*	104*	103	91*	104*
104*	104*	104*	104*	104*	79

Stars indicate censoring

With the random effect, this estimate becomes −0.93, with about the same change in deviance for treatment.

As in previous parts of this book, we see that, although it is important to check for heterogeneity, it will not always be present where we might expect it. At times, the experimental units selected may be very uniform.

In Section 8.4.2, we shall look at an example where the compounding normal distribution is introduced into the intensity function instead of the duration distribution.

8
Event histories

8.1 Repeated durations

As we have seen, the study of the durations of successive periods, or waiting times, between events occurring to a unit is often called a life or event history. In the simplest cases, events are accumulated without any fundamental change in the unit: a point process. In more complex situations, each event induces a change of state for the unit: a semi-Markov process (Section 2.10). As we saw in Section 7.2.1, in such longitudinal duration studies, if the successive measurements of duration for each unit are independent and identically distributed, then we have multiple renewal processes. All of the standard techniques of survival analysis may be used. The problem may be even simpler, because censoring is often not present. Our example there was one where times between successive failures of machines were compared (this would usually be under different conditions), where the machine is assumed to be 'as new' after repair. Note, however, that, in this context, a washout period is not usually possible, so that the renewal assumption may not be realistic, for example, when living subjects are involved. It may rather be necessary to model the stochastic dependence among successive responses.

In simple cases, we can model these durations between events directly, using an appropriate probability distribution. As we have seen in Section 2.2.1, for models based on the normal distribution, the mean and variance can vary independently of each other, because the variance function is $\tau^2 = 1$. For other distributions, including those which we might choose for durations, the mean and the variance are functionally related. If the covariances among responses are nonzero, these usually will also be functions of the means, as will any higher order stochastic dependencies. Two approaches are possible: 1. condition on previous responses or 2. try to model the covariance matrix directly. Usually, the latter is unfeasible for non-normal data, so that here we shall concentrate on the former.

The durations of event histories are not the only types of repeated measurement responses, concerning continuous variables, which might be non-normally distributed. Any other type of response which is constrained to be positive can be modelled by the same techniques, if an appropriate duration-type distribution can be found for it. Examples of such data will also be given below.

When the process is changing over time, and specifically between events, it can no longer be modelled directly by the duration distribution. In such cases, the intensity function must be used, although, of course, it can also be used in the simpler cases as well. However, it will not make sense for the other types of non-normally distributed responses just mentioned. For a point process, there will be one intensity function; for semi-Markov processes, there is one function for each possible transition between states. As we saw in Section 7.1.2, the simplest models are those for multiplicative intensities, because they form log linear or logistic models for discrete data.

If there are time-varying covariates, they may follow stochastic processes as well as the response being recorded. Only the case of external time-dependent covariates (Kalbfleisch and Prentice, 1980, pp. 122–127), which do not depend on previous durations, will be considered. Models with internal covariates are multivariate for two or more different types of responses, and are, thus, outside the scope of this book.

8.2   Distribution-based models

8.2.1   AUTOREGRESSION

In Section 2.7, we explored several ways in which analogues of the normal-theory autoregression models could be constructed. Thus, if it is thought that the present duration depends on previous such times, one possible approach is to fit lagged values of the response variable as covariates. When positive-valued responses, which do not form an event history, are being observed, those observations should be equally spaced in time. In some cases, it may be more suitable to transform the lagged response by the link function

$$g(\mu_{ij}) = \sum_l \beta_{il} x_{ijlt} + \sum_h \rho_h \left[g(y_{ij,t-h}) - \sum_l \beta_{il} x_{ijl,t-h} \right] \qquad (8.1)$$

because, for non-normal models, this function is not usually the identity. Thus, care may need to be taken with the form of the stochastic dependence in conditional models. For example, if the response times have a gamma distribution with canonical link (reciprocal), we might want to use

$$\frac{1}{\mu_t} = \sum_{l=1}^{P} \beta_l x_{lt} + \sum_{h=1}^{M} \frac{\rho_h}{y_{t-h}} \qquad (8.2)$$

as suggested by Zeger and Qaqish (1988).

As noted in Section 2.7, in contrast to normal regression models, the marginal distributions will have a different form from the corresponding

Table 8.1. Days to viral positivity in AIDS blood samples. (Wei, Lin, and Weissfeld, 1989)

Month								
1	2	3	1	2	3	1	2	3
Placebo			Low Dose			High Dose		
9	6	7	6	4	5	13	7	21*
4	5	10	16	17	21*	16	6	20
6	7	6	31	19*	21*	3	8	6
10	-	21*	27*	19*	-	21	-	25*
15	8	-	7	16	23*	7	19	3
3	-	6	28*	7	19*	11	13	21*
4	7	3	28*	3	16	27*	18*	9
9	12	12	15	12	16	14	14	6
9	19	19*	18	21*	22	8	11	15
6	5	6	8	4	7	8	4	7
9	-	18	4	21*	7	8	3	9
9	20*	17*	21	9	8*	19*	10	17*

Stars indicate censoring

conditional distributions, and neither will usually be tractable.

Let us look at data from a clinical trial to evaluate the effectiveness of the drug, ribavirin, on patients with AIDS (Wei, Lin, and Weissfeld, 1989), reproduced in Table 8.1. Three groups of 12 patients were assigned to placebo, low dose, and high dose. Blood samples, collected after four, eight and twelve weeks, were evaluated by determining the number of days until viral positivity was detected, as measured by a p24 antigen level greater than 100 picograms per millilitre. Although the responses are duration times, this is not a point process because they are not times between successive events. Because we shall be using autoregression, those patients with missing data at the second month will have no usable data with an AR(1) and must be dropped from all of the analyses; those with missing data at month three are used.

If we fit a generalized extreme value distribution, as in Equation (7.4), we obtain a value of ν close to zero, indicating that a Weibull distribution may be suitable. When we fit this distribution, the difference in deviance for treatment effects using a log link and all of the data, except that mentioned above, is 12.92 with 2 d.f. However, with an autoregression, we shall lose the first month's values. With this smaller set of observations, the effect of treatment now only gives a difference in deviance of 5.69. Adding the lagged response time of the previous measurement, taking the censored values as they stand, further reduces the deviance by 4.69 with 1 d.f. If we replace the lagged times by their logarithms, the link function being used here for the response, the final model is 0.74 smaller, indicating that the latter are slightly preferable. The parameter estimate, on a log scale, is 0.62 for the lagged log time, indicating longer times are followed by other long times, and conversely. The estimates of treatment effects, $(0.00, 0.73, 0.24)$, show that time to viral positivity is longest with the low dose. The high dose time is not much different from placebo. The Weibull power parameter has estimate 1.68, which is some distance from an exponential distribution.

This same procedure could have been used in this example, even if time-varying covariates were available. This would also be possible for event histories, as long as the covariates are constant over each duration period. If the covariates can change between events, the intensity function must be modelled directly.

8.2.2   CONDITIONAL EXPONENTIAL MODELS

The conditional exponential family (Feigin, 1981) of Section 2.7.2 provides a different generalization of the normal theory autoregression model. Here, we shall concentrate on the gamma distribution, which is most appropriate for duration data. In that model, given in Equation (2.76), the power or dispersion parameter is equated to the lagged response value. This is easily fitted as a generalized linear model by specifying the lagged values as prior weights.

The autoregression relationship, given by Equation (2.77), may be modelled in several ways. If we choose a log link, we obtain

$$\log(\mu_{jt}) = \log(y_{j,t-1}) + \log(\rho_j) \qquad (8.3)$$

Here, the log lagged response is an offset and the linear model is in terms of ρ_j. Thus, only changes in the autoregression coefficient, with covariates, can be introduced into the model to influence the mean. In most situations, this will be of little use, although an extension of this approach will be given in Section 8.2.3 below. A more promising alternative is to use the identity link in an additive location model, such as

$$\mu_{jt} = \rho_j y_{j,t-1} + \boldsymbol{\beta}_j^T \mathbf{z}_{jt} \qquad (8.4)$$

This is similar in conception to the autoregression model of Section 8.2.1 above, except that the power or dispersion parameter is changing in time, as well as the mean.

We shall fit this model to times, in months, between recurrences of tumours for 118 patients with bladder cancer, given in Table A7 (Andrews and Herzberg, 1985, pp. 254–259). Patients were randomized to three treatment groups: placebo, pyridoxine, and thiotepa. The 56 patients who did not suffer tumour recurrence will not be used in the present analysis. Although the placebo group had a lower percentage without tumour recurrence: 40% as compared to 53% in the two treatment groups, this difference is not significant (deviance 2.02 on 2 d.f.). Because we have a conditional model, those patients with only one recurrence will also be unusable, unless we were to model this initial distribution, as will be done with the model in the next subsection.

When we fit the conditional exponential model for the gamma distribution, we find that treatment has a significant effect, with a change in deviance of 20.53 on 2 d.f. If we add a variable to account for the number of previous recurrences, the deviance only decreases by 2.60 on 1 d.f. The placebo and thiotepa groups are similar with average recurrence times of about 7.5 months, while the pyridoxine group has a significantly *shorter* time of 5.7 months. The crude average recurrence times for the three groups are similar: respectively, 7.2, 5.3, and 7.1 . The estimate of the autoregression coefficient is $\hat{\rho} = -0.07$ (s.e. 0.013), which, although small, is significantly different from zero.

If we fit the more usual autoregression model, with a gamma distribution and identity link, but constant dispersion parameter, neither treatment nor number of previous recurrences shows a significant effect. The autoregression coefficient is estimated to have about the same value, but is now not significantly different from zero. The same type of autoregression model, but with log link and log lagged recurrence time gives a dependence on the number of previous recurrences (deviance 24.3) but no treatment effect (deviance 3.8). Here, the autoregression coefficient is estimated to be positive (0.08), but is not significantly different from zero.

Thus, when the dispersion parameter is held constant, treatment shows no effect but number of previous recurrences does, but the opposite is true when the dispersion parameter is equated to the lagged recurrence time. Once again, as in the previous chapter, different models lead to very different conclusions, pointing to the great care which must be taken in modelling the stochastic dependence among responses which do not have a normal distribution.

8.2.3 EXPONENTIAL DISPERSION MODELS

The exponential dispersion model (Jørgensen, 1986, 1987, 1992; see Section 2.5.3) is a generalization of the generalized linear model and also of the conditional exponential family. Consider a first order autoregression model with gamma conditional distributions. Suppose that

$$Y_1 \sim \text{ED}^{(2)}(\mu_1, \sigma^2)$$

$$Y_t | y_{t-1} \sim \text{ED}^{(2)} \left(\rho_t^* y_{t-1}, \frac{\sigma^2}{\rho y_{t-1}} \right) \qquad t = 2, \ldots, R \qquad (8.5)$$

where $\text{ED}^{(2)}(\cdot)$ signifies an exponential dispersion model with $\text{var}[Y_t] = \mu^2$, i.e. a gamma distribution. Thus, in contrast to the conditional exponential

Table 8.2. Rate of perceived exertion of people doing exercise. (Jørgensen, 1987)

\multicolumn{8}{c}{Minutes}							
6	12	20	40	6	12	20	40
\multicolumn{4}{c}{Female}				\multicolumn{4}{c}{Male}			
9	11	12	13	13	15	16	–
10	12	14	17	13	15	16	17
11	13	15	17	14	15	16	–
13	14	14	16	11	13	13	15
8	11	14	16	13	15	17	–
12	14	15	–	14	15	17	–
				14	15	17	–

Hyphens indicate censoring

model of Section 8.2.2, we, here, include both the autoregression parameter and a distinct dispersion parameter. In other words, the dispersion
parameter is proportional to the lagged response, instead of equal to it.
For this reason, we shall need to model the initial conditions, as well as
the subsequent conditional distribution in order to be able to estimate all
of the parameters.

The mean values, $\mu_t = \mathrm{E}[Y_t]$, $t = 1, \ldots, R$, satisfy $\mu_t = \rho_t^* \mu_{t-1}, t =$
$2, \ldots, R$. With a log link for the means, we have, for the conditional means,

$$\log(\mu_1) = \mathbf{z}_1^T \mathbf{B}$$
$$\log(\rho_t^* y_{t-1}) = \log(y_{t-1}) + (\mathbf{z}_t - \mathbf{z}_{t-1})^T \mathbf{B} \qquad t = 2, \ldots, R \qquad (8.6)$$

With ρ (a mean of the ρ_t^*) known, this is a generalized linear model with a
gamma distribution, log link, vector of offsets, $(0, \log(y_1), \ldots, \log(y_{R-1}))^T$,
prior weights, $(1, \rho y_1, \ldots, \rho y_{R-1})^T$, and design matrix with columns $\mathbf{z}_1^T$ and
$(\mathbf{z}_t - \mathbf{z}_{t-1})^T$, $t = 2, \ldots, R$. To estimate ρ, we fit separate models for the
marginal distribution of y_1 and the conditional distribution of $y_2, \ldots, y_R$
given y_1, as in Equations (8.5). As seen from these equations, the ratio of
dispersion parameters will give the required estimate of ρ.

As an example, we shall look at another data set which does not involve
an event history, although these models can also be applied in that case.
Consider subjects doing physical exercise who were instructed to express
their rate of perceived exertion on a scale from 6 to 20 at four times, 6, 12,
20, and 40 minutes after the start of the exercise period, or until exhaustion.
Data for six women and seven men are given in Table 8.2, from Jørgensen
(1987).

The conditional model with a common rate of change of perceived exertion over time (β) for all subjects fits about as well as that with different
rates for the two sexes (difference in deviance of 0.4 with 1 d.f.) or with

different rates for all subjects (further difference of 5.3 on 11 d.f.). The dependence on log time is estimated as $\hat{\beta} = 0.051$, showing that the perceived exertion is increasing over time. On the other hand, the marginal model for the initial conditions reveals a difference between the sexes; the intercept is lower for women: 2.260 as compared to 2.485, with a difference in deviance of 13.8 on 1 d.f. The estimates of the remaining parameters are $\hat{\sigma}^2 = 0.0186$ and $\hat{\rho} = 0.3174$, revealing a fairly strong correlation among successive responses.

If we fit the simpler conditional exponential model of Section 8.2.2, using the identity link and dispersion parameter equal to the lagged response, we find neither changing response over time nor differences between the sexes.

8.3 Intensity-based models

A *semi-Markov* or *Markov renewal process* (Section 2.10) has the intensity or risk for a transition between two states depending not only on the time lapsed because entry to the current state, as in a renewal process, but also on other factors over time, including its current state. Measurement of duration starts over again at zero after each transition, but the intensity function will have changed. A simpler special case is the *nonhomogeneous point process*, where an event occurs without transition to another state. Note, however, that the intensities may also depend on the total time since some zero point, as, for example, with the age of a unit. There may also be heterogeneity, which might be modelled as a *doubly stochastic point process*.

When the intensity function depends on time, we have a nonhomogeneous Poisson process. Models such as the Weibull distribution and the birth process are simple special cases. Because those which are members of the multiplicative intensity family can be modelled as log linear models for discrete data, they are both very flexible and especially easy to fit; we shall concentrate on them here. The cases which are best known are the simple Poisson process or exponential distribution, the Weibull distribution, the extreme value distribution, and the Cox proportional hazards model. For the Poisson process, the intensity is constant over time; for the Weibull, it is a power function of time; for the extreme value, $e^{\kappa t}$; and for the Cox model, it jumps at each event.

The situation is even more complex if the time-varying covariates are themselves random, as, for example, if the treatment of a patient varies as a function of previous response. An example of this, in another context, was the treatment of multiple sclerosis in Sections 4.3 and 4.4. In such cases, the total duration to an event cannot be predicted, and it is most evident that distribution-based models are not applicable.

The distribution of a point process is different depending on whether observations begin at a fixed point in time or with the first event. In the first case, we have both right and left censoring. In the examples, we shall

Table 8.3. Times (sec.) at initiations of mating between flies. (Aalen, 1978b)

Ebony Flies							
143	180	184	303	380	431	455	475
500	514	521	552	558	606	650	667
683	782	799	849	901	995	1131	1216
1591	1702	2212					
Oregon Flies							
555	742	746	795	934	967	982	1043
1055	1067	1081	1296	1353	1361	1462	1731
1985	2051	2292	2335	2514	2570	2970	

usually consider the process to start at an event, because ignoring the time previous to the first event does not bias the results and usually causes little loss of information.

8.3.1 POINT PROCESSES

In Section 6.5, we looked at point processes as categorical models by studying the numbers of events in fixed intervals of time. The renewal processes of Section 7.2.1 were also point processes, but where we looked directly at the intervals between events. A third approach will be used here, where we model the presence or absence of a single event in each interval of observation, as the counting process of Section 7.1.2. As stated there, this is the most general way to model point processes, because it allows variation in the intensity at any time, for example, as a function of time-varying covariates. It is also the simplest case of the type of stochastic process in which we are interested, because it leads on directly to semi-Markov processes where the event signals a change of state of the unit under observation.

Here, we shall look at a very simple point process, the times at which matings occur in a batch of flies. The series for two types of flies (Aalen, 1978b) are given in Table 8.3 and plotted in Figure 8.1. We see that the two curves seem to be fairly similar, except that the Oregon flies started mating about 400 seconds later than the ebony flies.

The deviances for the exponential, Weibull, extreme value, and Cox models, based on the approach of Equation (7.2), are given in Table 8.4. The full log linear models, with vectors of length 5182, are simply

$$\text{DURATION} * \text{SUBSPECIES}$$

using the Wilkinson and Rogers (1973) notation of Section 2.3.2, where DURATION is unity for the exponential model, log time since the previous mating for the Weibull, time for the extreme value, and the factor time variable for the Cox model.

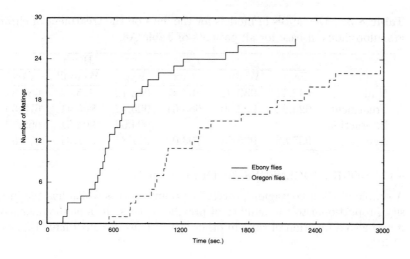

Fig. 8.1. Counting processes of two species of flies, from Table 8.3.

Table 8.4. Deviances of models for the fly mating times of Table 8.3.

	Exponential	Weibull	Extreme Value	Cox Model
Null	464.09	461.54	461.71	286.23
Subspecies	461.51	459.57	459.58	284.19
Interaction		459.57	458.17	227.18

Here, there is little difference among the exponential, Weibull, and extreme value models, where the latter two each have one more parameter than the first. They show no indication of difference between the two subspecies of flies. In the interaction models, a different power parameter is allowed in the probability distribution for the times for each type of fly. The Cox model, with 45 more parameters than the exponential, fits very much better, certainly, at least in part, because it can allow for the time lag before mating begins. The interaction model here is a stratified model, with a completely different form of intensity for each subspecies. This clearly illustrates the nonhomogeneous character of these two point processes.

If the delay of about 400 seconds in initiating mating for the Oregon flies is allowed for, the deviances for the parametric models are reduced by only from 6 to 8, while those for the Cox models remained unchanged. Thus, the differences are not primarily due to this delay.

Table 8.5. Deviances of models for the data on the treatment of chronic granulotomous disease for all patients of Table A8.

				Birth		
	Exp.	Weibull	Cox	Exp.	Weibull	Cox
Null	942.51	939.93	707.47	915.12	915.12	681.34
Treatment	923.92	922.50	688.61	905.42	905.41	670.55
Interaction				904.53	904.51	669.32
Age	937.86	935.53	703.02	912.35	912.34	678.52

8.3.2 NONHOMOGENEOUS BIRTH PROCESSES

A simple birth or contagion process is a point process which has the intensity proportional to the number of previous events. It is nonhomogeneous if it is also a function of time in other ways. Thus, we shall have Equation (2.101),

$$\omega(t|N_t) = N_t\omega(t)$$

where $\omega(t)$ is some arbitrary function, and may depend also on covariates. For example, for a Weibull model, $\omega(t) = t^{\kappa-1}$. If $\omega(t)$ is allowed to change at each event, we have a Cox proportional hazards model. In contrast, the exponential model yields a pure birth process.

Birth processes can be generalized by replacing N_t by some function of the number of previous events, for example,

$$\omega(t|N_t) = N_t^\nu\omega(t) \tag{8.7}$$

This is especially important if time is already measured before the first event, as in many medical studies, because then $N_t = 0$. Several approaches are possible. For example, all values of N_t could be augmented by one, as if time started at an event; N_t could be set to one in the first two periods, until there are two events; or N_t could be replaced by $e^{\nu N_t}$.

As an example, we shall look at data on the treatment of chronic granulotomous disease from a placebo controlled randomized clinical trial of gamma interferon, given in Table A8 of Appendix A (Fleming and Harrington, 1991, pp. 377–383). Only 44 of the 128 subjects had one or more serious infections. As shown in the table, a number of variables describing each patient are also available. We shall fit the exponential, Weibull, and Cox models. A few of the relevant deviances are given in the first three columns of Table 8.5, again from log linear models based on Equation (7.2). The full log linear model, with vectors of length 37477, is

TREATMENT + AGE + DURATION

where DURATION is unity for the exponential distribution, log time since

Table 8.6. Deviances of models for the data on the treatment of chronic granulotomous disease for patients with at least one infection in Table A8.

	Exp.	Weibull	Cox	Exp.	Birth Weibull	Cox
Null	789.13	787.74	547.60	789.10	787.72	547.34
Treatment	787.75	786.31	546.34	787.63	786.31	546.22
Interaction				786.64	786.23	545.35

the previous infection for the Weibull, and a factor time variable for the Cox model.

The simple exponential model fits about as well as the Weibull, but the Cox model is considerably better, although it has 74 extra parameters. Treatment is significant and age marginally so; no other variable changes the deviance very much at all. The log intensity is about 1.09 (s.e. 0.26) higher for the placebo group as compared to treatment, while it decreases by about 0.027 (s.e. 0.013) for each year of increase in age.

Let us now fit a birth model, taking into account the number of infections which have already occurred during the observation period; we have no information about previous infections. Here, we use the first approach mentioned above, with $N_t + 1$, so that a term, $\nu \log(N_t + 1)$, is included in the log linear model, because it gives a superior fit to the other two. This is equivalent to assuming that each patient has started with one infection; we have an ordinary exponential, Weibull, or Cox model until the first observed infection. The deviances are presented in the last three columns of Table 8.5. We see that these models fit much better than the corresponding simpler ones. As well, when the birth process is introduced, the Weibull model is even closer to the exponential and age shows less effect. When an interaction, between treatment and the variable describing the number of events, is introduced, the fit is not changed. The difference in log intensity for treatment is 0.87 (for the Cox model, with s.e. 0.27), an important reduction as compared to the models without a birth process. The corresponding estimate of ν for the event variable is 1.00 (s.e. 0.23). The intensity is higher under placebo, but increases at each infection to the same extent for both treatment and placebo.

Because the number of infections seems to play such an important role, let us now look only at those patients who had at least one infection. Most of the subjects with no infections were under treatment, so that we may expect that this analysis will produce different results from those for the complete data. They may indicate more clearly if gamma interferon is effective under repeated infection. The deviances are given in Table 8.6. The first thing which we discover is that none of the explanatory variables, including treatment and its interaction with the event variable, is significant here, even in the simpler models, without the birth effect.

And, the addition of this latter effect does not improve the fits of the models. Here, the estimate of ν for the event variable is virtually zero.

Thus, we conclude that the gamma interferon treatment has no apparent effect on patients with repeated serious infections. It reduces the risk of a first infection, but, if an infection does occur, the treatment may have little further effect. Note also that the average risk of infection is higher in this subgroup than among the patients in general.

Many other nonhomogeneous Poisson processes can also be fitted just as easily as in this example, especially when they are multiplicative intensity models. These will often include other time-varying covariates which also pose no problem.

8.3.3 SEMI-MARKOV PROCESSES

In a semi-Markov or Markov renewal process, the intensity function depends on the time since the last event, but the function will also be changing over time. The former can, for example, be as in the Weibull or Cox models. The latter may occur because the intensity is a function of some time-varying covariates. However, in contrast to a point process, here, there is also a time-changing function because the unit changes state at each event. In the simpler survival context, this is known as competing risks. If these changes take the multiplicative form with the intensity, the models may again be fitted by the log linear approach of Equation (7.2). We, then, wish to estimate the transition intensities between pairs of states. In the simpler cases, there may be a progression of irreversible states, such as in the advance of uncurable diseases, with a final, absorbing, death state. Then, there will be one less transition intensity than the number of states. When at least some of the state transitions are reversible, many more different intensity functions may need to be estimated.

Consider a study of chronic myeloid leukemia, where the patients may be in one of three states: stable, blast, or dead. Here, the state space is ordered, with only two types of transition possible. The data are presented in Table A9 of Appendix A, from Klein, Klotz, and Grever (1984). A time-varying covariate, the level of the enzyme, adenosine deaminase (ADA: 10^{-8} mole inosine/hour/10^6 blood cells), in leukemia cells, was measured periodically. We shall arbitrarily assume that the value at the beginning of a period holds over the whole period until the next measurement: for a more reasonable model which does not make this assumption, see Kay (1986).

As previously in this section, we shall base the analysis on exponential, Weibull, and Cox models, but with a binomial logistic model, instead of the Poisson log linear model of Equation (7.2), because each transition is irreversible and can only happen once. In the exponential model, only the constant intensity can change between states. However, in the Weibull model, the power parameter may also differ between states. For the Cox

Table 8.7. Deviances for various models fitted to the myeloid leukemia data of Appendix A9.

	Exp.	Weibull	Different Weibull	Cox	Stratified Cox
Null	492.66	492.02	488.15	268.99	248.05
ADA	490.67	490.37	486.12	266.56	245.45
ADA.State	486.08	485.20	478.82	265.54	242.50

model, the whole baseline intensity function may be different, yielding a stratified Cox model. The full model, with vectors of length 12176, is then

$$(\text{DURATION} + \text{ADA}) * \text{STATE}$$

where DURATION is unity for the exponential model, log time since the last event for the Weibull, and a factor time variable for the Cox model. Note that there are only 38 transition events in this fairly large data set. Patients in the blast state at the beginning of the study will be assumed to have just entered that state. If this is not true, the intensity for death will be overestimated.

The deviances for the five types of models, with and without the effect of ADA, are given in Table 8.7. As compared to the exponential models, the Weibull have one more parameter, the Weibull with different power parameter two more parameters, and the Cox model 36 more parameters. Again, the Cox models provides an improvement on the parametric models. The exponential models and the Weibull with common power parameter for the two states give almost identical fits, while, when distinct Weibull models are used for the two states, the fit is slightly better. Because there is also an interaction between ADA and state, the former must have a different meaning in the two states. In contrast, the Cox model shows little indication of a relationship between the intensities in different states and ADA.

It is useful to look at the parameters of the Weibull model to give a general idea of the changes in the intensity function. The log intensity of transition is, on average, 9.58 higher in the blast phase than in the stable phase. However, the Weibull power parameter is estimated as 2.04 in the first chronic or stable phase and as 0.97 in the accelerated or blast phase. This implies that the intensity of transition from stable to blast is increasing with time, while that from blast to death may be slightly decreasing, although close to the constant exponential intensity (for fixed level of ADA). The effect of ADA on the intensity is virtually zero in the blast phase, but increases the log intensity of transition from stable to blast by 0.104 (s.e. 0.0235) per unit increase in ADA. This is also true of the estimates from the stratified Cox model, although that from stable to blast is estimated to be 0.181 (s.e. 0.0893), larger than in the Weibull model, but with much less precision. Thus, an increased level of ADA in the stable state may be a warning sign of transition to the final accelerated phase, whereas such changes in the blast state do not indicate imminent death.

8.4 Stochastic processes with frailty

8.4.1 DYNAMIC GENERALIZED LINEAR MODELS

Dynamic generalized linear models can be applied to duration data, or at least to longitudinal data having positive response values which might follow a gamma, inverse Gaussian, or log normal distribution. Thus, the model is based on the distribution, not on the intensity function. Here, we continue with the gamma distribution, whose conjugate is also a form of gamma, allowing for frailty or heterogeneity among the units. The procedure for estimating the parameters is essentially the same as that described in Section 6.5.2, except for the change in distributions, and need not be repeated here.

We shall apply this model to measurements of plasma citrate concentration (μmol/l) for ten subjects over fourteen successive hourly observation points between eight in the morning and nine in the evening. The data, from Andersen, Jensen, and Schou (1981), are reproduced in Table 8.8. Because interest centres on daily rhythms, this dynamic generalized linear model with harmonics may be appropriate.

With short series, as in this example, fitting a different DGLM to each series is not reasonable; there would be too many parameters. Instead we fit 'parallel' and identical series, with 12 harmonics for a half day and a trend which might pick up a longer period. The resulting deviances are given in Table 8.9. The ten series are not identical, but have different levels, as already could be seen from Table 8.8. For example, subjects one and seven have consistently lower plasma citrate concentrations than the others. Because a trend does not appear, the model with only the harmonics is sufficient. In fact, this can be reduced from 12 to four harmonics with only a gain in deviance of 8.45. The observations are plotted, along with the fitted model, in Figure 8.2, arbitrarily separated into two plots to make them clearer. We see that the plasma citrate concentration is generally highest at about ten or eleven in the morning and lowest about four or five in the afternoon. There seems to be no relationship to the meal times of eight in the morning, noon, and five in the afternoon.

If these series were longer, another approach which might be considered for this type of data is the spectral analysis of Section 4.6.

8.4.2 INTENSITIES WITH FRAILTY

Because intensities can be modelled as log linear or logistic regression, the extension to random effects is straightforward. For example, the methods

Table 8.8.   Plasma citrate concentrations (μmol/l) for 10 subjects at 14 successive times during the day. (Andersen, Jensen, and Schou, 1981)

93	109	114	121	101	109	112	107	97	117
89	132	121	124						
116	116	111	135	107	115	114	106	92	98
116	105	135	83						
125	166	180	137	142	114	119	121	95	105
152	154	102	110						
144	157	161	173	158	138	148	147	133	124
122	133	122	130						
105	134	128	119	136	126	125	125	103	91
98	112	133	124						
109	121	100	83	87	110	109	100	93	80
98	100	104	97						
89	109	107	95	101	96	88	83	85	91
95	109	116	86						
116	138	138	128	102	116	122	100	123	107
117	120	119	99						
151	165	156	149	136	142	121	128	130	126
154	148	138	127						
137	155	145	139	150	141	125	109	118	109
112	102	107	107						

Table 8.9.   Deviances for several dynamic generalized linear models for the plasma citrate data of Table 8.8.

	Null	Trend	Harmonics	Both
Identical	1152.92	1149.57	1138.05	1134.96
'Parallel'	1079.68	1075.17	1053.87	1049.96

of Sections 7.3.3 and 7.3.4 can be directly applied. We shall illustrate this approach with an example similar to that in Section 8.3.2. Here we choose to use the normal compounding distribution with the intensity, because such a model was not presented in the previous chapter.

Let us look at data on the time to appearance of mammary cancer tumours in 48 female rats under two treatments, given in Table A10 of Appendix A, from Gail, Santner, and Brown (1980). A number of animals were injected with a carcinogen for mammary cancer and then given retinyl acetate to prevent the cancer. After 60 days, those still free of tumours were randomly assigned to the two treatments, one of which continued the retinoid prophylaxis and the other acted as a control. The rats were palpated for tumours twice weekly, with observation ending 182 days after the initial carcinogen injection. Because the presence of one or more tumours

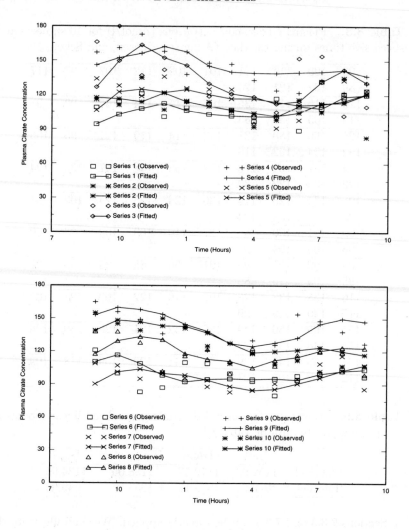

Fig. 8.2.   Hourly plasma citrate concentrations, from Table 8.8, with fitted DGLM.

may induce further tumours, a birth or contagion model may be appropriate. This 'snowball' effect can be checked by looking to see if the influence of the number of tumours is different under the two treatments.

As in previous examples using the intensity function, we shall consider the three multiplicative processes for $\omega(t)$, the exponential, Weibull, and Cox models. When we fit these as homogeneous processes, without and with treatment effects, we obtain the deviances in the first two lines of Table 8.10. The log linear model, with vectors of length 8736, contains

Table 8.10. Deviances for various continuous time models fitted to the tumour data of Appendix A10.

	Exponential	Weibull	Cox
Null	1443.4	1439.3	1118.0
Treatment	1424.9	1423.1	1100.3
Birth Process			
Null	1417.5	1412.1	1101.9
Treatment	1416.6	1409.6	1098.9
Generalized Birth Process			
Null	1400.2	1400.0	1086.5
Treatment	1392.9	1392.4	1077.4

Table 8.11. The parameter estimate for treatment effect in various models fitted to the tumour data of Appendix A10 (standard errors in parentheses).

Exponential	Weibull	Cox
0.649	0.618	0.657
(0.154)	(0.156)	(0.161)
Birth Process		
0.147	0.253	0.281
(0.156)	(0.161)	(0.165)
Generalized Birth Process		
0.433	0.443	0.497
(0.161)	(0.162)	(0.167)

TREATMENT + DURATION

with DURATION unity in the exponential model, log time since the previous new tumour for the Weibull, and a factor time variable for the Cox model. The Weibull models have one more parameter than the exponential, and the Cox model 181 more.

We see that there is an effect of treatment in all cases, that the exponential model fits about as well as the Weibull, and that the Cox model fits considerably better. The parameter estimates, and standard errors, for treatment effect are listed in the first two lines of Table 8.11. They are not very different for the three models, although the standard errors are necessarily smaller for the exponential and Weibull, because they contain many fewer parameters. The Weibull power parameter is estimated as $\hat{\kappa} = 0.92$ when treatment is present in the model, close to that for an exponential distribution, confirming the small difference from that model. There is no evidence of the need for a different value of this parameter under the two treatments.

We shall now fit a birth or contagion model, taking into account the number of tumours already present, but still without the power parameter, ν, of Equation (8.7), i.e. as an offset in the log linear model. However, as in Section 8.3.2, the problem is that there is no tumour present in the first period: $N_t = 0$ until the first event, although this time we know that there were no previous events. Several approaches were outlined in that section. Here, in contrast to that section, setting N_t to one in the first two periods, until there are two tumours gives a better fit than the other two methods.

With this model, we obtain the deviances in the third and fourth lines of Table 8.10. With no extra parameters added, these models fit much better. Indeed, the treatment effect has virtually disappeared, except perhaps for the Cox model. This is reflected in the corresponding parameter values in Table 8.11, which are much reduced in size. This result may not be surprising, because there are more tumours among the controls and this factor may be substituting for treatment effect.

When we generalize further, by adding the power parameter, ν, a term, $\nu \log(N_t)$, in the log linear model, the fit improves still more, as can be seen in the last two lines of Table 8.10. However, there is now, once again, a reasonably large difference in deviance for treatment effect. The parameter values are larger than in the ordinary birth process, but quite a bit smaller than in the homogeneous processes. The Weibull power parameter is now estimated as $\hat{\kappa} = 1.05$, while that for the log number of tumours (ν) ranges from 0.52 for the Cox model to 0.59 for the Weibull, all with treatment effect included. These are considerably different from the value of unity for the ordinary birth process.

There is no indication of interaction between the treatment effect and the number of tumours. For example, the difference in deviance for the Weibull model is 0.02 on 1 d.f. Thus, the influence of the number of tumours present seems to be the same under the two treatments, indicating that it may be a real effect, attenuating that of the treatment. Under the control, the accumulation of tumours has a 'snowball' effect on inducing further tumours. In other words, the retinoid treatment slows down the production of tumours, but the very fact that there are then fewer tumours further slows down the process compared to the controls.

Let us now introduce frailty or proneness. Up until now, we have used continuous time models. However, software is not readily available for a compound normal random effect in Poisson regression with large data sets, so that we are obliged to turn to that for logistic regression. This also gives us an opportunity to compare results for continuous and discrete time on the same data. We take the exponential and Weibull models and the corresponding generalized birth processes. The binomial deviances (not comparable with the Poisson deviances of Table 8.10) are given in Table 8.12. If we compare the first two columns of values in this table with the corresponding ones in Table 8.10, we see that their differences are

Table 8.12. Deviances for some discrete time models fitted to the tumour data of Appendix A10.

	Independence		Random Effect	
	Exponential	Weibull	Exponential	Weibull
Null	1815.3	1811.2	1796.6	1796.6
Treatment	1796.4	1794.5	1787.4	1787.4
	Generalized Birth Process			
Null	1770.8	1770.6	1770.8	1770.6
Treatment	1763.4	1762.8	1763.4	1762.8

Table 8.13. The parameter estimate for treatment effect in various models fitted to the tumour data of Appendix A10 (standard errors in parentheses).

Independence		Random Effect	
Exponential	Weibull	Exponential	Weibull
0.664	0.631	0.662	0.662
(0.157)	(0.159)	(0.210)	(0.230)
Generalized Birth Process			
0.444	0.454	0.429	0.443
(0.165)	(0.166)	(0.188)	(0.201)

similar. Comparing the first two with the last two columns in Table 8.12, we see that introduction of a random effect only improves the fit when the birth process is not allowed for. Thus, contagion and proneness are not necessary simultaneously in our model; the former alone gives the better fit. The subjects are relatively homogeneous, but the intensity of events is changing as their number increases.

The parameter estimates are given in Table 8.13. Again, these are similar to those for continuous time in Table 8.11. They do not change much when a random effect is added, although, as we have come to expect, the standard errors are greater. The estimate of the scale parameter in the normal compound model is 0.450 in the exponential and Weibull models and 0.002 in the generalized birth models.

If we introduced frailty into the example of Section 8.3.2, for patients with at least one infection, we would discover that it is not required at all, even in the poorly fitting models.

We have now fitted the exponential, Weibull, and Cox models to four different data sets. In each case, the Cox model provided a substantially better fit, but at the expense of a large number of parameters. In a testing situation, where we only wish to determine if some treatment has an effect, this is a reasonable solution. However, in a modelling context, where we seek to describe and explain the overall event history, it is not really satisfactory. With so many parameters, there is little chance that the same Cox model would be found acceptable (for the values of all parameters, not just treatment effect) if the study were repeated (see, for example, Dinse, 1988). In such a situation, it will usually be more satisfactory to explore the dependence of the intensity on various functions of time in place of, or in addition to, the $\log(t)$ of the Weibull model. Because we have a log linear regression model, such extensions can be looked at very easily.

8.5 Generalized estimating equations

As for categorical and count data, when the responses on a unit are dependent, generalized estimating equations (GEE) may also be applied. This approach has been described in detail in Sections 2.9 and 6.6. That explanation need not be repeated here. In the present context, one further handicap of this approach is that it applies only to the distribution, and not to the intensity function. Time-varying covariates may be used if they remain constant over a time duration, as for the distributional models already presented in this chapter. Because the GEE are applied to generalized linear models, the gamma and inverse Gaussian distributions will be of principal interest for durations.

Dependence among durations on a unit in terms of a covariance matrix is not clearly defined for event history data. The 'distance' between successive durations may depend on the durations themselves, for example, if time is used as distance. However, the simplest approach is to have an 'autocorrelation' which depends simply on the number of intervening durations. This would mean that the 'distance' is the number of events and not time. This may be reasonable in certain contexts. In a direct application of the GEE, this is what occurs.

As in Section 6.6, it will be useful to choose an example for which we have already constructed some models. The study of rate of perceived exertion of Section 8.2.3 is appropriate, because, as well, it is not an event history and does not have the problem of 'distance'. The response is not confounded with the passage of time. We use exactly the same procedure as for the GEE example of Section 6.6, however here using the gamma distribution. The main change with respect to that section is that the variance function is now the square of the mean.

The parameter estimates and standard errors are given in Table 8.14. The independence model, with $\hat{\phi} = 0.010$, gives basically the same results as the exponential dispersion model above. The interaction between sex and time is not significant, while the females have a lower intercept than the males. Here, the slope with time is smaller than before, 0.011 as compared to 0.051 above.

The uniform correlation GEE has $\hat{\phi} = 0.078$ and intra-unit correlation 0.700. Here, the interaction is significantly different from zero and the

Table 8.14. Parameter estimates, with standard errors, of various stochastic dependence 'models' for the data of Table 8.2.

	Independence	Uniform	AR(1)
Intercept	2.355	2.574	2.412
	(0.038)	(0.132)	(0.060)
Sex	0.227	0.695	0.356
	(0.055)	(0.174)	(0.078)
Time	0.011	0.006	0.009
	(0.002)	(0.002)	(0.002)
Sex.Time	−0.004	−0.013	−0.004
	(0.003)	(0.004)	(0.004)

difference in intercept at time zero is greater. The slope for females is smaller, and that for males is decreasing with time relative to the females.

The IWLS procedure did not converge for the AR(1) and the general covariance structure GEE for these data. For the AR(1), the procedure oscillated between two fixed sets of parameter values while, for the general GEE, it appeared to jump among arbitrary values. For the AR(1), that set of values closest to the other 'models' is presented in Table 8.14. For this solution, the autocorrelation was estimated as 0.72 (and 0.39 for the other), with $\hat{\phi} = 0.018$. The regression parameter values, and conclusions, are similar to those for the independence model.

Once again, we are presented with a series of results and no clear means of choosing among them. Given the additional restrictions of the GEE approach in the context of event histories, mentioned above, it can only be of limited usefulness.

For event history data, the reader should consult Lancaster (1990) and Blossfeld, Hamerle, and Mayer (1989); both treat the problem of heterogeneity as well as the more classical survival-type models for repeated events.

Appendix A
Data tables

A.1 Ultrafiltration data

	Centre 1			
TMP	160.0	265.0	365.0	454.0
UFR	600.0	1026.0	1470.0	1890.0
TMP	164.0	260.5	355.0	451.0
UFR	516.0	930.0	1380.0	1770.0
TMP	156.0	260.0	363.0	466.0
UFR	480.0	900.0	1380.0	1860.0
TMP	160.0	259.0	361.0	462.0
UFR	528.0	930.0	1410.0	1872.0
TMP	157.0	258.0	359.0	471.0
UFR	540.0	978.0	1410.0	1920.0
TMP	161.0	264.0	359.0	466.0
UFR	564.0	996.0	1422.0	1920.0
TMP	161.0	263.0	363.0	468.0
UFR	564.0	1062.0	1500.0	1980.0
TMP	158.0	255.0	360.0	461.0
UFR	492.0	900.0	1392.0	1860.0
TMP	161.0	263.0	361.0	462.0
UFR	516.0	960.0	1380.0	1800.0
TMP	155.0	255.0	355.0	455.0
UFR	528.0	930.0	1356.0	1860.0
TMP	158.0	267.0	360.0	464.0
UFR	564.0	1020.0	1380.0	1884.0
TMP	165.0	263.0	362.0	461.0
UFR	618.0	1056.0	1500.0	1920.0

TMP	158.0	263.0	367.0	464.0	
UFR	564.0	1038.0	1410.0	1770.0	
TMP	162.0	268.0	360.0	465.0	
UFR	552.0	1014.0	1440.0	1944.0	
TMP	171.0	256.0	357.0	466.0	
UFR	624.0	978.0	1440.0	1980.0	
TMP	158.5	263.0	361.0	460.0	
UFR	468.0	930.0	1332.0	1860.0	
TMP	162.0	263.0	356.0	463.0	
UFR	480.0	900.0	1272.0	1758.0	

Centre 2					
TMP	166.0	216.5	259.5	348.5	444.5
UFR	540.0	720.0	900.0	1320.0	1716.0
TMP	156.0	247.0	350.0	465.5	
UFR	426.0	810.0	1218.0	1692.0	
TMP	143.5	249.0	359.0	463.0	
UFR	390.0	822.0	1290.0	1716.0	
TMP	146.0	253.5	358.0	451.5	
UFR	570.0	900.0	1284.0	1728.0	
TMP	149.0	244.5	344.5	446.0	
UFR	360.0	870.0	1320.0	1800.0	
TMP	149.0	235.5	344.0	442.0	
UFR	480.0	840.0	1302.0	1752.0	
TMP	151.5	247.5	332.0	456.5	
UFR	450.0	810.0	1128.0	1560.0	
TMP	153.0	240.5	347.5	440.0	
UFR	480.0	870.0	1326.0	1716.0	
TMP	149.5	250.0	348.5	461.5	
UFR	378.0	840.0	1224.0	1608.0	
TMP	180.5	239.5	346.5	439.5	
UFR	594.0	762.0	1260.0	1620.0	
TMP	164.0	250.0	345.5	459.0	
UFR	516.0	870.0	1272.0	1704.0	
TMP	159.5	253.0	349.0	434.5	
UFR	486.0	852.0	1242.0	1620.0	
TMP	175.5	237.0	459.0		
UFR	570.0	774.0	1716.0		

	Centre 3			
TMP	183.0	249.0	345.5	449.5
UFR	600.0	900.0	1290.0	1710.0
TMP	160.5	244.5	348.5	424.5
UFR	480.0	810.0	1260.0	1560.0
TMP	149.5	254.5	353.0	403.5
UFR	510.0	930.0	1380.0	1560.0
TMP	188.5	243.0	339.5	462.5
UFR	600.0	840.0	1290.0	1710.0
TMP	208.0	248.5	350.0	448.0
UFR	720.0	900.0	1380.0	1740.0
TMP	182.5	245.5	350.5	443.5
UFR	600.0	840.0	1260.0	1560.0
TMP	184.0	250.0	350.0	446.0
UFR	600.0	840.0	1260.0	1620.0
TMP	192.0	251.5	342.5	452.0
UFR	600.0	900.0	1290.0	1710.0
TMP	181.5	251.5	344.0	479.0
UFR	480.0	780.0	1140.0	1710.0
TMP	174.5	254.0	345.0	455.0
UFR	540.0	840.0	1200.0	1650.0
TMP	213.5	235.5	341.0	454.0
UFR	780.0	780.0	1290.0	1680.0

Table A1: *In vivo* ultrafiltration characteristics of 41 hollow fibre dialyzers tested in three centres (Vonesh and Carter, 1987). TMP – transmembrane pressure (mm Hg), UFR – ultrafiltration rate (response, ml/hour).

A.2 Coronary sinus potassium data

Dog	1	3	5	7	9	11	13
				1. Control			
1	4.0	4.0	4.1	3.6	3.6	3.8	3.1
2	4.2	4.3	3.7	3.7	4.8	5.0	5.2
3	4.3	4.2	4.3	4.3	4.5	5.8	5.4
4	4.2	4.4	4.6	4.9	5.3	5.6	4.9
5	4.6	4.4	5.3	5.6	5.9	5.9	5.3
6	3.1	3.6	4.9	5.2	5.3	4.2	4.1
7	3.7	3.9	3.9	4.8	5.2	5.4	4.2
8	4.3	4.2	4.4	5.2	5.6	5.4	4.7
9	4.6	4.6	4.4	4.6	5.4	5.9	5.6
		2. Extrinsic cardiac denervation					
			immediately prior				
10	3.2	3.3	3.8	3.8	4.4	4.2	3.7
11	3.3	3.4	3.4	3.7	3.7	3.6	3.7
12	3.1	3.3	3.2	3.1	3.2	3.1	3.1
13	3.6	3.4	3.5	4.6	4.9	5.2	4.4
14	4.5	4.5	5.4	5.7	4.9	4.0	4.0
15	3.7	4.0	4.4	4.2	4.6	4.8	5.4
16	3.5	3.9	5.8	5.4	4.9	5.3	5.6
17	3.9	4.0	4.1	5.0	5.4	4.4	3.9

Dog	1	3	5	7	9	11	13
3. Bilateral thoracic sympathectomy							
and stellectomy 3 weeks prior							
18	3.1	3.5	3.5	3.2	3.0	3.0	3.2
19	3.3	3.2	3.6	3.7	3.7	4.2	4.4
20	3.5	3.9	4.7	4.3	3.9	3.4	3.5
21	3.4	3.4	3.5	3.3	3.4	3.2	3.4
22	3.7	3.8	4.2	4.3	3.6	3.8	3.7
23	4.0	4.6	4.8	4.9	5.4	5.6	4.8
24	4.2	3.9	4.5	4.7	3.9	3.8	3.7
25	4.1	4.1	3.7	4.0	4.1	4.6	4.7
26	3.5	3.6	3.6	4.2	4.8	4.9	5.0
4. Extrinsic cardiac denervation							
3 weeks prior							
27	3.4	3.4	3.5	3.1	3.1	3.7	3.3
28	3.0	3.2	3.0	3.0	3.1	3.2	3.1
29	3.0	3.1	3.2	3.0	3.3	3.0	3.0
30	3.1	3.2	3.2	3.2	3.3	3.1	3.1
31	3.8	3.9	4.0	2.9	3.5	3.5	3.4
32	3.0	3.6	3.2	3.1	3.0	3.0	3.0
33	3.3	3.3	3.3	3.4	3.6	3.1	3.1
34	4.2	4.0	4.2	4.1	4.2	4.0	4.0
35	4.1	4.2	4.3	4.3	4.2	4.0	4.2
36	4.5	4.4	4.3	4.5	5.3	4.4	4.4

Table A2: Coronary sinus potassium (mil equivalents per litre) in dogs measured at 2 minute intervals after occlusion (Grizzle and Allen, 1969).

A.3 Azams data — AFCR

					P + P				
1	-27	-13	28	56	84	168	259	331	427
	561	334	157	374	191	465	125	212	232
	504	672	771	834	945	1008	1092	1289	1306
	177	98	207	127	202	143	174	216	245
	1351								
	237								
2	-14	-6	58	253	358	508	574	672	855
	429	587	446	269	131	50	145	634	273
	924								
	144								
3	-14	-7	0	56	84	168	336	420	504
	231	312	123	127	297	337	225	312	178
	672	756	840	924	1000	1135	1260	1280	1337
	111	97	133	239	151	115	297	141	113
4	-14	-7	0	28	56	83	169	252	330
	351	369	177	107	273	263	266	221	246
	420	504	588	678	756	840	924	1008	1120
	226	323	157	260	234	283	211	210	174
	1176	1260	1298						
	335	216	90						
5	-22	-8	0	27	83	167	405	504	588
	347	77	120	379	330	620	798	386	211
	672	756	861	924	1008	1092	1176	1197	1281
	176	506	239	312	235	206	430	308	251
	1309								
	151								
6	-42	-7	0	28	84	246	343	392	483
	23	126	108	350	222	216	312	154	148
	644	819	896	1001	1091	1148	1232	1256	1284
	132	217	151	102	304	85	462	141	107
7	-14	-7	0	28	56	84	168	261	392
	273	269	248	345	366	414	341	291	226
	477	561	644	736	812	897	980	1155	1184
	331	114	87	100	339	278	133	121	40
	1198	1232	1254						
	116	112	182						

8	-7	0	48	85	168	252	335	426	504
	339	431	277	330	602	264	207	165	65
	588	693	797	923	986	1190			
	133	202	242	228	147	147			
9	-77	-7	0	28	56	83	168	251	342
	187	429	465	504	187	199	444	153	489
	371	426	518	584	609	672	755	853	924
	248	286	120	206	77	43	243	212	171
	1008	1091	1176	1214	1317				
	137	23	283	143	122				
10	-14	-7	28	56	112	175	251	336	419
	952	572	776	541	622	900	744	940	707
	510	595	672	756	924	979	1092	1176	1186
	1175	367	353	704	449	459	701	178	644
	1214	1305							
	150	1153							
11	-14	-7	20	56	84	168	259	336	489
	90	126	316	245	358	143	148	226	227
	588	693	784	840	924	1029	1113	1134	1176
	118	129	148	298	246	177	156	144	232
12	-14	-7	0	28	56	85	168	335	426
	544	120	209	418	262	251	351	282	69
	496	595	672	763	840	924	1092	1274	
	156	228	211	83	236	175	146	262	
13	-21	-14	0	28	56	84	167	336	426
	94	475	240	139	382	326	176	133	252
	505	603	679	762	854	909			
	137	95	87	212	75	212			
14	-80	-73	-31	-3	25	53	137	221	305
	288	184	276	483	308	324	173	318	112
	389	473	641	725	809	936			
	196	184	314	217	118	223			

15	-7	0	84	252	329	427	497	580	664
	141	149	250	168	189	111	109	162	112
	756	960	1121	1130	1140	1175			
	175	89	110	218	127	71			
16	-21	-7	0	28	167	252	335	447	504
	426	418	404	297	287	60	175	197	289
	587	672	840	1092	1098				
	194	482	315	261	149				
17	-175	-168	-162	83	168	252	440	504	713
	291	920	144	187	256	91	31	33	144
	756	811	917	1056	1064	1137	1151		
	114	30	113	90	82	146	211		

	AZ + P								
1	-15	-7	0	28	56	84	252	308	405
	537	341	328	168	208	228	48	123	19
	518	588	671	756	860	945	1008	1112	1196
	34	41	18	12	80	14	60	70	11
	1259	1281	1309	1333	1358	1400			
	40	25	138	102	136	88			
2	-20	-13	-6	21	50	78	161	246	323
	272	207	245	151	123	142	12	26	42
	420	497	581	666	750	834	918	1002	1086
	46	34	17	10	13	20	77	25	16
	1170	1288	1302	1338	1365				
	32	26	50	238	47				
3	-14	-7	0	28	56	84	168	252	323
	158	114	78	64	65	63	67	44	23
	419	514	588	700	791	868	945	1035	1134
	81	76	18	52	32	70	12	10	12
	1260	1288	1319	1340					
	57	42	143	41					
4	-14	-7	0	28	56	84	175	252	343
	259	213	96	60	48	82	36	58	31
	357	504	588	770	847	924	1008	1273	1297
	27	54	22	37	80	31	78	58	38
	1344								
	52								
5	-14	-7	0	28	84	169	336	420	504
	221	130	207	176	380	139	130	33	21
	588	673	756	840	924	995	1108	1183	1260
	20	115	40	14	11	0	90	18	25
	1299								
	117								
6	-14	-7	168	252	336	420	504	658	
	316	229	145	37	104	16	34	60	
7	-14	-7	0	84	168	239	336		
	442	288	148	201	223	50	95		

8	-69	-62	21	84	168	239	420	504	609
	748	454	438	230	617	295	195	154	138
	778	869	938	1022	1156	1177	1206		
	87	134	20	60	30	57	34		
9	-14	0	28	56	84	253	421	511	609
	296	485	165	279	278	163	82	49	88
	672	841	924	1008	1100	1176	1331		
	127	65	32	65	34	41	36		
10	-14	-7	0	28	56	84	168	253	336
	423	518	324	578	391	420	285	96	221
	420	504	588	686	756	840	931	1009	1092
	164	130	84	185	132	88	154	285	215
	1148	1176	1208	1241					
	319	111	84	150					
11	-20	0	56	86	168	267	336	421	511
	438	435	204	98	149	78	27	16	17
	672	785	841	980	1093	1176			
	35	168	23	89	29	64			
12	-14	-7	0	56	84	174	258	342	420
	190	250	306	591	263	41	102	23	32
	504	588	672	756	924				
	39	20	16	17	26				
13	-14	0	29	56	85	169	238	336	420
	266	347	316	221	200	40	30	49	22
	504	588	672	756	840	945	1029	1092	1120
	28	54	46	55	79	48	84	51	57
14	-35	-21	56	84	168	252	343	420	505
	224	428	134	171	47	258	100	178	159
	603	680	848						
	203	211	117						
15	-14	-7	0	28	84	168	252	336	504
	216	388	253	88	139	139	52	78	108

AZ + MP								
1	-125	-13	0	28	56	168	239		
	515	814	469	274	486	230	216		
2	-14	28	56	84	168	252	343		
	267	71	73	80	238	0	0		
3	-7	0	28	58	84	175	252	336	504
	408	613	374	273	48	358	84	29	120
	568	672	770	841	946	1022	1092	1296	1302
	293	503	533	205	298	153	131	129	173
	1373								
	304								
4	-20	-13	10	28	56	84	168	252	420
	408	263	119	515	177	194	351	112	113
	504	673	756	840	924	1015	1120		
	47	40	18	14	60	80	13		
5	-14	-7	0	84	156	252	338	428	581
	206	373	262	126	151	162	125	91	99
	672	840	1128	1191	1260				
	60	27	34	120	43				
6	-14	-7	0	30	56	84	168	252	420
	344	468	441	162	485	509	200	66	99
	497	588	672	749	854	932	1029	1100	1177
	115	69	27	80	20	33	11	70	48
	1281	1306	1330						
	177	84	69						
7	-22	-15	-8	28	56	84	162	252	321
	378	349	209	150	470	116	94	37	80
	419	503	587	671	728	755	853	937	1234
	125	56	82	29	65	135	72	18	80
	1241	1287	1315						
	38	59	96						
8	-14	-6	0	28	56	84	169	252	316
	137	198	219	303	499	182	123	156	114
	420	504	581	674	688	875	1017	1094	1212
	167	280	406	213	147	178	144	167	389
	1239	1267							
	105	119							

9	-14	-7	10	56	78	168	266	420	498
	350	576	364	268	438	296	128	98	17
	588	673	756	840	924	1043	1157	1352	
	287	52	74	114	119	123	89	71	
10	-14	-7	0	28	56	84	168	259	365
	372	412	310	457	815	779	597	372	257
	449	533	617	701	1205				
	228	51	296	40	212				
11	-21	-19	0	28	56	84	168	252	336
	480	626	184	290	462	176	317	109	140
	420	504	594	672	756	840	938	1008	1105
	91	62	61	50	106	19	60	13	22
	1176	1186	1225						
	56	133	143						
12	-14	0	28	56	86	168	420	504	581
	436	143	311	277	156	82	40	13	27
	756								
	25								
13	-42	-35	0	28	56	86	340		
	295	315	176	317	194	108	212		
14	-14	-7	0	56	169	240	298	329	
	88	117	272	305	281	60	189	139	
15	-28	0	56	168	252	336	427	518	588
	228	201	375	143	50	22	31	37	50
	671	756	833	973	1063	1071	1130	1176	
	34	21	29	37	33	59	105	290	
16	-43	0	56	84	168	259	336	434	510
	343	286	159	163	37	45	55	37	59
	602	672	756	840	945	1091	1101	1119	1151
	60	34	33	70	42	26	52	34	121

Table A3: Evolution of AFCR (number of lymphocytes bearing a F_c receptor in one mm^3 of peripheral blood) for three treatments of multiple sclerosis (Heitjan, 1991b). For each patient, times and AFCR are on pairs of lines. Negative times are before treatment began.

A.4 Azams data — doses

	P + P							
1	0	28	58	85	113	159	203	333
	1.000	1.167	1.333	1.500	1.667	1.833	2.000	1.833
	375	585	591	1306				
	2.000	0.000	2.000	0.000				
2	0	218	312	352	403	406	973	
	1.000	1.200	1.400	1.600	1.800	2.000	0.000	
3	0	29	57	87	119	164	203	241
	1.000	1.200	1.400	1.600	1.800	2.000	2.200	2.400
	287	818	835	1280				
	2.600	0.000	2.400	0.000				
4	0	34	56	85	122	150	178	205
	1.000	1.250	1.500	1.750	2.000	2.250	2.500	0.000
	221	248	357	379	392	414	422	974
	2.500	2.750	0.000	1.750	2.000	2.250	2.500	0.000
	1002	1018	1266					
	2.000	2.500	0.000					
5	0	17	58	85	127	169	198	225
	1.000	0.000	0.667	0.833	1.000	1.167	1.333	1.500
	254	297	317	437	505	541	577	589
	1.667	1.833	2.000	1.833	2.000	1.833	2.000	1.833
	645	1259						
	2.000	0.000						
6	0	30	58	88	120	155	205	218
	1.000	1.167	1.333	1.500	1.667	1.833	1.667	2.000
	248	1284						
	2.167	0.000						
7	0	31	60	88	114	141	162	190
	1.000	1.125	1.250	1.375	1.500	1.625	1.750	1.875
	218	900	907	1061	1069	1198		
	2.000	0.000	2.000	0.000	2.000	0.000		
8	0	55	87	122	148	170	229	265
	1.000	1.286	1.429	1.571	1.714	1.857	2.000	2.143
	590	676	679	1077	1095	1171	1190	
	2.000	0.000	2.000	0.000	1.714	2.000	0.000	

9	0	30	58	85	114	142	174	204
	1.000	1.200	1.400	1.600	1.800	2.000	2.200	2.400
	1231							
	0.000							
10	0	31	59	97	129	185	210	253
	1.000	1.100	1.200	1.300	1.400	1.500	1.600	1.700
	272	296	338	385	423	883	890	1246
	1.600	1.700	1.800	1.900	2.000	0.000	2.000	0.000
	1408	1427						
	1.000	0.000						
11	0	38	77	129	149	169	213	236
	1.000	1.167	1.333	1.167	0.833	0.500	0.333	0.500
	260	330	337	489	504	674	745	786
	0.333	0.167	0.000	0.083	0.000	0.167	0.083	0.000
	1002	1025						
	0.083	0.000						
12	0	31	58	86	114	122	128	139
	1.000	1.200	1.400	1.600	0.000	0.800	1.200	1.400
	163	191	220	1225				
	1.600	1.800	2.000	0.000				
13	0	30	59	86	112	142	169	599
	1.000	1.167	1.333	1.500	1.667	1.833	2.000	0.000
	604	611	619	1057				
	1.333	1.500	2.000	0.000				
14	0	28	139	186	222	969		
	1.000	1.143	1.429	1.571	1.714	0.000		
15	0	29	59	71	129	175	206	233
	1.000	1.200	1.000	1.200	1.400	1.600	1.800	2.000
	255	1144						
	2.200	0.000						
16	0	30	56	87	129	190	226	254
	1.000	1.167	1.333	1.167	1.333	1.500	1.667	1.833
	283	674	679	1009	1063	1094	1116	
	2.000	0.000	2.000	0.000	1.000	1.667	0.000	
17	0	30	63	105	145	164	198	232
	1.000	1.143	1.286	1.429	1.571	1.714	1.857	2.000
	409	441	458	470	472	479	1064	
	0.000	0.143	1.143	0.000	0.571	2.000	0.000	

				AZ + P			
1	0	29	56	86	133	168	196	227
	1.000	1.143	1.286	1.429	1.571	1.714	1.857	2.000
	260	282	292	406	451	457	485	513
	1.714	1.857	2.000	1.714	1.571	1.714	1.857	1.714
	534	592	674	764	1112	1149	1309	
	1.857	1.714	1.857	1.714	1.429	1.571	0.000	
2	0	23	51	142	219	326	828	1213
	1.000	1.143	1.286	1.429	1.571	1.429	1.286	1.429
	1302							
	0.000							
3	0	30	64	169	188	625	632	701
	1.000	1.250	1.000	0.750	1.000	0.000	0.750	0.500
	1288							
	0.000							
4	0	7	14	15	56	86	114	149
	1.000	0.000	0.000	1.000	1.167	1.333	1.500	1.667
	211	234	358	463	506	563	589	597
	1.833	1.667	1.833	1.667	1.500	1.667	1.500	1.667
	653	689	864	882	954	1094	1181	1297
	1.833	2.000	1.833	1.667	1.833	1.667	2.000	0.000
5	0	35	36	51	106	155	212	249
	1.000	1.125	1.000	0.500	0.625	0.750	0.875	1.000
	295	326	368	401	421	591	610	639
	1.125	1.250	1.375	1.500	1.250	1.000	1.125	1.000
	665	787	1324					
	1.125	1.000	0.000					
6	0	13	62	68	70	74	84	94
	1.000	0.000	0.375	0.750	1.000	0.000	0.750	0.500
	104	134	162	189	219	246	359	393
	0.750	0.875	1.000	1.125	1.250	1.375	1.500	1.625
	423	428	435	511	525	542	569	690
	1.500	1.625	1.500	1.375	1.125	1.000	1.125	0.000
7	0	20	57	85	140	202	233	272
	1.000	0.600	0.800	1.000	1.200	1.400	1.600	1.800
	358	371	396	437	455	479		
	2.000	2.200	1.800	2.000	2.200	0.000		

8	0	23	30	60	154	479	733	790
	1.000	0.000	0.857	1.000	0.857	1.000	1.143	1.286
	871							
	0.000							
9	0	32	59	87	115	142	197	294
	1.000	1.200	1.400	1.600	1.800	2.000	2.200	2.000
	477	486	504	597	609	632	721	760
	2.200	2.200	2.000	0.000	2.200	0.000	2.200	0.000
	779	1176	1224	1226				
	2.000	2.200	2.400	0.000				
10	0	1	21	58	87	116	142	170
	1.000	0.000	1.000	1.143	1.286	1.429	1.571	1.714
	207	337	758	765	1176			
	1.857	2.000	0.000	1.857	0.000			
11	0	58	88	121	157	452	493	616
	1.000	1.125	1.250	1.375	1.500	1.000	1.250	1.000
	624	787	816	864	956	1005	1029	1065
	0.000	0.750	0.875	1.000	1.125	0.000	1.000	0.000
	1071	1178	1261					
	0.750	0.875	0.000					
12	0	31	59	87	113	513	610	960
	1.000	1.143	1.286	1.429	1.571	1.429	1.143	0.000
13	0	32	58	88	116	165	200	228
	1.000	1.143	1.286	1.429	1.571	1.714	1.857	2.000
	266	275	338	353	361	416	421	437
	2.143	2.000	0.000	1.714	1.857	1.714	0.000	1.714
	728	737	750	759	786	975	1061	1127
	0.000	1.571	0.000	1.429	1.571	1.714	1.857	0.000
14	0	30	58	149	155	203	239	315
	1.000	1.167	1.000	0.833	0.667	0.833	1.000	0.000
	327	385						
	0.833	0.000						
15	0	45	57	158	293	398	426	538
	1.000	1.200	1.400	1.600	1.200	1.400	1.600	0.000

AZ + MP							
1	0	20	25	26	27	39	59	141
	1.000	0.000	1.000	0.000	1.000	1.250	0.000	0.250
	149	155	191	234				
	0.500	0.750	1.000	0.000				
2	0	28	58	109	116	170	210	263
	1.000	1.200	1.400	1.600	1.400	1.600	1.800	1.400
	298	344	352	368	373			
	1.200	0.600	0.800	1.000	0.000			
3	0	29	60	92	127	163	193	254
	1.000	1.143	1.286	1.429	1.571	1.714	1.857	2.000
	505	547	590	619	655	681	717	753
	1.714	1.857	1.429	1.714	1.857	2.000	2.143	2.286
	823	843	1251	1257	1282	1299	1303	
	2.429	2.286	0.000	2.286	0.000	2.286	0.000	
4	0	30	56	91	144	185	190	476
	1.000	1.200	1.200	1.400	1.600	1.800	1.600	1.800
	626	662	709	757	778	808	843	899
	2.000	2.200	2.400	0.000	1.600	1.800	1.600	1.800
	956	997	1033	1085	1130	1175		
	2.000	2.200	0.000	2.200	2.000	0.000		
5	0	30	60	86	114	134	156	166
	1.000	1.250	1.500	1.750	2.250	0.000	1.500	0.000
	187	252	278	469	473	490	494	536
	1.000	0.000	1.000	1.250	1.000	0.000	1.000	0.000
	542	597	652	667	673	680	687	732
	1.000	0.750	0.500	0.000	0.500	0.250	0.500	0.000
	758	827	892	893	899	928	935	959
	0.250	0.500	0.000	0.500	0.250	0.000	0.125	0.000
	1284	1296						
	0.500	0.000						

6	0	32	59	86	113	141	170	198
	1.000	1.143	1.286	1.429	1.571	1.714	1.857	2.000
	226	267	310	324	337	394	408	449
	2.143	2.286	2.000	1.143	0.000	1.714	2.000	0.000
	504	540	568	587	590	673	739	751
	1.143	1.429	1.571	1.714	1.714	1.857	1.714	1.571
	778	787	794	820	857	878	1145	1212
	0.000	1.143	1.429	1.571	1.429	1.571	1.571	1.714
	1306							
	0.000							

7	0	29	62	129	192	211	227	373
	1.000	1.250	1.500	1.750	0.000	1.250	1.500	1.250
	484	569	587	688	730	765	804	909
	1.000	1.250	1.500	1.250	1.500	1.250	1.500	0.000
	926	997	1107	1232	1241			
	1.250	1.500	1.000	0.500	0.000			

8	0	31	59	85	122	150	176	222
	1.000	1.125	1.250	1.375	1.500	1.625	1.750	1.625
	235	239	268	291	403	424	528	547
	1.500	1.250	1.000	1.125	0.000	0.875	0.750	0.875
	583	641	695	758	770	905	943	952
	0.750	0.625	0.500	0.625	0.500	0.625	0.000	0.500
	976	1148	1239					
	0.625	0.750	0.000					

9	0	28	58	84	115	150	203	254
	1.000	1.167	1.333	1.500	1.667	1.833	2.000	1.000
	297	331	500	654	680	704	1000	1033
	1.667	1.333	1.500	1.667	1.500	1.667	0.000	1.667
	1150	1212	1229	1260				
	1.833	0.000	1.667	0.000				

10	0	29	59	86	120	189	199	224
	1.000	1.143	1.286	1.429	1.571	1.286	1.143	1.286
	261	310	344	372	422	515	588	592
	1.429	1.571	1.714	1.857	2.000	2.143	2.286	1.714
	632	645	654	689	704	722	1108	1150
	2.000	0.000	2.000	2.143	2.000	0.000	2.000	2.143
	1185	1319	1337	1435				
	2.000	0.000	2.000	0.000				
11	0	30	58	86	114	143	174	218
	1.000	1.111	1.222	1.333	1.444	1.556	1.667	1.778
	283	729	799	842	877	905	969	1094
	1.667	1.778	1.889	2.000	2.111	2.222	2.333	2.444
	1130	1196						
	2.556	0.000						
12	0	29	58	92	121	148	171	277
	1.000	1.167	1.333	1.500	1.667	1.833	1.667	1.333
	452	757	980					
	1.500	1.333	0.000					
13	0	31	50	66	95	98	99	
	1.000	1.167	0.667	0.833	1.000	0.667	0.000	
14	0	30	60	88	114	171	200	205
	1.000	1.250	1.500	1.750	2.000	2.250	2.500	0.000
	217	239						
	2.500	0.000						
15	0	63	150	471	519	741	1140	
	1.000	1.143	1.429	1.571	1.429	1.571	0.000	
16	0	48	87	141	1031	1101		
	1.000	1.143	1.286	1.000	1.143	0.000		

Table A4: Treatment doses (one unit equals 2.2 mg/kg daily) for multiple sclerosis patients (Heitjan, 1991b). For each patient, times and doses are on pairs of lines.

A.5   Blood sugar level data

Group 1 (ABAB)					
Rabbit	Period	0.0	1.5	3.0	4.5	6.0
1	1	77	52	35	56	64
	2	90	47	52	68	90
	3	85	52	35	39	60
	4	94	60	60	77	94
2	1	77	56	43	56	64
	2	85	52	60	81	94
	3	103	26	68	30	30
	4	107	60	68	90	94
3	1	77	43	64	39	64
	2	90	30	30	47	60
	3	99	39	39	73	90
	4	103	60	60	90	99
4	1	81	56	22	35	39
	2	107	47	47	47	73
	3	77	26	56	64	64
	4	116	52	26	68	120
5	1	90	73	52	60	64
	2	94	60	60	77	90
	3	90	60	18	35	64
	4	94	56	47	81	99
6	1	103	47	22	60	99
	2	90	47	30	43	68
	3	85	56	56	77	94
	4	111	73	85	103	111
7	1	99	47	52	47	81
	2	99	26	64	52	90
	3	90	43	18	30	39
	4	111	56	43	90	111
8	1	90	35	22	26	22
	2	94	8	30	26	43
	3	103	39	22	39	94
	4	90	18	26	18	22

Rabbit	Period	0.0	1.5	3.0	4.5	6.0
9	1	90	64	47	47	60
	2	81	39	47	77	94
	3	77	56	26	64	94
	4	94	52	47	81	90
10	1	90	35	12	26	68
	2	81	26	35	30	85
	3	81	56	60	81	103
	4	90	22	52	85	90
11	1	85	68	56	68	73
	2	94	56	68	73	90
	3	105	73	47	77	90
	4	97	73	52	81	94

Group 2 (BABA)						
Rabbit	Period	0.0	1.5	3.0	4.5	6.0
12	1	103	26	22	39	68
	2	90	35	30	39	94
	3	103	26	26	56	103
	4	111	68	52	77	103
13	1	85	68	56	64	81
	2	94	52	43	47	56
	3	85	43	22	68	103
	4	111	52	99	103	120
14	1	85	35	8	64	99
	2	90	35	30	43	99
	3	73	22	12	56	103
	4	94	52	47	73	101

Rabbit	Period	0.0	1.5	3.0	4.5	6.0
15	1	85	35	47	77	68
	2	85	39	52	52	64
	3	90	12	43	85	99
	4	92	52	73	99	101
16	1	85	35	39	85	77
	2	85	47	56	77	81
	3	90	43	35	64	81
	4	85	52	52	77	99
17	1	103	56	47	64	56
	2	103	35	39	68	99
	3	90	35	18	64	94
	4	107	47	43	85	101
18	1	90	52	35	35	39
	2	90	43	30	47	77
	3	77	56	64	81	99
	4	111	60	68	107	111
19	1	81	22	12	22	77
	2	81	12	2	30	77
	3	68	12	8	8	39
	4	85	26	12	18	39
20	1	85	30	22	52	94
	2	90	30	26	52	90
	3	85	39	22	56	90
	4	105	68	64	81	81
21	1	85	47	33	77	107
	2	85	39	39	43	60
	3	85	35	26	43	73
	4	90	43	35	81	90
22	1	94	39	35	52	85
	2	94	26	30	39	81
	3	90	22	26	73	99
	4	103	47	35	64	94

Table A5: Blood sugar levels (mg %) of rabbits treated with two insulin mixtures (Ciminera and Wolfe, 1953).

A.6 Epileptic fit data

	Placebo				
	Baseline	Week			
Age	Count	2	4	6	8
31	11	5	3	3	3
30	11	3	5	3	3
25	6	2	4	0	5
36	8	4	4	1	4
22	66	7	18	9	21
29	27	5	2	8	7
31	12	6	4	0	2
42	52	40	20	23	12
37	23	5	6	6	5
28	10	14	13	6	0
36	52	26	12	6	22
24	33	12	6	8	4
23	18	4	4	6	2
36	42	7	9	12	14
26	87	16	24	10	9
26	50	11	0	0	5
28	18	0	0	3	3
31	111	37	29	28	29
32	18	3	5	2	5
21	20	3	0	6	7
29	12	3	4	3	4
21	9	3	4	3	4
32	17	2	3	3	5
25	28	8	12	2	8
30	55	18	24	76	25
40	9	2	1	2	1
19	10	3	1	4	2
22	47	13	15	13	12

| Progabide | | | | | |
| | Baseline | Week | | | |
Age	Count	2	4	6	8
18	76	11	14	9	8
32	38	8	7	9	4
20	19	0	4	3	0
30	10	3	6	1	3
18	19	2	6	7	4
24	24	4	3	1	3
30	31	22	17	19	16
35	14	5	4	7	4
27	11	2	4	0	4
20	67	3	7	7	7
22	41	4	18	2	5
28	7	2	1	1	0
23	22	0	2	4	0
40	13	5	4	0	3
33	46	11	14	25	15
21	36	10	5	3	8
35	38	19	7	6	7
25	7	1	1	2	3
26	36	6	10	8	8
25	11	2	1	0	0
22	151	102	65	72	63
32	22	4	3	2	4
25	41	8	6	5	7
35	32	1	3	1	5
21	56	18	11	28	13
41	24	6	3	4	0
32	16	3	5	4	3
26	22	1	23	19	8
21	25	2	3	0	1
36	13	0	0	0	0
37	12	1	4	3	2

Table A6: Numbers of epileptic fits in four two week periods (Thall and Vail, 1990).

A.7 Tumour recurrence data

	Placebo								
Id	Recurrence Times								
6	6								
9	5								
10	12	4							
12	10	5							
13	3	13	7						
14	3	6	12						
15	7	3	6	8					
16	3	12	10						
18	1								
19	2	24							
20	25								
24	28	2							
25	2	15	5						
26	3	3	2	4	14				
27	12	3	9						
31	29								
33	9	8	5	2					
34	16	3	4	6	5	6			
36	3								
37	6								
38	3	3	3						
39	9	2	9	6	4				
40	18								
42	35								
43	17								
44	3	12	31	5	2				
46	2	13	9	6	4	5	4	6	3
47	5	9	5	8	14				
48	2	6	4	1	4	4	12	16	

		Pyridoxine Recurrence Times							
Id									
51	3	1							
53	2	1							
56	4								
57	3								
61	5								
64	3	7	12	4	8				
65	3	6	6	4	6				
67	3	4	5	4	3	9	6	2	3
70	2	4	4	6	7	4	9	3	3
72	10								
73	6	14							
74	8	7	3	2	2	3	13	2	
75	42								
77	44	3							
78	8	6	6	5	4	4	15	1	

		Thiotepa					
83	5						
87	3						
88	1	2	2	2	3		
90	17						
91	2						
92	17	2					
97	6	6	1				
98	6						
99	2						
100	26	9					
102	22	1	4	5			
103	4	12	7	4	6	3	1
104	24	2	3	11			
107	1	26					
109	2	18	3	4	11		
111	2						
115	4	20	23				
117	38						

Table A7: Times (months) between recurrences of tumours for patients with bladder cancer (Andrews and Herzberg, 1985, pp. 254–259).

A.8   Chronic granulotomous disease data

1	2	38	152.2	66.7	2	1	2	2	0
293									
2	1	14	144.0	32.8	2	1	1	2	0
255									
2	1	26	81.3	55.0	2	1	1	2	0
213									
2	1	26	178.5	69.3	2	1	1	2	0
203									
1	2	12	147.0	62.0	2	2	2	2	0
219	373	414							
2	2	15	159.0	47.5	2	1	1	2	0
8	26	152	241	249	322	350	439		
1	1	19	171.0	72.7	2	1	1	2	0
382									
1	1	12	142.0	34.0	2	1	1	2	0
388									
2	1	1	79.0	10.5	2	1	1	2	0
211	260	265	269	307	363				
1	2	9	134.5	32.7	2	1	1	2	0
82	114	337	367						
2	1	1	79.0	11.5	2	1	1	2	0
18	362								
1	1	5	102.0	18.0	2	1	1	2	0
267	360								
1	2	22	169.0	52.2	2	1	1	2	0
337									
2	2	19	159.0	46.0	2	2	2	2	0
270									
2	1	7	115.5	19.5	2	1	1	2	0
274									
1	1	25	185.0	58.4	2	1	1	2	0
271									
2	1	31	170.0	80.5	2	1	1	2	0
252									

1	1	37	155.0	67.5	2	1	2	2	0
243									
2	2	6	130.0	21.6	2	1	2	2	0
104	227								
2	2	3	96.0	13.1	2	1	2	2	0
227									
2	2	22	163.0	49.1	2	1	2	4	0
198									
2	1	17	169.0	63.5	2	1	1	4	0
207									
2	1	19	182.0	63.9	2	1	1	4	0
168	200								
1	2	36	167.0	60.8	2	1	2	4	0
197									
2	1	17	162.5	52.7	2	1	1	1	0
246	253	383							
2	2	27	176.0	82.8	2	1	1	1	0
294	349								
1	1	5	113.0	19.5	2	1	1	1	0
371									
2	1	2	93.0	13.2	2	1	1	1	0
19	102								
1	1	8	124.0	25.4	2	1	1	1	0
373	388								
1	1	12	144.0	36.9	2	1	1	1	0
388									
2	1	27	174.0	67.8	2	1	1	1	0
365									
2	1	14	143.5	33.4	2	1	1	1	0
334	370	382							
1	2	11	149.0	50.9	2	1	1	1	0
373									
2	2	29	175.0	73.1	2	1	1	1	0
280	385								

1	2	31	167.0	51.8	2	1	2	1	0
376									
1	1	7	121.0	19.9	2	1	1	1	0
360									
2	2	26	153.0	46.9	2	1	2	1	0
306									
1	1	13	145.2	36.2	2	1	1	1	0
118	240	251							
1	2	25	168.0	68.9	2	1	2	1	0
187	356								
1	2	9	140.0	36.0	2	1	2	1	0
339									
2	1	28	174.0	63.7	2	1	1	1	0
6	301								
1	2	13	139.0	34.8	2	1	2	1	0
334									
1	2	24	177.0	78.4	2	1	1	1	0
269									
2	1	11	123.0	24.3	2	1	1	1	0
273									
1	1	4	103.8	16.8	2	1	1	1	0
113	273								
2	1	19	170.0	71.2	2	1	1	1	1
99	306								
1	1	18	166.0	58.1	2	1	1	1	0
263									
1	2	7	135.0	42.9	2	1	1	1	0
167	240								
2	2	12	166.0	51.9	2	1	2	1	0
271									
1	1	10	129.0	27.4	2	1	1	1	0
254									
2	1	9	129.4	28.7	2	1	1	2	0
278									

1	1	5	112.3	20.7	2	1	1	2	0
265									
1	1	1	76.3	11.3	2	1	1	2	0
217									
1	1	7	119.0	20.6	2	1	1	2	0
165	210								
2	2	11	137.5	40.3	2	1	1	2	0
192									
2	2	4	98.3	14.4	2	1	1	2	0
11	22	169	195						
2	2	7	113.0	20.4	2	1	2	2	0
120	203								
1	1	15	178.7	60.5	2	1	1	2	0
197									
2	1	1	79.0	12.2	2	1	1	2	0
206	347								
1	1	7	116.6	23.3	2	1	1	2	0
335									
1	1	17	170.2	47.9	2	1	1	2	0
274	361	365							
1	1	8	125.4	27.7	2	1	1	2	0
336									
2	1	20	173.0	68.4	2	1	1	2	0
52	65	255	270						
2	1	5	114.5	23.0	2	2	1	2	0
67	248	250	284	347					
1	1	6	105.5	19.5	1	1	1	2	0
318									
2	1	9	129.6	29.3	2	2	1	2	0
318	359								
1	1	5	110.4	19.1	2	1	1	2	0
259									
1	2	44	153.3	45.0	2	2	2	2	0
364									

2	1	22	175.0	59.7	2	1	1	2	0
292	364								
1	1	7	111.0	17.4	2	1	1	2	0
363									
1	1	19	173.6	61.4	2	2	1	2	0
350									
1	1	34	182.6	94.8	2	2	1	2	0
269									
1	1	32	177.9	63.4	2	2	1	2	0
185									
2	1	25	185.0	74.6	2	2	1	2	0
91									
2	1	21	189.0	101.5	2	2	1	2	0
91									
2	1	7	109.0	14.7	2	1	1	2	0
357									
2	1	5	101.0	15.4	2	1	1	2	0
175	280	353							
2	1	24	169.7	59.0	2	1	1	2	0
343									
1	1	12	138.0	28.1	2	1	1	2	0
265	303								
2	1	5	97.1	15.3	2	1	1	2	0
226	322								
1	1	24	171.0	55.1	2	1	1	2	0
255									
1	1	26	176.8	66.0	2	1	1	4	0
65	343								
2	1	6	104.4	13.8	2	1	1	4	0
294									
1	2	9	122.8	20.1	2	1	2	4	0
303									
2	1	25	176.9	73.5	2	2	1	4	0
23	270								

1	1	2	93.5	14.2	2	1	1	4	0
270									
2	1	8	121.6	22.4	2	1	1	4	0
245									
2	1	10	125.9	29.7	2	1	1	4	0
261									
1	1	20	169.8	50.1	2	1	1	4	0
284									
2	2	34	166.5	58.2	2	1	1	4	0
276									
1	1	6	119.9	26.2	2	1	1	4	0
294									
1	1	11	139.2	34.9	2	1	1	4	0
277									
2	2	3	91.7	14.4	2	1	1	4	0
4	159	213	287						
1	1	9	131.3	24.0	2	1	1	4	0
331									
2	2	11	138.6	36.1	2	1	2	4	0
288									
2	1	17	156.7	36.8	2	1	1	4	0
269									
2	1	10	143.0	31.3	2	1	1	4	0
269									
2	2	17	171.0	46.1	2	1	2	2	0
330									
2	2	8	115.0	19.0	2	1	2	2	0
57	121	351							
1	1	6	111.9	17.9	2	1	1	2	0
297									
2	2	7	116.2	22.2	2	1	1	2	0
281									
1	1	7	119.7	21.5	2	1	1	2	0
276									

1	1	8	116.8	20.0	2	1	1	2	0
279									
2	1	11	141.5	36.0	2	1	1	2	0
14	278								
1	1	4	108.1	17.3	2	1	1	2	0
199									
2	2	18	179.0	67.0	2	1	1	3	0
308									
1	1	13	151.0	49.0	2	1	1	3	0
327									
2	1	11	136.5	31.2	2	2	1	3	0
329									
1	1	2	86.0	13.5	2	1	1	3	0
318									
1	1	17	180.0	68.0	2	1	1	3	0
304									
2	2	35	181.5	80.0	2	2	1	3	0
316									
2	1	25	172.5	50.0	2	1	1	3	0
300									
1	1	14	145.0	29.0	2	2	1	3	0
146	188	300							
2	2	25	187.5	74.0	2	1	1	3	0
304	312								
2	1	27	169.5	65.0	1	2	2	3	0
91	121	203	287						
2	2	32	185.0	95.0	2	1	1	3	0
293									
1	1	6	120.0	23.0	2	1	1	3	0
293									
2	2	8	138.0	31.0	2	2	1	3	0
264	286								
1	2	9	144.0	34.4	2	1	1	3	0
286									

1	2	23	170.0	49.0	2	2	2	3	0
273									
2	2	23	122.0	31.2	1	1	2	3	0
236	273								
1	2	17	127.0	27.0	2	1	2	3	0
273									
1	2	21	158.0	49.0	2	1	1	3	0
207	273								
2	1	1	81.0	10.4	2	1	1	3	0
264									
1	1	12	136.5	30.0	2	1	1	2	0
160									
2	1	7	120.0	23.0	2	1	1	2	0
146	316								
2	1	1	79.0	10.6	2	1	1	2	0
316									
1	1	3	97.8	13.0	2	1	1	2	0
315									

First line for each individual:

col. 1 — treatment (1: interferon, 2: placebo)

col. 2 — inheritance (1: x-linked, 2: autosomal recessive)

col. 3 — age

col. 4 — height (cm.)

col. 5 — weight (kg.)

col. 6 — using corticosteroids at entry (1: yes, 2: no)

col. 7 — using antibiotics at entry (1: yes, 2: no)

col. 8 — sex (1: male, 2 female)

col. 9 — hospital (1: US-NIH, 2: US-other,
                      3: Europe- Amsterdam, 4: Europe-other)

col. 10 — censor of last time (1: uncensored)

Second line for each individual: times

Table A8: Placebo controlled randomized trial of gamma interferon for chronic granulotomous disease (Fleming and Harrington, 1991, pp. 377–383).

A.9 Chronic myeloid leukemia data

Id	Day	State	ADA	Id	Day	State	ADA
1	0	1	6.3	2	785	1	5.9
1	51	2	9.9	2	798	1	10.2
1	63	2	11.7	2	812	1	6.7
1	107	2	21.8	2	840	1	14.0
1	124	3	0.0	2	875	1	13.1
2	0	1	4.7	2	896	1	11.3
2	28	1	8.7	2	918	1	8.0
2	56	1	7.6	2	945	1	10.1
2	77	1	7.8	2	980	1	13.4
2	105	1	7.3	2	995	1	19.4
2	112	1	6.2	2	1008	1	22.4
2	140	1	12.5	2	1036	1	43.0
2	168	1	7.1	2	1051	2	60.0
2	210	1	11.3	2	1057	2	75.1
2	245	1	6.0	2	1066	2	76.0
2	259	1	6.5	2	1073	2	84.8
2	280	1	8.6	3	0	2	24.1
2	301	1	5.0	3	41	2	22.2
2	329	1	7.9	3	52	2	15.4
2	343	1	8.0	3	59	2	14.7
2	371	1	6.2	3	81	3	0.0
2	455	1	5.7	4	0	1	6.0
2	483	1	9.1	4	26	1	6.1
2	491	1	7.2	4	147	1	13.9
2	497	1	5.2	5	0	1	6.7
2	511	1	7.9	5	2	1	5.3
2	539	1	2.9	5	863	1	7.1
2	644	1	8.2	6	0	2	20.7
2	658	1	11.2	6	21	2	30.7
2	735	1	8.3	6	63	3	0.0
2	742	1	7.8	7	0	1	5.4
2	749	1	9.8	7	63	1	14.3
2	763	1	7.6	7	92	1	8.7
2	770	1	8.2	7	147	1	5.9
2	777	1	6.9	7	196	1	8.8
7	255	1	15.7	11	28	1	8.8
7	280	1	9.0	11	35	1	13.1
7	308	1	14.0	11	63	1	7.5

State: 1 — stable, 2 — blast, 3 — dead

Id	Day	State	ADA	Id	Day	State	ADA
7	371	1	8.7	11	77	1	6.3
7	486	1	10.2	11	91	1	6.5
7	539	2	18.3	11	155	1	17.1
7	553	2	14.7	11	182	1	3.7
7	560	2	16.4	11	224	1	21.3
7	567	2	16.4	11	238	2	13.3
7	574	2	15.7	11	252	2	14.9
7	581	2	12.7	11	268	2	8.6
7	595	2	16.3	11	294	2	7.7
7	609	2	12.6	11	383	2	6.7
7	679	2	32.7	11	386	2	13.6
7	693	2	21.3	11	454	3	0.0
7	694	2	19.8	12	0	2	8.7
7	700	2	17.8	12	77	3	0.0
7	728	2	24.3	13	0	2	7.4
7	744	2	35.9	13	10	3	0.0
7	746	3	0.0	14	0	2	15.9
8	0	2	16.3	14	67	2	11.6
8	7	2	10.2	14	77	2	7.1
8	82	3	0.0	14	84	2	39.4
9	0	2	23.1	14	107	3	0.0
9	362	3	0.0	15	0	1	4.2
10	0	2	16.2	15	34	1	5.4
10	18	2	13.4	15	148	1	12.7
10	25	2	13.9	15	204	1	5.9
10	44	2	11.7	15	351	1	5.9
10	60	2	28.0	15	399	1	5.4
10	66	2	12.6	15	521	1	4.7
10	94	2	9.4	15	542	1	4.7
10	266	2	27.6	15	605	1	4.7
10	321	3	0.0	15	806	1	6.5
11	0	1	9.0	16	0	1	4.0
16	266	1	6.1	25	0	2	9.9
17	0	1	9.8	25	14	2	9.9
17	273	1	8.2	25	28	2	12.7
17	329	1	9.5	25	44	2	15.9
17	343	1	8.0	25	86	3	0.0
17	385	1	12.8	26	0	1	9.0

State: 1 — stable, 2 — blast, 3 — dead

Id	Day	State	ADA	Id	Day	State	ADA
17	476	1	6.1	26	133	1	11.5
17	644	1	9.0	26	266	1	8.6
17	759	1	11.6	26	378	1	11.4
17	882	1	9.0	26	441	1	7.8
18	0	2	14.2	26	570	1	10.0
18	252	3	0.0	27	0	2	26.2
19	0	1	3.8	27	9	2	30.9
19	113	1	5.8	27	15	3	0.0
19	136	1	9.2	28	0	2	19.9
19	226	1	12.2	28	56	2	21.1
19	318	1	4.3	28	64	3	0.0
20	0	1	3.4	29	0	2	9.3
20	2	1	4.4	29	14	3	0.0
20	56	1	2.6	30	0	1	9.9
21	0	2	12.8	30	14	1	8.6
21	30	3	0.0	30	35	1	7.6
22	0	2	31.0	30	42	1	7.4
22	85	2	38.2	30	56	1	8.4
22	88	3	0.0	30	63	1	9.9
23	0	1	9.9	30	87	1	6.1
23	90	1	7.8	30	98	1	9.1
23	174	1	8.5	30	203	1	10.5
24	0	1	8.5	30	287	1	12.3
24	140	1	9.7	31	0	2	32.0
24	182	1	7.7	31	130	2	17.0
24	232	1	8.1	31	151	2	20.0
24	329	1	13.3	31	179	2	6.9
24	693	1	10.2	31	207	2	70.2
24	938	2	12.9	31	319	2	8.0
31	340	2	28.3	35	217	2	6.8
31	373	2	29.7	35	231	2	14.7
31	402	2	70.0	35	243	2	8.7
31	409	3	0.0	35	273	2	8.6
32	0	1	4.0	35	348	2	39.5
32	26	1	9.1	35	371	2	24.2
32	89	1	11.1	35	375	2	22.5
32	103	1	6.4	35	376	2	18.1
32	124	1	9.9	35	377	2	20.7

State: 1 — stable, 2 — blast, 3 — dead

Id	Day	State	ADA	Id	Day	State	ADA
32	180	1	6.1	35	386	3	0.0
32	194	1	9.0	36	0	1	9.4
32	201	1	11.9	36	21	1	13.0
32	215	1	8.7	36	98	1	6.9
32	223	1	8.6	36	231	1	10.6
32	236	1	25.6	36	378	1	5.8
32	257	1	8.4	36	539	1	10.3
32	271	1	5.7	36	791	1	5.9
32	278	1	5.9	36	805	1	14.3
32	642	1	6.9	36	819	1	8.1
33	0	1	4.5	36	854	1	4.7
33	35	1	5.2	36	868	1	8.0
33	147	1	3.9	36	882	1	8.6
33	206	1	4.6	36	945	1	11.6
33	343	1	5.1	37	0	1	4.0
33	385	1	5.0	37	28	1	4.0
33	451	1	3.0	37	56	1	5.5
33	537	1	5.7	37	84	1	6.8
34	0	1	6.3	37	126	1	5.4
34	21	1	5.3	37	154	1	6.7
35	0	2	15.0	37	182	1	7.8
35	130	2	24.7	37	210	1	6.1
35	143	2	5.3	37	238	1	7.9
35	160	2	6.1	37	267	1	7.6
35	180	2	11.3	37	329	1	10.3
35	195	2	6.4	37	357	1	6.5
37	413	1	5.3	37	949	2	6.4
37	497	1	5.2	37	956	2	7.2
37	525	1	9.9	37	973	2	8.5
37	553	1	8.0	37	980	2	6.3
37	581	1	9.7	37	994	2	9.6
37	609	1	8.8	37	1001	2	20.0
37	637	1	3.2	37	1008	2	7.8
37	693	1	4.8	37	1012	2	6.2
37	721	1	11.3	37	1015	2	6.4
37	726	1	8.4	37	1022	2	7.2
37	749	1	14.4	37	1026	2	7.4
37	756	1	11.2	37	1033	2	21.6

State: 1 — stable, 2 — blast, 3 — dead

Id	Day	State	ADA	Id	Day	State	ADA
37	763	1	16.8	37	1043	3	0.0
37	770	1	14.5	38	0	2	31.4
37	777	1	9.8	38	6	3	0.0
37	784	1	9.1	39	0	2	22.0
37	791	1	7.3	39	15	3	0.0
37	798	1	7.3	40	0	2	6.1
37	807	1	7.6	40	59	2	6.2
37	812	1	6.5	40	99	2	2.9
37	819	1	5.7	40	107	2	4.6
37	840	1	9.6	40	116	3	0.0
37	848	1	10.7	41	0	2	17.8
37	861	1	7.1	41	5	3	0.0
37	868	1	9.1	42	0	1	7.5
37	875	2	7.7	42	336	1	5.4
37	882	2	7.4	42	371	1	5.1
37	896	2	4.3	42	567	1	6.6
37	910	2	7.7	43	0	1	7.6
37	918	2	7.5	43	21	1	7.1
37	924	2	7.3	43	49	1	7.8
37	931	2	7.2	43	63	1	4.5
37	938	2	20.5	43	119	1	5.6
37	942	2	9.0	43	147	1	9.3
37	945	2	9.9	43	160	1	7.7
43	168	1	6.0	47	168	2	28.3
43	287	1	10.4	47	178	3	0.0
43	365	1	4.5	48	0	1	6.1
43	392	1	9.7	48	128	1	6.6
43	406	1	7.1	48	162	1	4.2
43	420	1	12.0	48	169	1	8.0
43	483	1	11.5	48	204	1	5.7
43	598	1	6.4	48	232	1	9.2
44	0	2	24.6	48	414	1	9.4
44	19	2	65.8	48	442	1	11.3
44	33	2	14.1	48	527	1	9.4
44	133	2	113.4	49	0	1	8.4
44	164	2	52.8	49	398	2	16.9
44	173	2	73.5	49	438	2	21.6
44	175	2	105.4	49	454	3	0.0

State: 1 — stable, 2 — blast, 3 — dead

Id	Day	State	ADA	Id	Day	State	ADA
44	230	2	73.2	50	0	1	12.4
44	263	3	0.0	50	51	1	11.9
45	0	1	8.6	51	0	2	22.0
45	364	1	8.5	51	92	2	7.3
45	427	1	7.9	51	107	2	6.1
46	0	1	7.1	51	149	2	16.7
46	14	1	7.3	51	438	3	0.0
46	35	1	7.3	52	0	2	16.8
46	42	1	6.6	52	76	3	0.0
46	63	1	5.0	53	0	1	11.6
46	91	1	5.6	53	36	1	6.7
46	119	1	6.3	53	199	1	9.3
46	161	1	6.6	53	218	1	4.9
46	175	1	6.7	53	225	1	10.3
46	196	1	7.0	53	232	1	19.9
46	224	1	6.4	53	239	1	7.3
46	273	1	6.1	53	248	1	10.2
47	0	2	8.7	53	265	1	10.0
47	87	2	25.6	53	274	1	8.4
47	131	2	35.3	53	290	1	7.8
53	316	1	11.3	53	672	2	23.6
53	344	1	14.4	53	717	2	26.6
53	386	1	7.6	54	0	1	7.4
53	406	1	14.3	54	20	1	6.8
53	434	1	10.0	54	62	1	7.2
53	472	1	15.2	54	76	1	11.7
53	528	1	9.5	55	0	1	8.1
53	542	1	9.6	55	77	1	5.8
53	547	1	9.3	55	105	1	8.6
53	643	1	15.1	55	133	1	5.8
53	654	2	18.3				

State: 1 — stable, 2 — blast, 3 — dead

Table A9: Event histories of 55 patients with chronic myeloid leukemia (Klein, Klotz, and Grever, 1984).

A.10 Mammary tumour data

Rat	Retinoid						
1	182						
2	182*						
3	63	68	182*				
4	152	182*					
5	130	134	145	152	182*		
6	98	152	182				
7	88	95	105	130	137	167	182*
8	152	182*					
9	81	182*					
10	71	84	126	134	152	182*	
11	116	130	182*				
12	91	182*					
13	63	68	84	95	152	182*	
14	105	152	182*				
15	63	102	152	182*			
16	63	77	112	140	182*		
17	77	119	152	161	167	182*	
18	105	112	145	161	182		
19	152	182*					
20	81	95	182*				
21	84	91	102	108	130	134	182*
22	182*						
23	91	182*					

Stars indicate censoring

Rat	Control								
24	63	102	119	161	161	172	179	182*	
25	88	91	95	105	112	119	119	137	145
		167	172	182*					
26	91	98	108	112	134	137	161	161	179
		182*							
27	71	174	182*						
28	95	105	134	134	137	140	145	150	150
		182*							
29	68	68	130	137	182*				
30	77	95	112	137	161	174	182*		
31	81	84	126	134	161	161	174	182*	
32	68	77	98	102	102	102	182*		
33	112	182*							
34	88	88	91	98	112	134	134	137	137
		140	140	152	152	182*			
35	77	179	182*						
36	112	182*							
37	71	71	74	77	112	116	116	140	140
		167	182*						
38	77	95	126	150	182*				
39	88	126	130	130	134	182*			
40	63	74	84	84	88	91	95	108	134
		137	179	182*					
41	81	88	105	116	123	140	145	152	161
		161	179	182*					
42	88	95	112	119	126	126	150	157	179
		182*							
43	68	68	84	102	105	119	123	123	137
		161	179	182					
44	140	182*							
45	152	182	182						
46	81	182*							
47	63	88	134	182*					
48	84	134	182						

Stars indicate censoring

Table A10: Times of appearance of mammary tumours in female rats under two treatments (Gail, Santner, and Brown, 1980).

Appendix B
Programs used

Chapter 3
Section 3.1 – MatLab
Section 3.2 – GLIM
Section 3.3 – GLIM
Section 3.4 – CARMA, GLIM
Chapter 4
Section 4.1 – FORTRAN, GLIM, MatLab
Section 4.2 – CARMA
Section 4.3 – CARMA
Section 4.4 – GROWTH
Section 4.5 – CARMA
Section 4.6 – GLIM
Chapter 5
Section 5.1 – GLIM
Section 5.2 – GLIM
Section 5.3 – FORTRAN, GLIM
Section 5.4 – GLIM
Chapter 6
Section 6.1 – GLIM
Section 6.2 – GLIM
Section 6.3 – GLIM
Section 6.4 – GLIM
Section 6.5 – DISCOUNT, GLIM
Section 6.6 – MatLab
Chapter 7
Section 7.2 – GLIM
Section 7.3 – GLIM, MIXTURE
Chapter 8
Section 8.2 – GLIM
Section 8.3 – GLIM
Section 8.4 – DISCOUNT, GLIM, SABRE
Section 8.5 – MatLab

Bibliographic codes

Although the bibliography is fairly extensive, I can make no claim to its exhaustiveness. These are the books and articles which I have been able to consult (i.e. find and read), although several hundred of those I did look at are not included below, since they were deemed less pertinent. Indeed, a complete list would surely be several times longer and is growing every day. As well, the topics of classical normal, discrete data, and survival models are only touched in so far as they directly concern repeated measures.

An attempt to provide an indication of the contents of each article is given by the following codes:

AR	–	autoregression
CD	–	compound distribution
CO	–	crossover trial
DES	–	experimental design
DGLM	–	dynamic generalized linear model, Kalman filter
DIST	–	distribution
DD	–	discrete data model
EH	–	event histories
GC	–	growth curve
GEE	–	generalized estimating equations, quasi-likelihood
GLM	–	generalized linear model
MC	–	Markov chain
MP	–	Markov process, semi-Markov process
MULT	–	general multivariate methods and distributions
NPAR	–	nonparametric methods
OD	–	overdispersion
PP	–	point process
RP	–	renewal process
SURV	–	survival model
TS	–	longitudinal data, time series
VC	–	variance components, random effects, heterogeneity

Some especially clear introductory references for various subjects are marked with an asterix.

Bibliography

1. Aalen, O.O. (1976) Nonparametric inference in connection with multiple decrement models. *Scandinavian Journal of Statistics* **3**, 15–27.
   DD MP NPAR SURV TS

2. Aalen, O.O. (1978a) Nonparametric estimation of partial transition probabilities in multiple decrement models. *Annals of Statistics* **6**, 534–555.
   DD EH MP NPAR SURV TS

3. Aalen, O.O. (1978b) Nonparametric inference for a family of counting processes. *Annals of Statistics* **6**, 701–726.
   DD EH MP NPAR SURV TS

4. Aalen, O.O. (1987a) Two examples of modelling heterogeneity in survival analysis. *Scandinavian Journal of Statistics* **14**, 19–25.
   SURV VC

5. Aalen, O.O. (1987b) Mixing distributions on a Markov chain. *Scandinavian Journal of Statistics* **14**, 281–289.
   CD MC TS

6. Aalen, O.O. (1988) Heterogeneity in survival analysis. *Statistics in Medicine* **7**, 1121–1137.
   SURV VC

7. Aalen, O.O. (1989) A linear regression model for the analysis of life times. *Statistics in Medicine* **8**, 907–925.
   MP SURV TS

8. Aalen, O.O. (1991) Modelling the influence of risk factors on familial aggregation of disease. *Biometrics* **47**, 933–945.
   VC

9. *Aalen, O.O., Borgan, Ø., Keiding, N., and Thormann, J. (1980) Interaction between life history events: nonparametric analysis for prospective and retrospective data in the presence of censoring. *Scandinavian Journal of Statistics* **7**, 161–171.
   EH MP NPAR SURV TS

10. *Aalen, O.O. and Husebye, E. (1991) Statistical analysis of repeated events forming renewal processes. *Statistics in Medicine* **10**, 1227–1240.
    EH RP TS

11. Aalen, O.O. and Johansen, S. (1978) An empirical transition matrix for nonhomogeneous Markov chains based on censored observations. *Scandinavian Journal of Statistics* **5**, 141–150.
    DD MC TS

12. Abeyasekera, S. and Curnow, R.N. (1984) The desirability of adjusting

for residual effects in a crossover design. *Biometrics* **40**, 1071–1078.
CO

13. Abramson, I. (1988) A recursive regression for high-dimensional models, with application to growth curves and repeated measures. *Journal of the American Statistical Association* **83**, 1073–1077.
GC TS

14. Abu-Libdeh, H., Turnbull, B.W., and Clark, L.C. (1990) Analysis of multi-type recurrent events in longitudinal studies: application to a skin cancer prevention trial. *Biometrics* **46**, 1017–1034.
DD VC

15. Adke, S.R. and Balakrishna, N. (1992) Renewal counting process induced by a discrete Markov chain. *Australian Journal of Statistics* **34**, 115–121.
DD EH RP TS

16. Afsarinejad, K. (1990) Repeated measurements designs — a review. *Communication in Statistics* **A19**, 3985–4028.
DES

17. Agresti, A. (1988) A model for agreement between ratings on an ordinal scale. *Biometrics* **44**, 539–548.
DD VC

18. Agresti, A. (1989a) An agreement model with kappa as parameter. *Statistics and Probability Letters* **7**, 271–273.
DD VC

19. Agresti, A. (1989b) A survey of models for repeated ordered categorical response data. *Statistics in Medicine* **8**, 1209–1224.
DD

20. *Agresti, A. (1990) *Categorical Data Analysis.* New York: John Wiley.
DD

21. Agresti, A. (1992a) Modelling patterns of agreement and disagreement. *Statistical Methods in Medical Research* **1**, 201–218.
DD VC

22. Agresti, A. (1992b) Analysis of ordinal paired comparison data. *Journal of the Royal Statistical Society* **C41**, 287–297.
DD

23. Aitkin, M. (1981) Regression models for repeated measurements. *Biometrics* **37**, 831–832.
VC

24. Aitkin, M. (1987) Modelling variance heterogeneity in normal regression using GLIM. *Journal of the Royal Statistical Society* **C36**, 332–339.
VC

25. Aitkin, M., Anderson, D., Francis, B. and Hinde, J. (1989) *Statistical Modelling with GLIM.* Oxford: Oxford University Press.
GLM

26. Aitkin, M., Anderson, D., and Hinde, J. (1981) Statistical modelling of data on teaching styles. *Journal of the Royal Statistical Society* **A14**, 419–461.
VC

27. Aitkin, M., Laird, N., and Francis, B. (1983) A reanalysis of the Stanford heart transplant data. *Journal of the American Statistical Association* **78**, 264–292.
    SURV

28. Albert, P.S. (1991) A two-state Markov model for a time series of epileptic seizure counts. *Biometrics* **47**, 1371–1381.
    DD MC OD TS

29. Alho, J.M. (1992) On prevalence, incidence, and duration in general stable populations. *Biometrics* **48**, 587–592.
    SURV

30. Allen, D.J., Reimers, H.J., Feuerstein, I.A., and Mustard, J.F. (1975) The use and analysis of multiple responses in multicompartment cellular systems. *Biometrics* **31**, 921–929.
    MP SURV TS

31. Alling, D.W. (1958) The after-history of pulmonary tuberculosis a stochastic model. *Biometrics* **14**, 527–547.
    DD PP TS

32. Allison, P.D. (1984) *Event History Analysis. Regression For Longitudinal Event Data.* Newbury Park: Sage.
    EH MP SURV TS

33. Allison, P.D. (1985) Survival analysis of backward recurrence times. *Journal of the American Statistical Association* **80**, 315–322.
    EH SURV

34. Altham, P.M.E. (1976) Discrete variable analysis for individuals grouped into families. *Biometrika* **63**, 263–269.
    DD OD

35. Altham, P.M.E. (1978) Two generalizations of the binomial distribution. *Journal of the Royal Statistical Society* **C27**, 162–167.
    DD OD

36. Altham, P.M.E. (1979) Detecting relationships between categorical variables observed over time: a problem of deflating a Chi-squared statistic. *Journal of the Royal Statistical Society* **C28**, 115–125.
    DD TS

37. Altham, P.M.E. (1984) Improving the precision of estimation by fitting a model. *Journal of the Royal Statistical Society* **B46**, 118–119.
    GLM

38. Altham, P.M.E. and Porteous, B.T. (1986) Markov chains, reversibility, equilibrium and GLIM. *GLIM Newsletter* **11**, 46–50.
    DD MC TS

39. Amemiya, T. (1971) The estimation of the variances in a variance components model. *International Economic Review* **12**, 1–13.
    VC

40. Andersen, A.H., Jensen, E.B., and Schou, G. (1981) Two-way analysis of variance with correlated errors. *International Statistical Review* **49**, 153–167.
    AR TS

41. Andersen, P.K. (1985) Statistical models for longitudinal labour market data based on counting processes. In Heckman and Singer (1985), pp. 294–307.
    EH MP SURV TS VC

42. Andersen, P.K. (1988) Multistate models in survival analysis: a study of nephropathy and mortality in diabetes. *Statistics in Medicine* **7**, 661–670.
    EH MP SURV TS

43. Andersen, P.K. (1992) Repeated assessment of risk factors in survival analysis. *Statistical Methods in Medical Research* **1**, 297–315.
    EH MP SURV TS

44. Andersen, P.K., Borch-Johnsen, K., Deckert, T., Green, A., Hougaard, P., Keiding, N., and Kreiner, S. (1985) A Cox regression model for the relative mortality and its application to Diabetes mellitus survival data. *Biometrics* **41**, 921–932.
    SURV

45. *Andersen, P.K. and Borgan, Ø. (1985) Counting process models for life history data: a review. *Scandinavian Journal of Statistics* **12**, 97–158.
    EH MP SURV TS

46. Andersen, P.K., Borgan, Ø., Gill, R.D., and Keiding, N. (1985) Linear nonparametric tests for comparison of counting processes with applications to censored survival data. *International Statistical Review* **50**, 219–258, **52**, 225.
    MP NPAR SURV TS

47. Andersen, P.K. and Gill, R.D. (1982) Cox's regression model for counting processes: a large sample study. *Annals of Statistics* **10**, 1100–1120.
    MP SURV TS

48. Andersen, P.K. and Green, A. (1985) Evaluation of estimating bias in an illness–death–emigration model. *Scandinavian Journal of Statistics* **12**, 63–68.
    EH MP SURV TS

49. Andersen, P.K., Hansen, L.S., and Keiding, N. (1991a) Non- and semi-parametric estimation of transition probabilities from censored observation of a non-homogeneous Markov process. *Scandinavian Journal of Statistics* **18**, 153–167.
    EH MP NPAR SURV TS

50. Andersen, P.K., Hansen, L.S., and Keiding, N. (1991b) Assessing the influence of reversible disease indicators on survival. *Statistics in Medicine* **10**, 1061–1067.
    EH MP NPAR SURV TS

51. Andersen, P.K. and Rasmussen, N.K.R. (1986) Psychiatric admissions and choice of abortion. *Statistics in Medicine* **5**, 243–253.
    EH MP SURV TS

52. Anderson, D.A. (1988) Some models for overdispersed binomial data. *Australian Journal of Statistics* **30**, 125–148.
    DD OD

53. Anderson, D.A. and Aitkin, M. (1985) Variance component models with

binary response: interviewer variability. *Journal of the Royal Statistical Society* **B47**, 203–210.
DD VC

54. Anderson, D.A. and Hinde, J.P. (1988) Random effects in generalized linear models and the EM algorithm. *Communications in Statistics* **A17**, 3847–3856.
GLM VC

55. Anderson, J.E., Louis, T.A., Holm, N.V., and Harvald, B. (1992) Time-dependent association measures for bivariate survival distributions. *Journal of the American Statistical Association* **87**, 641–650.
MP SURV TS VC

56. Anderson, T.W. (1978) Repeated measures on autoregressive processes. *Journal of the American Statistical Association* **73**, 371–378.
AR TS

57. Anderson, T.W. and Goodman, L.A. (1957) Statistical inference about Markov chains. *Annals of Mathematical Statistics* **28**, 89–109.
DD MC TS

58. Anderson, T.W. and Hsiao, C. (1981) Estimation of dynamic models with error components. *Journal of the American Statistical Association* **76**, 598-606.
AR TS VC

59. *Anderson, T.W. and Hsiao, C. (1982) Formulation and estimation of dynamic models using panel data. *Journal of Econometrics* **18**, 47–82.
AR TS VC

60. Andrews, D.F. and Herzberg, A.M. (1985) *Data. A Collection of Problems from Many Fields for the Student and Research Worker.* Berlin: Springer Verlag.

61. Anscombe, F.J. (1949) The statistical analysis of insect counts based on the negative binomial distribution. *Biometrics* **5**, 165–173.
CD DD

62. Anscombe, F.J. (1950) Sampling theory of the negative binomial and logarithmic series distributions. *Biometrika* **37**, 358–382.
CD

63. Ansell, J.I. and Phillips, M.J. (1989) Practical problems in the statistical analysis of reliability data. *Journal of the Royal Statistical Society* **C38**, 205–247.
RP TS

64. Antelman, G.R. (1972) Interrelated Bernouilli processes. *Journal of the American Statistical Association* **67**, 831–841.
DD OD

65. Arbous, A.G. and Kerrich, J.E. (1951) Accident statistics and the concept of accident-proneness. I: A critical evaluation. II: The mathematical background. *Biometrics* **7**, 340–432.
CD

66. Arbous, A.G. and Sichel, H.S. (1954) New techniques for the analysis of absenteeism data. *Biometrika* **41**, 77–90.

CD DD MC TS

67. *Arjas, E. (1989) Survival models and martingale dynamics. *Scandinavian Journal of Statistics* **16**, 177–225.
DD PP SURV TS VC

68. Arjas, E. and Greenwood, P. (1981) Competing risks and independent minima: a marked point process approach. *Advances in Applied Probability* **13**, 669–680.
DD PP SURV TS

69. Arjas, E. and Haara, P. (1984) A marked point process approach to censored failure data with complicated covariates. *Scandinavian Journal of Statistics* **11**, 193–209.
DD PP SURV TS

70. Arjas, E. and Haara, P. (1987) A logistic regression model for hazard: asymptotic results. *Scandinavian Journal of Statistics* **14**, 1–18.
DD PP SURV TS

71. Arjas, E. and Haara, P. (1988) A note on the exponentiality of total hazards before failure. *Journal of Multivariate Analysis* **26**, 207–218.
DD PP SURV TS

72. Arjas, E. and Haara, P. (1992) Observation scheme and likelihood. *Scandinavian Journal of Statistics* **19**, 111–132.
DD PP SURV TS

73. Arnold, B.C. (1967) A note on multivariate distributions with specified marginals. *Journal of the American Statistical Association* **62**, 1460–1461.
DIST MULT

74. Arnold, B.C. (1968) Parameter estimation for a multivariate exponential distribution. *Journal of the American Statistical Association* **63**, 848–852.
DIST MULT SURV

75. Arnold, B.C., Castillo, E., and Sarabia, J.M. (1992) *Conditionally Specified Distributions.* Berlin: Springer Verlag.
DIST MULT

76. Arnold, B.C. and Press, S.J. (1989) Compatible conditional distributions. *Journal of the American Statistical Association* **84**, 152–156.
DIST MULT

77. Arnold, B.C. and Strauss, D.J. (1988) Bivariate distributions with exponential conditionals. *Journal of the American Statistical Association* **83**, 522–527.
DIST MULT SURV

78. Arnold, B.C. and Strauss, D.J. (1991) Bivariate distributions with conditionals in prescribed exponential families. *Journal of the Royal Statistical Society* **B53**, 365–375.
DIST MULT

79. Ashby, M., Neuhaus, J.M., Hauck, W.W., Bacchetti, P., Heilbron, D.C., Jewell, N.P., Segal, M.R., and Fusaro, R.E. (1992) An annotated bibliography of methods for analyzing correlated categorical data. *Statistics in Medicine* **11**, 67–99.
DD OD TS VC

80. Atkinson, A.C. and Yeh, L. (1982) Inference for Sichel's compound Poisson distribution. *Journal of the American Statistical Association* **77**, 153–158.
    CD

81. Aven, T. (1986) Bayesian inference in a parametric counting process model. *Scandinavian Journal of Statistics* **13**, 87–97.
    DD PP SURV TS

82. Avery, R.B., Hansen, L.P., and Hotz, J.V. (1983) Multiperiod probit models and orthogonality condition estimation. *International Economic Review* **24**, 21–35.
    DD TS

83. Azzalini, A. (1983) Maximum likelihood estimation of order m for stationary stochastic processes. *Biometrika* **70**, 381–387.
    MP TS

84. Azzalini, A. (1984) Estimation and hypothesis testing for collections of autoregressive time series. *Biometrika* **71**, 85–90.
    AR GC TS

85. Azzalini, A. (1989) An analysis of variance table for repeated measurements with unknown autoregressive parameter (AS 246). *Journal of the Royal Statistical Society* **C38**, 402–411.
    AR GC TS

86. Azzalini, A. (1991) An explicit nearly unbiased estimate of the AR(1) parameter for repeated measurements. *Journal of Time Series Analysis* **12**, 273–281.
    AR GC TS

87. Bai, D.S. (1975) Efficient estimation of transition probabilities in a Markov chain. *Annals of Statistics* **3**, 1305–1317.
    DD MC TS

88. Bailey, B.J.R. (1990) A model for function word counts. *Journal of the Royal Statistical Society* **C39**, 107–114.
    DD MC TS

89. Bailey, N.T.J. (1951) On estimating the size of mobile populations from recapture data. *Biometrika* **38**, 293–306.
    DD TS

90. Bailey, N.T.J. (1964) *The Elements of Stochastic Processes with Applications to the Natural Sciences.* New York: John Wiley.
    DD MC PP RP TS

91. Bain, L.J., Englehardt, M., and Wright, F.T. (1985) Tests for an increasing trend in the intensity of a Poisson process: a power study. *Journal of the American Statistical Association* **80**, 419–422.
    DD PP TS

92. Baker, S.G., Freedman, L.S., and Parmar, M.K.B. (1991) Using replicate observations in observer agreement studies with binary assessments. *Biometrics* **47**, 1327–1338.
    DD

93. Baksalary, R.S., Corsten, L.C.A., and Kala, R. (1978) Reconciliation of two different views on estimation of growth curve parameters. *Biometrika* **65**, 661–665.
    GC MULT TS

94. Balestra, P. and Nerlove, M. (1966) Pooling cross section and time series data in the estimation of a dynamic model: the demand for natural gas. *Econometrica* **34**, 585–612.
    AR TS VC

95. Banerjee, A.K. and Bhattacharyya, G.K. (1976) A purchase incidences model with inverse Gaussian interpurchase times. *Journal of the American Statistical Association* **71**, 823–829.
    EH RP TS VC

96. Barcikowski, R.S. and Robey, R.R. (1984) Decisions in single group repeated measures analysis: statistical tests and three computer packages. *American Statistician* **38**, 148–150.
    VC

97. Barnard, G.A. (1953) Time intervals between accidents — a note on Maguire, Pearson, and Wynn's paper. *Biometrika* **40**, 212–213.
    EH RP SURV TS

98. *Barnard, G.A., Jenkins, G.M., and Winsten, C.B. (1962) Likelihood inference and time series. *Journal of the Royal Statistical Society* **A125**, 321–352.
    AR TS

99. Barndorff-Nielsen, O.E., Blaesild, P., and Seshadri, V. (1992) Multivariate distributions with generalized inverse Gaussian marginals, and associated Poisson mixtures. *Canadian Journal of Statistics* **20**, 109–120.
    CD DIST MULT SURV VC

100. Barndorff-Nielsen, O.E. and Cox, D.R. (1984) The effect of the sampling rules on likelihood statistics. *International Statistical Review* **52**, 309–326.
     MP SURV TS

101. Barndorff-Nielsen, O.E. and Yeo, G.F. (1969) Negative binomial processes. *Journal of Applied Probability* **6**, 633–647.
     CD DD PP TS

102. Barnwall, R.K. and Paul, S.R. (1988) Analysis of one-way layout of count data with negative binomial variation. *Biometrika* **75**, 215–222.
     CD DD OD

103. Barron, D.N. and Hannan, M.T. (1991) Autocorrelation and density dependence in organizational founding rates. *Sociological Methods and Research* **20**, 218–241.
     AR DD GEE OD SURV TS

104. Barry, J.T., Francis, B.J., and Davies, R.B. (1989) SABRE: software for the analysis of binary recurrent events. In Decarli *et al.*, pp. 56–63.
     DD VC

105. Bar-Shalom, Y. (1971) On the asymptotic properties of the maximum likelihood estimate obtained from dependent observations. *Journal of the*

Royal Statistical Society **B33**, 72–77.
MP TS

106. Bartholomew, D.J. (1956) Tests for randomness in a series of events when the alternative is a trend. *Journal of the Royal Statistical Society* **B18**, 234–239.
DD TS

107. Bartholomew, D.J. (1975) Errors of prediction for Markov chain models. *Journal of the Royal Statistical Society* **B37**, 444–456.
DD MC TS

108. Bartholomew, D.J. (1982) *Stochastic Models for Social Processes*. New York: John Wiley.
DD MC TS

109. Bartholomew, D.J. (1983) Some recent developments in social statistics. *International Statistical Review* **51**, 1–9.
DD EH MC MP TS

110. Bartholomew, D.J. (1984) Recent developments in nonlinear stochastic modelling of social processes. *Canadian Journal of Statistics* **12**, 39–52.
DD EH MC MP TS

111. Bartlett, M.S. (1937) Some examples of statistical methods of research in agriculture and applied biology. *Journal of the Royal Statistical Society (Suppl.)* **4**, 137–170.
VC

112. Bartlett, M.S. (1949) Some evolutionary stochastic processes. *Journal of the Royal Statistical Society* **B11**, 211–229.
DD EH GC PP RP TS

113. Bartlett, M.S. (1953) Stochastic processes or the statistics of change. *Journal of the Royal Statistical Society* **C2**, 44–64.
DD EH PP RP TS

114. Bartlett, M.S. (1963) The spectral analysis of point processes. *Journal of the Royal Statistical Society* **B25**, 264–296.
DD PP TS

115. Bartlett, M.S., Brennan, J.M., and Pollock, J.N. (1971) Stochastic analysis of some experiments on the mating of blowflies. *Biometrics* **27**, 725–729.
DD PP RP TS

116. Barton, D.E., David, F.N., and Fix, E. (1962) Persistence in a chain of multiple events when there is simple dependence. *Biometrika* **49**, 351–357.
DD MC TS

117. Bartoszynksi, R., Brown, B.W., McBride, C.M., and Thompson, J.R. (1981) Some nonparametric techniques for estimating the intensity function of a cancer related nonstationary Poisson process. *Annals of Statistics* **9**, 1050–1060.
MP NPAR SURV TS

118. Basawa, I.V. (1971) Some models based on the interaction of two independent Markovian point processes. *Journal of Applied Probability* **8**,

193–197.
DD MP PP TS

119. Basawa, I.V. (1974) Maximum likelihood estimation of parameters in renewal and Markov-renewal processes. *Australian Journal of Statistics* **16**, 33–43.
EH MP RP TS

120. Basawa, I.V., Feigin, P.D., and Heyde, C.C. (1976) Asymptotic properties of maximum likelihood estimators for stochastic processes. *Sankhya* **A38**, 259–270.
MP TS

121. Basu, A.P. (1971) Bivariate failure rate. *Journal of the American Statistical Association* **66**, 103–104.
MULT SURV

122. Bateman, G.I. (1950) The power of the χ^2 index of dispersion when Neyman's contagious distribution is the alternate hypothesis. *Biometrika* **37**, 59–63.
CD OD

123. Bather, J.A. (1965) Invariant conditional distributions. *Annals of Mathematical Statistics* **36**, 829–846.
MP TS

124. Battese, G.E., Harter, R.M., and Fuller, W.A. (1988) An error-components model for prediction of county crop areas using survey and satellite data. *Journal of the American Statistical Association* **83**, 28–36.
VC

125. Baum, L.E., Petrie, T., Soules, G., and Weis, N. (1970) A maximization technique occurring in the statistical analysis of probabilistic functions of Markov chains. *Annals of Mathematical Statistics* **41**, 164–171.
MC TS

126. Bayo Lawal, H.B. and Upton, G.J.G. (1990) Alternative interaction structures in square contingency tables having ordered classificatory variables. *Quality and Quantity* **24**, 107-127.
DD MC TS

127. Beal, S.L. (1991) Computing initial estimates with mixed effects models: a general method of moments. *Biometrika* **78**, 217–220.
VC

128. Becker, M.P. (1989) Using association models to analyze agreement data: two examples. *Statistics in Medicine* **8**, 1199–1207.
DD VC

129. Becker, M.P. (1990) Quasisymmetric models for the analysis of square contingency tables. *Journal of the Royal Statistical Society* **B52**, 369–378.
DD MC TS

130. Becker, M.P. and Agresti, A. (1992) Log-linear modelling of pairwise interobserver agreement on a categorical scale. *Statistics in Medicine* **11**, 101–114.
DD VC

131. Begg, C.B. and Larson, M. (1982) A study of the use of the probability-of-being-in-response function as a summary of tumor response data. *Biometrics* **38**, 59–66.
    MP SURV TS

132. Beitler, P.J. and Landis, J.R. (1985) A mixed-effects model for categorical data. *Biometrics* **41**, 991–1000.
    DD VC

133. Bellhouse, D.R. (1990) On the equivalence of marginal and approximate conditional likelihoods for correlation parameters under a normal model. *Biometrika* **77**, 743–746.
    DGLM TS

134. Bemis, B.M., Bain, I.J., and Higgins, J.J. (1972) Estimation and hypothesis testing for the parameters of a bivariate exponential distribution. *Journal of the American Statistical Association* **67**, 927–929.
    DIST MULT SURV

135. Bennett, S. (1988) An extension of Williams' method for overdispersion models. *GLIM Newsletter* **17**, 12–18.
    DD OD

136. Berk, K. (1987) Computing for incomplete repeated measures. *Biometrics* **43**, 385–398.
    GC TS VC

137. Berkey, C.S. (1982a) Comparison of two longitudinal growth models for pre-school children. *Biometrics* **38**, 221–234.
    GC

138. Berkey, C.S. (1982b) Bayesian approach for a nonlinear growth model. *Biometrics* **38**, 953–961.
    GC

139. Berkey, C.S., Laird, N.M., Valadian, I., and Gardner, J. (1991) Modelling adolescent blood pressure patterns and their prediction of adult pressures. *Biometrics* **47**, 1005–1018.
    VC

140. Berlin, B., Brodsky, J., and Clifford, P. (1979) Testing disease dependence in survival experiments with serial sacrifice. *Journal of the American Statistical Association* **74**, 5–14.
    SURV

141. Berman, S.M. (1990) A stochastic model for the distribution of HIV latency time based on T4 counts. *Biometrika* **77**, 733–741.
    MP TS

142. Berman, M. and Turner, T.R. (1992) Approximating point process likelihoods with GLIM. *Journal of the Royal Statistical Society* **C41**, 31–38.
    DD PP TS

143. Berzeg, K. (1979) The error component model. Conditions for the existence of the maximum likelihood estimates. *Journal of Econometrics* **10**, 99–102.
    VC

144. Bhargava, A. and Sargan, J.D. (1983) Estimating dynamic random effects models from panel data covering short time periods. *Econometrica* **51**, 1635–1659.
     AR TS VC

145. Bhattacharyya, G.K. and Johnson, R.A. (1973) On a test of independence in a bivariate exponential distribution. *Journal of the American Statistical Association* **68**, 704–706.
     DIST MULT SURV

146. Bildikar, S. and Patil, G.P. (1968) Multivariate exponential-type distributions. *Annals of Mathematical Statistics* **39**, 1316–1326.
     DIST MULT

147. Billingsley, P. (1961) Statistical methods in Markov chains. *Annals of Mathematical Statistics* **32**, 12–40.
     MC TS

148. Birnbaum, A. (1954) Statistical methods for Poisson processes and exponential populations. *Journal of the American Statistical Association* **49**, 254–266.
     DD PP RP TS

149. Bishop, Y.M.M., Fienberg, S.E., and Holland, P.W. (1975) *Discrete Multivariate Analysis: Theory and Practice.* Cambridge: MIT Press.
     DD

150. Bissell, A.F. (1972) A negative binomial model with varying element sizes. *Biometrika* **59**, 435–441.
     CD DD OD

151. Bjornsson, H. (1978) Analysis of a series of long-term grassland experiments with autocorrelated errors. *Biometrics* **34**, 645–651.
     AR TS

152. Blackith, R.E. (1960) A synthesis of multivariate techniques to distinguish patterns of growth in grasshoppers. *Biometrics* **16**, 28–40.
     GC TS

153. Blackwell, L.M. and Singpurwalla, N.D. (1988) Inference from accelerated life tests using filtering in coloured noise. *Journal of the Royal Statistical Society* **B50**, 281–292.
     DGLM SURV

154. Blackwood, L. (1988) Latent variable models for the analysis of medical data with repeated measures of binary variables. *Statistics in Medicine* **7**, 975–981.
     DD VC

155. Bliss, C.I. and Owen, A.R.G. (1958) Negative binomial distributions with a common k. *Biometrics* **45**, 37–50.
     CD

156. Blomqvist, N. (1977) On the relation between change and initial value. *Journal of the American Statistical Association* **72**, 746–749.
     VC

157. Bloomfield, P. (1976) *Fourier Analysis of Time Series: An Introduction.* New York: John Wiley.

TS

158. Blossfeld, H.P. and Hamerle, A. (1989) Unobserved heterogeneity in hazard rate models a test and an illustration from a study of career mobility. *Quality and Quantity* **23**, 129–141.
SURV VC

159. Blossfeld, H.P. and Hamerle, A. (1992) Unobserved heterogeneity in event history models. *Quality and Quantity* **26**, 157–168.
EH SURV TS VC

160. *Blossfeld, H.P., Hamerle, A., and Mayer, K.U. (1989) *Event History Analysis*. Hillsdale: Lawrence Erlbaum Associates.
EH SURV TS VC

161. Boik, R.J. (1991) Scheffé's mixed model for multivariate repeated measures a relative efficiency evaluation. *Communication in Statistics* **A20**, 1233–1255.
VC

162. Bonney, G.E. (1986) Regressive logistic models for familial disease and other binary traits. *Biometrics* **42**, 611–625.
DD VC

163. Bonney, G.E. (1987) Logistic regression for dependent binary observations. *Biometrics* **43**, 951–973.
DD TS

164. Borgan, Ø. (1984) Maximum likelihood estimation in parametric counting process models, with applications to censored failure time data. *Scandinavian Journal of Statistics* **11**, 1–16, 275.
EH MP SURV TS

165. Borgan, Ø. and Ramlau-Hansen, H. (1985) Demographic incidence rates and estimation of intensities with incomplete information. *Annals of Statistics* **13**, 564–582.
MP SURV TS

166. Box, G.E.P. (1950) Problems in the analysis of growth and wear curves. *Biometrics* **6**, 362–389.
GC TS VC

167. Box, G.E.P. and Cox, D.R. (1964) An analysis of transformations. *Journal of the Royal Statistical Society* **B26**, 211–252.
DIST

168. Boyle, K.E. and Starr, T.B. (1985) Survival models for fertility evaluation. *Journal of the American Statistical Association* **80**, 823–827.
SURV

169. Brand, R. and Kragt, H. (1992) Importance of trends in the interpretation of an overall odds ratio in the meta-analysis of clinical trials. *Statistics in Medicine* **11**, 2077–2082.
DD

170. Brännäs, K. (1986) On heterogeneity in econometric duration models. *Sankhya* **B48**, 284–295.
EH SURV VC

171. Brass, W. (1958) Simplified methods of fitting the truncated negative binomial distribution. *Biometrika* **45**, 59–68.
     CD

172. Brass, W. (1974) Perspectives in population prediction: illustrated by the statistics of England and Wales. *Journal of the Royal Statistical Society* **A137**, 532–583.
     GC TS

173. Brecht, L. (1992) A new approach for the analysis of multi-episode duration data. In Fahrmeir *et al.*, pp. 34–39.
     EH MP SURV TS

174. Breslow, N.E. (1984) Extra-Poisson variation in log-linear models. *Journal of the Royal Statistical Society* **C33**, 38–44.
     DD OD

175. Breslow, N.E. (1989) Score tests in overdispersed GLM's. In Decarli *et al.*, pp. 64–74.
     DD GLM OD

176. Breslow, N.E. (1990a) Tests of hypotheses in overdispersed Poisson regression and other quasi-likelihood models. *Journal of the American Statistical Association* **85**, 565–571.
     DD OD

177. Breslow, N.E. (1990b) Biostatistics and Bayes. *Statistical Science* **5**, 269–298.
     DD OD

178. *Breslow, N.E. and Day, N.E. (1987) *The Design and Analysis of Cohort Studies.* Lyon: International Agency for Research on Cancer.
     DD OD SURV TS

179. Breslow, N.E., Lubin, J.H., Marek, P., and Langholz, B. (1989) Multiplicative models and cohort analysis. *Journal of the American Statistical Association* **78**, 1–12.
     DD SURV

180. Brier, S.S. (1980) Analysis of contingency tables under cluster sampling. *Biometrika* **67**, 591–596.
     DD VC

181. Brillinger, D.R. (1966) Estimation of the second-order intensities of a bivariate stationary point process. *Journal of the Royal Statistical Society* **B38**, 60–66.
     DD PP TS

182. Brillinger, D.R. (1972) The spectral analysis of stationary interval functions. *Proceedings of the Sixth Berkeley Symposium* **1**, 483–513.
     DD PP SURV TS

183. Brillinger, D.R. (1980) Analysis of variance and problems under time series models. *Handbook of Statistics* **1**, 237–278.
     GC TS

184. Brillinger, D.R. (1986) The natural variability of vital rates and associated statistics. *Biometrics* **42**, 693–734.
     DD OD

185. Brillinger, D.R. (1992) Nerve cell spike train data analysis: a progression of technique. *Journal of the American Statistical Association* **87**, 260–271.
DD TS

186. Brillinger, D.R. and Preisler, H.K. (1983) Maximum likelihood estimation in a latent variable problem. In Karlin S., Amemiya, T., and Goodman, L.A. (ed.) *Studies of Econometrics, Time Series and Multivariate Statistics.* New York: Academic Press, pp. 31–65.
DD VC

187. Brindley, E.C. and Thompson, W.A. (1972) Dependence and aging aspects of multivariate survival. *Journal of the American Statistical Association* **67**, 822–830.
DIST MULT SURV

188. Bristol, D.R. and Patel, H.I. (1990) A Markovian model for comparing incidences of side effects. *Statistics in Medicine* **9**, 803–809.
DD MC TS

189. Brookmeyer, R. and Gail, M.H. (1988) A method for obtaining short-term projections and lower bounds on the size of the AIDS epidemic. *Journal of the American Statistical Association* **83**, 301–308.
GC TS

190. Brooks, R.J. (1984) Approximate likelihood ratio tests in the analysis of beta-binomial data. *Journal of the Royal Statistical Society* **C33**, 285–289.
CD DD OD

191. Brooks, R.J., James, W.H., and Gray, E. (1991) Modelling sub-binomial variation in the frequency of sex combinations in litters of pigs. *Biometrics* **47**, 403–417.
DD OD

192. Broström, G. (1987) The influence of mother's death on infant mortality: a case study in matched data survival analysis. *Scandinavian Journal of Statistics* **14**, 113–123.
SURV VC

193. Brown, B.W. (1980) The crossover experiment for clinical trials. *Biometrics* **36**, 69–79.
CO

194. Brown, C.C. (1975) On the use of indicator variables for studying the time-dependence of parameters in a response-time model. *Biometrics* **31**, 863–872.
DD PP SURV TS VC

195. Brown, C.H. (1990) Protecting against nonrandomly missing data in longitudinal studies. *Biometrics* **46**, 143–155.
MULT

196. Brown, G.H. and Donnelly, J.B. (1988) A note on comparing growth rates of animals between groups. *Biometrics* **44**, 985–993.
GC TS VC

197. Buist, A.S. and Vollmer, W.M. (1988) The use of lung function tests in identifying factors that affect lung growth and aging. *Statistics in*

Medicine **7**, 11–18.
VC

198. Burdett, K., Kiefer, N.M., Mortensen, D.T., and Neumann, G.R. (1984) Wages, unemployment and the allocation of time over time. *Review of Economic Studies* **51**, 559–578.
EH MP SURV TS

199. Burdett, K., Kiefer, N.M., and Sharma, S. (1985) Layoffs and duration dependence in a model of turnover. *Journal of Econometrics* **28**, 51–69.
EH MP SURV TS

200. Burnett, T.D. and Guthrie, D. (1970) Estimation of stationary stochastic regression parameters. *Journal of the American Statistical Association* **65**, 1547–1553.
VC

201. Butler, S.M. and Louis, T.A. (1992) Random effects models with non-parametric priors. *Statistics in Medicine* **11**, 1981–2000.
VC

202. Bye, B.V. and Schechler, E.S. (1986) A latent Markov model approach to the estimation of response errors in multiwave panel data. *Journal of the American Statistical Association* **81**, 375–380.
DD MC TS

203. Byrne, P.J. and Arnold, S.F. (1983) Inference about multivariate means for a nonstationary autoregressive model. *Journal of the American Statistical Association* **78**, 850–855.
AR TS

204. Cacoullos, T. and Papageorgiou, H. (1980) On some bivariate probability models applicable to traffic accidents and fatalities. *International Statistical Review* **48**, 345–356.
DIST MULT

205. Calvin, J.A. and Dysktra, ·R.L. (1991) Least squares estimation of co-variance matrices in balanced multivariate variance components models. *Journal of the American Statistical Association* **86**, 388–395.
VC

206. Carlin, B.P., Polson, N.G., and Stoffer, D.S. (1992) A Monte Carlo approach to nonnormal and nonlinear state-space modeling. *Journal of the American Statistical Association* **87**, 493–500.
DGLM TS

207. Cameron, A.C. and Trivedi, P.K. (1986) Econometric models based on count data: comparisons and applications of some estimators and tests. *Journal of Applied Econometrics* **1**, 29–53.
DD OD

208. Cameron, A.C. and Trivedi, P.K. (1990) Regression-based tests for overdispersion in the Poisson model. *Journal of Econometrics* **46**, 347–364.
DD OD

209. Campbell, M.J., Machin, D., and D'Arcangues, C. (1991) Coping with extra Poisson variability in the analysis of factors influencing vaginal ring

expulsions. *Statistics in Medicine* **10**, 241–254.
DD OD

210. Cane, V.R. (1959) Behaviour sequences as semi-Markov chains. *Journal of the Royal Statistical Society* **B21**, 36–58.
DD MC TS

211. Capéraà, P. and Genest, C. (1990) Concepts de dépendence et ordres stochastiques pour des lois bidimensionnelles. *Canadian Journal of Statistics* **18**, 315–326.
DIST MULT

212. Carr, G.J. and Chi, E.M. (1992) Analysis of variance for repeated measures data: a generalized estimating equations approach. *Statistics in Medicine* **11**, 1033–1040.
GEE

213. Carr, G.J., Hafner, K.B., and Koch, G.G. (1989) Analysis of rank measures of association for ordinal data from longitudinal studies. *Journal of the American Statistical Association* **84**, 797–804.
DD NPAR VC

214. Carter, E.M. and Hubert, J.J. (1984) A growth-curve model approach to multivariate quantal bio-assay. *Biometrics* **40**, 699–706.
DD GC TS

215. Carter, R.L., Resnick, M.B., Ariet, M., Shieh, G., and Vonesh, E.F. (1992) A random coefficient growth curve analysis of mental development in low-birth-weight infants. *Statistics in Medicine* **11**, 243–256.
GC TS VC

216. Catalano, P.J. and Ryan, L.M. (1992) Bivariate latent variable models for clustered discrete and continuous outcomes. *Journal of the American Statistical Association* **87**, 651–658.
DD OD

217. *Chamberlain, G. (1980) Analysis of covariance with qualitative data. *Review of Economic Studies* **47**, 225–238.
DD VC

218. Chamberlain, G. (1985) Heterogeneity, duration dependence and omitted variable bias. In Heckman and Singer (1985), pp. 3–38.
SURV VC

219. Chassan, J.B. (1964) On the analysis of simple cross-overs with unequal numbers of replicates. *Biometrics* **20**, 205–208.
CO

220. Chatfield, C. (1973) Statistical inference regarding Markov chain models. *Journal of the Royal Statistical Society* **C22**, 7–20.
DD MC TS

221. Chatfield, C., Ehrenberg, A.S.C., and Goodhardt, G.J. (1966) Progress on a simplified model of stationary purchasing behaviour. *Journal of the Royal Statistical Society* **A129**, 317–367.
DD OD

222. Chatfield, C. and Goodhardt, G.J. (1970) The beta-binomial model for consumer purchasing behaviour. *Journal of the Royal Statistical Society*

C19, 240–250.
CD DD OD

223. Cheke, R.A. (1985) Winter feeding assemblies, wing lengths and weights
of British dunnocks. In Morgan, B.J.T. and North, P.M. (ed.) *Statistics
in Ornithology.* Berlin: Springer Verlag, pp. 13–24.
DD TS

224. Chen, J.J. and Kodell, R.L. (1989) Quantitative risk assessment for teratological effects. *Journal of the American Statistical Association* **84**, 966–
971.
DD OD

225. Chen, J.J., Kodell, R.L., Howe, R.B., and Gaylor, D.W. (1991) Analysis
of trinomial responses from reproductive and developmental toxicity experiments. *Biometrics* **47**, 1049–1058.
DD OD

226. Chesher, A. (1984) Testing for neglected heterogeneity. *Econometrica* **52**,
865–872.
OD VC

227. Cheuvart, B. (1988) A nonparametric model for multiple recurrences.
Journal of the Royal Statistical Society **C37**, 157–168.
DD MC NPAR TS

228. Chi, E.M. (1991) Recovery of inter-block information in cross-over trials.
Statistics in Medicine **10**, 1115–1122.
CO

229. Chi, E.M. and Reinsel, G.C. (1989) Models for longitudinal data with
random effects and AR(1) errors. *Journal of the American Statistical Association* **84**, 452–459.
AR TS

230. Chiang, C.L. (1968) *Introduction to Stochastic Processes in Biostatistics.*
New York: John Wiley.
DD MC PP RP SURV TS

231. Chinchilli, V.M. and Carter, W.H. (1984) A likelihood ratio test for a
patterned covariance matrix in a multivariate growth-curve model. *Biometrics* **40**, 151–156.
GC TS VC

232. Ciminera, J.L., Bolognese, J.A., and Gregg, M.H. (1987) The statistical evaluation of a three-period two-treatment crossover pharmacokinetic
drug interaction study. *Biometrics* **43**, 713–718.
CO

233. Ciminera, J.L. and Wolfe, E.K. (1953) An example of the use of extended
cross-over designs in the comparison of NPH insulin mixtures. *Biometrics*
9, 431–446.
CO

234. Çinclar, E. (1969) Markov renewal theory. *Advances in Applied Probability*
1, 123–187.
EH MP TS

235. Clason, D.L. and Murray, L.W. (1992) Exponential families and variance component models. *American Statistician* **46**, 29–31.
VC

236. Clayton, D.G. (1978) A model for association in bivariate life-tables and its application in epidemiological studies of familial tendency in chronic disease incidence. *Biometrika* **65**, 141–151.
MULT SURV VC

237. Clayton, D.G. (1988) The analysis of event history data: a review of progress and outstanding problems. *Statistics in Medicine* **7**, 819–841.
EH DD PP SURV TS VC

238. Clayton, D.G. (1991) A Monte Carlo method for Bayesian inference in frailty models. *Biometrics* **47**, 467–485.
SURV VC

239. Clayton, D.G. and Cuzick, J. (1985) Multivariate generalizations of the proportional hazards model. *Journal of the Royal Statistical Society* **A148**, 82–117.
DIST MULT SURV

240. Clayton, D.G. and Schifflers, E. (1987) Models for temporal variation in cancer rates. I. Age–period and age–cohort models. II. Age–period–cohort models. *Statistics in Medicine* **6**, 449–481.
DD

241. Coffey, M. (1988) A random effects model for binary data from dependent samples. *Biometrics* **44**, 787–801.
DD VC

242. Cohen, J.E. (1968) On estimating the equilibrium and transition probabilities of a finite-state Markov chain from the same data. *Biometrics* **26**, 185–187.
MC TS

243. Cohen, J.E. (1976) The distribution of the chi-squared statistic under cluster sampling from contingency tables. *Journal of the American Statistical Association* **71**, 665–670.
DD VC

244. Cole, J.W.L. and Grizzle, J.E. (1966) Applications of multivariate analysis of variance to repeated measurements experiments. *Biometrics* **22**, 810–828.
MULT

245. Coles, S.G. and Tawn, J.A. (1991) Modelling extreme multivariate events. *Journal of the Royal Statistical Society* **B53**, 377–392.
MULT PP SURV

246. *Collet, D. (1991) *Modelling Binary Data*. London: Chapman and Hall.
DD

247. Collings, B.J. and Margolin, B.H. (1985) Testing goodness of fit for the Poisson assumption when observations are not identically distributed. *Journal of the American Statistical Association* **74**, 411–418.
DD OD

248. Conaway, M.R. (1989) Analysis of repeated categorical measurements with conditional likelihood methods. *Journal of the American Statistical Association* **84**, 53–62.
     DD TS

249. Conaway, M.R. (1990) A random effects model for binary data. *Biometrics* **46**, 317–328.
     DD VC

250. Conaway, M.R. (1992) The analysis of repeated categorical measurements subject to nonignorable nonresponse. *Journal of the American Statistical Association* **87**, 817–824.
     DD VC

251. Connolly, M. and Liang, K.Y. (1988) Conditional logistic regression models for correlated binary data. *Biometrika* **75**, 501–506.
     DD VC

252. Consul, P.C. (1990) On some properties and applications of quasi-binomial distribution. *Communications in Statistics* **A19**, 477–504.
     DD OD

253. Consul, P.C. and Famoye, F. (1992) Generalized Poisson regression model. *Communications in Statistics* **A21**, 89–109.
     DD OD

254. Cooil, B. (1991) Using medical malpractice data to predict the frequency of claims: a study of Poisson process models with random effects. *Journal of the American Statistical Association* **86**, 285–295.
     DD OD PP TS

255. Cook, R.D. and Johnson, M.E. (1981) A family of distributions for modelling non-elliptically symmetric multivariate data. *Journal of the Royal Statistical Society* **B43**, 210–218.
     DIST MULT

256. Cook, R.D. and Johnson, M.E. (1986) Generalized Burr-Pareto-logistic distributions with applications to a uranium exploration data set. *Technometrics* **28**, 123–131.
     DIST MULT

257. Coons, I. (1957) The analysis of covariance as a missing plot technique. *Biometrics* **13**, 387–405.
     VC

258. Cormack, R.M. (1985) Examples of the use of GLIM to analyze capture–recapture studies. In Morgan, B.J.T. and North, P.M. (ed.) *Statistics in Ornithology*. Berlin: Springer Verlag, pp. 243–273.
     DD TS

259. Cormack, R.M. (1989) Log-linear models for capture–recapture. *Biometrics* **45**, 395–413.
     DD TS

260. Cormack, R.M. (1992) Interval estimation for mark–recapture studies of closed populations. *Biometrics* **48**, 567–576.
     DD TS

261. Cormack, R.M. and Jupp, P.E. (1991) Inference for Poisson and multinomial models for capture–recapture experiments. *Biometrika* **78**, 911–916.
    DD TS

262. Cox, D.R. (1955) Some statistical methods connected with series of events. *Journal of the Royal Statistical Society* **B17**, 129–164.
    DD TS

263. Cox, D.R. (1958) The regression analysis of binary sequences. *Journal of the Royal Statistical Society* **B20**, 215–242.
    DD TS

264. *Cox, D.R. (1962) *Renewal Theory*. London: Methuen.
    RP SURV TS

265. Cox, D.R. (1965) On the estimation of the intensity function of a stationary point process. *Journal of the Royal Statistical Society* **B27**, 332–337.
    DD PP TS

266. Cox, D.R. (1972a) Regression models and life-tables. *Journal of the Royal Statistical Society* **B34**, 187–220.
    SURV

267. Cox, D.R. (1972b) The analysis of multivariate binary data. *Journal of the Royal Statistical Society* **C21**, 113–120.
    DD TS

268. Cox, D.R. (1975) A note on partially Bayes inference and the linear model. *Biometrika* **62**, 651–654.
    VC

269. Cox, D.R. (1983) Some remarks on overdispersion. *Biometrika* **70**, 269–274.
    OD

270. Cox, D.R., Fitzpatrick, R., Fletcher, A.E., Gore, S.M., Spiegelhalter, D.J., and Jones, D.R. (1992) Quality-of-life assessment: can we keep it simple? *Journal of the Royal Statistical Society* **A155**, 353–393.
    MP SURV TS VC

271. Cox, D.R. and Isham, V. (1980) *Point Processes*. London: Chapman and Hall.
    DD PP TS

272. *Cox, D.R. and Lewis, P.A.W. (1966) *The Statistical Analysis of Series of Events*. London: Methuen.
    DD TS

273. Cox, D.R. and Lewis, P.A.W. (1972) Multivariate point processes. *Proceedings of the Sixth Berkeley Symposium* **3**, 401–448.
    DD PP TS

274. Cox, D.R. and Miller, H.D. (1965) *The Theory of Stochastic Processes*. London: Methuen.
    DD TS

275. Cox, D.R. and Oakes, D. (1984) *Analysis of Survival Data*. London: Chapman and Hall.
    SURV

276. Cox, D.R. and Smith, W.L. (1953) The superposition of several strictly periodic sequences of events. *Biometrika* **40**, 1–11.
     DD PP RP TS

277. Cox, D.R. and Smith, W.L. (1954) On the superposition of renewal processes. *Biometrika* **41**, 91–99.
     DD PP RP TS

278. Cox, D.R. and Solomon, P.J. (1986a) On testing for serial correlation in large numbers of small samples. *Biometrika* **75**, 145–148.
     AR TS

279. Cox, D.R. and Solomon, P.J. (1986b) Analysis of variability with large numbers of small samples. *Biometrika* **75**, 543–554.
     OD VC

280. Crépeau, H., Koziol, J., Reid, N., and Yuh, Y.S. (1985) Analysis of incomplete multivariate data from repeated measurement experiments. *Biometrics* **41**, 505–514.
     MULT

281. Crouch, E.A.C. and Spiegelman, D. (1990) The evaluation of integrals of the form $\int_{-\infty}^{+\infty} f(t) \exp(-t^2) dt$: application to logistic-normal models. *Journal of the American Statistical Association* **85**, 464–469.
     DD VC

282. Crouchley, R., Davies, R.B., and Pickles, A.R. (1984) Methods for the identification of lexian, Poisson, and Markovian variations in the secondary sex ratio. *Biometrics* **40**, 165–175.
     MP TS VC

283. Crowder, M.J. (1976) Maximum likelihood estimation for dependent observations. *Journal of the Royal Statistical Society* **B38**, 45–53.
     GLM TS

284. Crowder, M.J. (1978) Beta-binomial ANOVA for proportions. *Journal of the Royal Statistical Society* **C27**, 34–37.
     DD OD

285. Crowder, M.J. (1979) Inference about the intraclass correlation coefficient in the beta-binomial ANOVA for proportions. *Journal of the Royal Statistical Society(B)* **14**, 230–234.
     CD DD OD

286. Crowder, M.J. (1980) Proportional linear models. *Journal of the Royal Statistical Society* **C29**, 299–303.
     GC TS

287. Crowder, M.J. (1983) A growth curve analysis for EDP curves. *Journal of the Royal Statistical Society* **C32**, 15–18.
     AR GC TS

288. Crowder, M.J. (1985a) Gaussian estimation for correlated binomial data. *Journal of the Royal Statistical Society* **B47**, 229–237.
     DD OD

289. Crowder, M.J. (1985b) A distributional model for repeated failure time measurements. *Journal of the Royal Statistical Society* **B47**, 447–452.
     DIST MULT SURV

290. Crowder, M.J. (1987) On linear and quadratic estimating functions. *Biometrika* **74**, 591–597.
GEE

291. Crowder, M.J. (1989) A multivariate distribution with Weibull connections. *Journal of the Royal Statistical Society* **B51**, 93–107.
DIST MULT SURV

292. Crowder, M.J. (1991) On the identifiability crisis in competing risks analysis. *Scandinavian Journal of Statistics* **18**, 223–233.
SURV VC

293. Crowder, M.J. (1992) Interlaboratory comparisons: round robins with random effects. *Journal of the Royal Statistical Society* **C41**, 409–425.
DD VC

294. Crowder, M.J. and Hand, D.J. (1990) *Analysis of Repeated Measures*. London: Chapman and Hall.
VC

295. Crowder, M.J. and Tredger, J.A. (1981) The use of exponentially damped polynomials for biological recovery data. *Journal of the Royal Statistical Society* **C30**, 147–152.
AR GC TS

296. Crowley, J. and Hu, M. (1977) Covariance analysis of heart transplant survival data. *Journal of the American Statistical Association* **72**, 27–36.
SURV

297. Cullis, B.R. and McGilchrist, C.A. (1990) A model for the analysis of growth data from designed experiments. *Biometrics* **46**, 131–142.
GC TS VC

298. Cullis, B.R., McGilchrist, C.A., and Gleeson, A.C. (1990) Error model diagnostics in the general linear model relevant to the analysis of repeated measurements and field experiments. *Journal of the Royal Statistical Society* **B53**, 409–416.
GC TS VC

299. Cullis, B.R. and Verbyla, A.P. (1992) Nonlinear regression modelling and time dependent covariates in repeated measures experiments. *Australian Journal of Statistics* **34**, 145–160.
TS VC

300. Cummings, W.B. and Gaylor, D.W. (1974) Variance components testing in unbalanced nested designs. *Journal of the American Statistical Association* **69**, 765–771.
VC

301. Cupples, L.A., D'Agostino, R.B., Anderson, K., and Kannel, W.B. (1988) Comparison of baseline and repeated measure covariate techniques in the Framingham heart study. *Statistics in Medicine* **7**, 205–222.
SURV

302. Curnow, R.N. and Kirkwood, T.B.L. (1989) Statistical analysis of deoxyribonucleic acid sequence data — a review. *Journal of the Royal Statistical Society* **A152**, 199–220.
DD MC

303. Dale, J.R. (1984) Local versus global association for bivariate ordered responses. *Biometrika* **71**, 507–514.
     DD MULT

304. Dale, J.R. (1986) Global cross-ratio models for bivariate, discrete, ordered responses. *Biometrics* **42**, 909–917.
     DD MULT

305. Dallal, G.E. (1988) Paired Bernoulli trials. *Biometrics* **44**, 253–257.
     DD VC

306. Danaher, P.J. (1988) Parameter estimation and applications for a generalization of the beta-binomial distribution. *Australian Journal of Statistics* **30**, 263–275.
     CD DD OD

307. Danaher, P.J. (1989) A Markov mixture model for magazine exposure. *Journal of the American Statistical Association* **84**, 922–926.
     CD DD MC OD TS

308. Danaher, P.J. (1990) A Markov chain model for magazine exposure. *Australian Journal of Statistics* **32**, 163–176.
     DD MC OD TS

309. Danford, M.B., Hughes, H.M., and McNee, R.C. (1960) On the analysis of repeated-measurements experiments. *Biometrics* **16**, 547–565.
     AR GC TS

310. Darroch, J.N. (1981) The Mantel-Haenszel test and tests of marginal symmetry: fixed-effects and mixed models for categorical responses. *International Statistical Review* **49**, 285–307.
     DD VC

311. Darroch, J.N. and McCloud, P.I. (1986) Category distinguishability and observer agreement. *Australian Journal of Statistics* **28**, 371–388.
     DD VC

312. Darroch, J.N. and McCloud, P.I. (1990) Separating two sources of dependence in repeated influenza outbreaks. *Biometrika* **77**, 237–243.
     DD TS VC

313. Darwin, J.H. (1957) The power of the Poisson index of dispersion. *Biometrika* **44**, 286–289.
     DD OD

314. Davidian, M. and Carroll, R.J. (1987) Variance function estimation. *Journal of the American Statistical Association* **82**, 1079–1091.
     GEE

315. Davidian, M. and Carroll, R.J. (1988) A note on extended quasi-likelihood. *Journal of the Royal Statistical Society* **B50**, 74–82.
     GEE

316. Davies, M. and Fleiss, J.L. (1982) Measuring agreement for multinomial data. *Biometrics* **38**, 1047–1051.
     DD VC

317. Davis, C.S. (1991) Semi-parametric and non-parametric methods for the analysis of repeated measurements with applications to clinical trials.

Statistics in Medicine **10**, 1959–1980.
DD NPAR

318. Davis, C.S. and Wei, L.J. (1988) Nonparametric methods for analyzing incomplete nondecreasing repeated measurements. *Biometrics* **44**, 1005–1018.
GC NPAR TS

319. Davis, M.H.A. (1984) Piecewise-deterministic Markov processes: a general class of non-diffusion stochastic models. *Journal of the Royal Statistical Society* **B46**, 353–388.
MP TS

320. Davison, A.C. (1992) Treatment effect heterogeneity in paired data. *Biometrika* **79**, 463–474.
DD OD VC

321. Day, N.E. (1966) Fitting curves to longitudinal data. *Biometrics* **22**, 276–291.
GC

322. de Jong, P. (1988) The likelihood of a state space model. *Biometrika* **75**, 165–169.
DGLM TS

323. de Jong, P. and Grieg, M. (1985) Models and methods for pairing data. *Canadian Journal of Statistics* **13**, 233–241.
DD VC

324. de Stavola, B.L. (1988) Testing departures from time homogeneity in multistate Markov processes. *Journal of the Royal Statistical Society* **C37**, 242–250.
MP SURV TS

325. Dean, C. (1991) Estimating equations for mixed Poisson models. In Godambe, V.P. (ed.) *Estimating Functions*. Oxford: Oxford University Press, pp. 35–46.
CD GEE OD VC

326. Dean, C. (1992) Testing for overdispersion in Poisson and binomial regression models. *Journal of the American Statistical Association* **87**, 451–457.
DD OD

327. Dean, C. and Lawless, J.F. (1989) Tests for detecting overdispersion in Poisson regression models. *Journal of the American Statistical Association* **84**, 467–471.
DD OD

328. Dean, C., Lawless, J.F., and Willmot, G.E. (1989) A mixed Poisson-inverse-Gaussian regression model. *Canadian Journal of Statistics* **17**, 171–181.
CD DD OD

329. Decarli, A., Francis, B.J., Gilchrist, R., and Seeber, G.U.H. (1989, ed.) *Statistical Modelling*. Berlin: Springer Verlag.
GLM

330. DeGruttola, V. and Lagakos, S. (1989) Analysis of doubly censored survival data, with application to AIDS. *Biometrics* **45**, 1–11.

MP SURV TS

331. DeGruttola, V., Lange, N. and Dafni, U. (1991) Modeling the progression of HIV infection. *Journal of the American Statistical Association* **86**, 569–577.
    VC

332. Deheuvels, P. (1983) Point processes and multivariate extreme values. *Journal of Multivariate Analysis* **13**, 257–272.
    DD MULT PP SURV TS

333. del Pino, G. (1989) The unifying role of iterative generalized least squares in statistical algorithms. *Statistical Science* **4**, 394–408.
    GLM

334. Dempster, A.P., Rubin, D.B., and Tsutakawa, R.K. (1981) Estimation in covariance components models. *Journal of the American Statistical Association* **76**, 341–353.
    VC

335. Dempster, A.P., Selwyn, M.R., Patel, C.M. and Roth, A.J. (1984) Statistical and computational aspects of mixed model analysis. *Journal of the Royal Statistical Society* **C33**, 203-214.
    VC

336. Dempster, A.P., Selwyn, M.R., and Weeks, B.J. (1983) Combining historical and randomized controls for assessing trends in proportions. *Journal of the American Statistical Association* **78**, 221–227.
    VC

337. Denby, L. and Vardi, Y. (1985) A short-cut method for estimation of renewal processes. *Technometrics* **27**, 361–373.
    RP TS

338. Denny, J. and Yakowitz, S. (1978) Admissible run-contingency type tests for independence and Markov dependence. *Journal of the American Statistical Association* **73**, 177–181.
    DD MC TS

339. Derman, C. (1961) Remark concerning two-state semi-Markov processes. *Annals of Mathematical Statistics* **32**, 615–616.
    MP TS

340. Derman, C., Ileser, L.J., and Olkin, I. (1973) *A Guide to Probability Theory and Applications.* New York: Holt, Rinehart and Winston.
    DIST

341. Dersimonian, R. and Baker, S.G. (1988) Two-process models for discrete-time serial categorical response. *Statistics in Medicine* **7**, 965–974.
    DD TS

342. Desmond, A. (1985) Stochastic models of failure in random environments. *Canadian Journal of Statistics* **13**, 171–183.
    DIST SURV

343. Diaconis, P. and Ylvisaker, D. (1979) Conjugate priors for exponential families. *Annals of Statistics* **7**, 269–281.
    CD

344. Diaz, J. (1982) Bayesian detection of a change of scale parameter in sequences of independent gamma random variables. *Journal of Econometrics* **19**, 23–29.
RP TS

345. Diekmann, A. and Mitter, P. (1984, ed.) *Stochastic Modelling of Social Processes.* Orlando: Academic Press.
GLM EH SURV TS VC

346. Diem, J.E. and Liukkonen, J.R. (1988) A comparative study of three methods for analyzing longitudinal pulmonary function data. *Statistics in Medicine* **7**, 19–28.
AR TS VC

347. Dietz, E. (1992) Estimation of heterogeneity — a GLM approach. In Fahrmeir *et al.*, pp. 66–71.
GLM SURV VC

348. Diggle, P.J. (1985) Comparing estimated spectral densities using GLIM. In Gilchrist *et al.*, pp. 34–43.
TS

349. Diggle, P.J. (1988) An approach to the analysis of repeated measurements. *Biometrics* **44**, 959–971.
AR GC TS VC

350. Diggle, P.J. (1989) Testing for random dropouts in repeated measurement data. *Biometrics* **45**, 1255–1258.
MULT

351. *Diggle, P.J. (1990) *Time Series. A Biostatistical Introduction.* Oxford: Oxford University Press.
AR GC TS VC

352. Diggle, P.J. and Donnelly, J.B. (1989) A selected bibliography on the analysis of repeated measurements and related areas. *Australian Journal of Statistics* **31**, 183–193.
AR CO GC TS VC

353. Diggle, P.J. and Fisher, N.I. (1991) Nonparametric comparison of cumulative periodograms. *Journal of the Royal Statistical Society* **C40**, 423–434.
NPAR TS

354. Diggle, P.J. and Hutchinson, M.F. (1989) On spline smoothing with autocorrelated errors. *Australian Journal of Statistics* **31**, 166–182.
AR GC TS

355. Diggle, P.J., Lange, N., and Beneš, F.M. (1991) Analysis of variance for replicated spatial point patterns in clinical neuroanatomy. *Journal of the American Statistical Association* **86**, 618–625.
DD OD PP TS

356. Diggle, P.J. and Milne, R.K. (1983) Bivariate Cox processes: some models for bivariate spatial patterns. *Journal of the Royal Statistical Society* **B45**, 11–21.
DD PP TS VC

357. Diggle, P.J. and Zeger, S.L. (1987) Modelling endocrinological time-series. *Bulletin of the International Statistical Institute* **46**, 123–134.

TS

358. Dinse, G.E. (1988) Simple parametric analysis of animal tumorigenicity data. *Journal of the American Statistical Association* **83**, 638–649.
SURV

359. Dinse, G.E. and Larson, M.G. (1986) A note on semi-Markov models for partially censored data. *Biometrika* **73**, 379–386.
MP SURV TS

360. *Dobson, A.J. (1990) *An Introduction to Generalized Linear Models.* London: Chapman and Hall.
GLM

361. Dolby, G.R. and Lipton, S. (1972) Maximum likelihood estimation of the general nonlinear functional relationship with replicated observations and correlated errors. *Biometrika* **59**, 121–129.
VC

362. Donald, A. and Donner, A. (1987) Adjustments to the Mantel–Haenszel chi-square statistic and odds ratio variance estimator when the data are clustered. *Statistics in Medicine* **6**, 491–499.
DD VC

363. Donald, A. and Donner, A. (1990) A simulation study of the analysis of sets of 2x2 contingency tables under cluster sampling: estimation of a common odds ratio. *Journal of the American Statistical Association* **85**, 537–543.
DD VC

364. Donner, A. (1986) A review of inference procedures for the intraclass correlation coefficient in the one-way random effects model. *International Statistical Review* **54**, 67–82.
VC

365. Donner, A. and Koval, J.J. (1980) The large sample variance of an intraclass correlation. *Biometrika* **67**, 719–722.
VC

366. Donner, A. and Koval, J.J. (1983) Variance-component estimation from human sibship data. *Biometrics* **39**, 599–605.
VC

367. Donner, A., Koval, J.J., and Bull, S. (1984) Testing the effect of sex differences on sib–sib correlations. *Biometrics* **40**, 349–356.
VC

368. Donner, A. and Rosner, B. (1980) On inferences concerning a common correlation coefficient. *Journal of the Royal Statistical Society* **C29**, 69–76.
VC

369. Doob, J.L. (1953) *Stochastic Processes.* New York: John Wiley.
MC MP TS

370. Downton, F. (1970) Bivariate exponential distributions in reliability theory. *Journal of the Royal Statistical Society* **B32**, 408–417.
DIST MULT SURV

371. Duffy, J.C. and Latcham, R.W. (1986) Liver cirrhosis mortality in England and Wales compared to Scotland: an age–period–cohort analysis. *Journal of the Royal Statistical Society* **A149**, 45–59.
DD

372. Duncan, D.B. and Horn, S.D. (1972) Linear dynamic recursive estimation from the viewpoint of regression analysis. *Journal of the American Statistical Association* **67**, 815–821.
DGLM TS VC

373. Duncan, O.D. (1979) How destination depends on origin in the occupational mobility table. *American Journal of Sociology* **84**, 793–803.
DD MC TS

374. Dunsmore, I.R. (1981a) Growth curves in two-period change over models. *Journal of the Royal Statistical Society* **C30**, 575–578.
CO GC TS

375. Dunsmore, I.R. (1981b) Analysis of preferences in two-period cross-over designs. *Biometrics* **37**, 575–578.
CO DD

376. Efron, B. (1986) Double exponential families and their use in generalized linear regression. *Journal of the American Statistical Association* **81**, 709–721.
DIST GLM OD

377. Efron, B. (1992) Poisson overdispersion estimates based on the method of asymmetric maximum likelihood. *Journal of the American Statistical Association* **87**, 98–107.
DD OD

378. Elashoff, J.D. (1981) Repeated-measures bioassay with correlated errors and heterogeneous variances: a Monte Carlo study. *Biometrics* **37**, 475–482.
TS VC

379. Elbers, C. and Ridder, G. (1982) True and spurious duration dependence: the identifiability of the proportional hazard model. *Review of Economic Studies* **49**, 403–409.
SURV VC

380. Eliasziw, M. and Donner, A. (1990) Comparison of recent estimators of interclass correlation from familial data. *Biometrics* **46**, 391–398.
VC

381. Eliasziw, M. and Donner, A. (1991) Application of the McNemar test to non-independent matched pair data. *Statistics in Medicine* **10**, 1981–1991.
DD VC

382. Elston, R.C. (1964) On estimating time–response curves. *Biometrics* **20**, 643–647.
GC TS VC

383. Elston, R.C. and Grizzle, J.F. (1962) Estimation of time response curves and their confidence bands. *Biometrics* **18**, 148–159.
GC TS VC

384. Enberg, J., Gottschalk, P., and Wolf, D. (1990) A random-effects logit
     model of work–welfare transitions. *Journal of Econometrics* **43**, 63–75.
     DD TS

385. Engel, B. (1990) The analysis of unbalanced linear models with variance
     components. *Statistica Neerlandica* **44**, 195–219.
     VC

386. Engel, J. (1984) Models for response data showing extra-Poisson varia-
     tion. *Statistica Neerlandica* **38**, 159–167.
     DD OD

387. Engel, J. (1985) On the analysis of variance for beta binomial responses.
     Statistica Neerlandica **39**, 27–34.
     CD DD OD

388. Engel, J. (1986) Split-plot design: model and analysis for count data.
     Statistica Neerlandica **40**, 21–33.
     DD VC

389. Engel, J. (1988) Rank tests and random blocking of classified data. *Sta-
     tistica Neerlandica* **42**, 17–27.
     DD VC

390. Engel, J. (1990) Quasi-likelihood inference in a generalized linear mixed
     model for balanced data. *Statistica Neerlandica* **44**, 221–239.
     GEE GLM VC

391. England, P.D. and Verrall, R.J. (1992) Modelling excess mortality of di-
     abetics: generalized linear models and dynamic estimation. In Fahrmeir
     et al., pp. 78–84.
     DD DGLM TS

392. Esary, J.D. and Proschan, F. (1972) Relationships among some concepts
     of bivariate dependence. *Annals of Mathematical Statistics* **43**, 651–655.
     MULT

393. Esary, J.D., Proschan, F., and Walkup, D.W. (1967) Association of ran-
     dom variables, with applications. *Annals of Mathematical Statistics* **38**,
     1466–1474.
     MULT

394. Espeland, M.A., Platt, O.S., and Galagher, D. (1989) Joint estimation
     of incidence and diagnostic error rates from irregular longitudinal data.
     Journal of the American Statistical Association **84**, 972–979.
     DD TS

395. Evans, J.C. and Roberts, E.A. (1979) Analysis of sequential observations
     with applications to experiments on grazing animals and perennial plants.
     Biometrics **35**, 687–693.
     GC MULT TS

396. Ezzet, F. and Whitehead, J. (1989) Models for nested binary and ordinal
     data. In Decarli *et al.*, pp. 144–150.
     DD VC

397. Ezzet, F. and Whitehead, J. (1991) A random effects model for ordinal
     responses from a crossover trial. *Statistics in Medicine* **10**, 901–907.
     CO VC

398. Ezzet, F. and Whitehead, J. (1992) A random effects model for binary data from crossover clinical trials. *Journal of the Royal Statistical Society* **C41**, 117–128.
CO DD VC

399. Faddy, M.J. (1976) A note on the general time-dependent stochastic compartmental model. *Biometrics* **32**, 443–448.
MP SURV TS

400. Faddy, M.J. (1977) Stochastic compartmental models as approximations to more general stochastic systems with the general stochastic epidemic as an example. *Advances in Applied Probability* **9**, 448–461.
MP SURV TS

401. Fahrmeir, L. (1989) Extended Kalman filtering for nonnormal longitudinal data. In Decarli *et al.*, pp. 151–156.
DGLM TS

402. Fahrmeir, L. (1992) Posterior mode estimation by extended Kalman filtering for multivariate dynamic generalized linear models. *Journal of the American Statistical Association* **87**, 501–509.
DGLM TS

403. Fahrmeir, L., Francis, B., Gilchrist, R., and Tutz, G. (1992) *Advances in GLIM and Statistical Modelling*. Berlin: Springer Verlag.
GLM

404. Fahrmeir, L., Hennevogl, W., and Klemme, K. (1992) Smoothing in dynamic generalized linear models by Gibbs sampling. In Fahrmeir *et al.*, pp. 85–90.
DGLM TS

405. Farewell, V.T. (1979) An application of Cox's proportional hazard model to multiple infection data. *Journal of the Royal Statistical Society* **C28**, 136–143.
SURV

406. Farewell, V.T. (1985) Some remarks on the analysis of cross-over trials with a binary response. *Journal of the Royal Statistical Society* **C34**, 121–128.
CO DD

407. Farewell, V.T. and Sprott, D.A. (1988) The use of a mixture model in the analysis of count data. *Biometrics* **44**, 1191–1194.
CD DD

408. Fearn, T. (1975) A Bayesian approach to growth curves. *Biometrika* **62**, 89–100.
GC TS VC

409. Fearn, T. (1977) A two-stage model for growth curves which leads to Rao's covariance adjusted estimators. *Biometrika* **64**, 141–143.
GC TS VC

410. Federer, W.T. and Meredith, M.P. (1992) Covariance analysis for split-plot and split-block designs. *American Statistician* **46**, 155–162.
VC

411. Feigin, P.D. (1976) Maximum likelihood estimation for continuous-time stochastic processes. *Advances in Applied Probability* **8**, 712–736.
AR GLM TS

412. Feigin, P.D. (1978) The efficiency criteria problem for stochastic processes. *Stochastic Processes and their Applications* **6**, 115–127.
AR GLM TS

413. Feigin, P.D. (1981) Conditional exponential families and a representation theorem for asymptotic inference. *Annals of Statistics* **9**, 597–603.
AR GLM TS

414. Fellegi, I.P. (1964) Response variance and its estimation. *Journal of the American Statistical Association* **59**, 1016–1041.
DD VC

415. Fellegi, I.P. (1974) An improved method of estimating the correlated response variance. *Journal of the American Statistical Association* **69**, 496–501.
DD VC

416. Feller, W. (1968, 3rd ed.) *An Introduction to Probability Theory and its Applications*. Vol. 1. New York: John Wiley.
DD DIST

417. Feller, W. (1970, 2nd ed.) *An Introduction to Probability Theory and its Applications*. Vol. 2. New York: John Wiley.
DIST

418. Fenech, A.P. and Harville, D.A. (1991) Exact confidence sets for variance components in unbalanced mixed linear models. *Annals of Statistics* **19**, 1771–1785.
VC

419. Feuer, E.J., Hankey, B.F., Gaynor, J.J., Wesley, M.N., Baker, S.G., and Meyer, J.S. (1992) Graphical representation of survival curves associated with a binary non-reversible time dependent covariate. *Statistics in Medicine* **11**, 455–474.
SURV

420. Fidler, V. (1984) Change-over clinical trial with binary data: mixed-model-based comparison of tests. *Biometrics* **56**, 1063–1070.
CO DD VC

421. Fidler, V. (1986) Change-over trials with binary data: estimation. *Statistica Neerlandica* **40**, 81–86.
CO DD

422. Fienberg, S.E. (1972) The multiple recapture census for closed populations and incomplete 2^k contingency tables. *Biometrika* **59**, 591–603.
DD TS

423. Fienberg, S.E. (1977) *The Analysis of Cross-Classified Categorical Data*. Cambridge: MIT Press.
DD

424. Finch, P.D. (1982) Difficulties of interpretation with Markov chain models. *Australian Journal of Statistics* **24**, 343–349.
DD MC TS

425. *Fingleton, B. (1984) *Models of Category Counts*. Cambridge: Cambridge University Press.
DD

426. Finney, D.J. (1990) Repeated measurements: what is measured and what repeats? *Statistics in Medicine* **9**, 639–644.
CO TS

427. Finney, D.J. and Phillips, P. (1977) The form and estimation of a variance function, with particular reference to radioimmunoassay. *Journal of the Royal Statistical Society* **C26**, 312–320.
DD OD

428. Firth, D. (1987) On the efficiency of quasi-likelihood estimation. *Biometrika* **74**, 233–245.
GEE

429. Firth, D. and Harris, I.R. (1991) Quasi-likelihood for multiplicative random effects. *Biometrika* **78**, 545–555.
GEE GLM VC

430. Fisher, R.A. (1949) A biological assay of tuberculins. *Biometrics* **5**, 300–316.
VC

431. Fleiss, J.L. (1975) Measuring agreement between two judges on the presence or absence of a trait. *Biometrics* **31**, 651–659.
DD VC

432. Fleiss, J.L. (1986) On multiperiod crossover studies. *Biometrics* **42**, 449–450.
CO

433. Fleming, T.R. (1978) Nonparametric estimation for nonhomogeneous Markov processes in the problem of competing risks. *Annals of Statistics* **6**, 1057–1070.
MP SURV TS

434. Fleming, T.R. and Harrington, D.P. (1978) Estimation for discrete time nonhomogeneous Markov chains. *Stochastic Processes and their Applications* **7**, 131-139.
DD MC TS

435. Fleming, T.R. and Harrington, D.P. (1991) *Counting Processes and Survival Analysis*. New York: John Wiley.
SURV

436. Flinn, C. and Heckman, J.J. (1982) New methods for analyzing structural models of labor force dynamics. *Journal of Econometrics* **18**, 115–168.
SURV VC

437. Follmann, D.A. (1990) Modelling failures of intermittently used machines. *Journal of the Royal Statistical Society* **C39**, 115–123.
EH RP TS

438. Follmann, D.A. and Goldberg, M.S. (1988) Distinguishing heterogeneity from decreasing hazard rates. *Technometrics* **30**, 389–396.
SURV VC

439. Follmann, D.A. and Lambert, D. (1989) Generalizing logistic regression by nonparametric mixing. *Journal of the American Statistical Association* **84**, 295–300.
CD DD NPAR OD

440. Forcina, A. (1989) Correlated GLMs with non-normal error and the multinomial density. *GLIM Newsletter* **17**, 19–23.
GLM VC

441. Forcina, A. (1992) Modelling balanced longitudinal data: maximum likelihood estimation and analysis of variance. *Biometrics* **48**, 743–750.
AR TS VC

442. Forcina, A. and Marchetti, G.M. (1989) Modelling transition probabilities in the analysis of aggregated data. In Decarli *et al.*, pp. 157–164.
DD MC OD TS

443. Foulkes, M.A. and Davis, C.E. (1981) An index of tracking of longitudinal data. *Biometrics* **37**, 439–446.
GC TS VC

444. Foutz, R.V., Jensen, D.R., and Anderson, G.W. (1985) Multiple comparisons in the randomization analysis of designed experiments with growth curve responses. *Biometrics* **41**, 29–37.
GC TS

445. France, L.A., Lewis, J.A., and Kay, R. (1991) The analysis of failure time data in crossover studies. *Statistics in Medicine* **10**, 1099–1113.
CO SURV

446. Francom, S.F., Chuang-Stein, C., and Landis, J.R. (1989) A log-linear model for ordinal data to characterize differential change among treatments. *Statistics in Medicine* **8**, 571–582.
DD

447. Freeman, P.R. (1989) The performance of the two-stage analysis of two-treatment, two-period crossover trials. *Statistics in Medicine* **8**, 1421–1432.
CO

448. Freund, J.E. (1961) A bivariate extension of the exponential distribution. *Journal of the American Statistical Association* **56**, 971–977.
DIST MULT SURV

449. Frey, C.M. and Muller, K.E. (1992) Analysis methods for nonlinear models with compound symmetric covariance. *Communications in Statistics* **A21**, 1163–1182.
VC

450. Frison, L. and Pocock, S.J. (1992) Repeated measures in clinical trials: analysis using mean summary statistics and its implications for design. *Statistics in Medicine* **11**, 1685–1704.
VC

451. Frühwirth-Schnatter, S. (1992) Approximate predictive integrals for dynamic generalized linear models. In Fahrmeir *et al.*, pp. 101–106.
DGLM TS

452. Frydman, H. (1992) A nonparametric estimation procedure for a periodically observed three-state Markov process, with application to AIDS. *Journal of the Royal Statistical Society* **B54**, 853–866.
   MP NPAR SURV TS

453. Fujikoshi, Y. and Rao, C.R. (1991) Selection of covariables in the growth curve model. *Biometrika* **78**, 779–785.
   GC MULT TS

454. Gabriel, K.R. (1954) The distribution of the number of successes in a sequence of dependent trials. *Biometrika* **46**, 454–460.
   DD TS

455. Gabriel, K.R. (1962) Ante-dependence analysis of an ordered set of variables. *Annals of Mathematical Statistics* **33**, 201–212.
   AR TS

456. Gail, M.H. (1978) The analysis of heterogeneity for indirect standardized mortality ratios. *Journal of the Royal Statistical Society* **A141**, 224–234.
   DD VC

457. Gail, M.H. (1981) Evaluating serial cancer marker studies in patients at risk of recurrent disease. *Biometrics* **37**, 67–78.
   DD SURV TS

458. Gail, M.H., Santner, T.J., and Brown, C.C. (1980) An analysis of comparative carcinogenesis experiments based on multiple times to tumor. *Biometrics* **36**, 255–266.
   MP SURV TS

459. Gallo, J. and Khuri, A.I. (1990) Exact tests for the random and fixed effects in an unbalanced mixed two-way cross-classification model. *Biometrics* **46**, 1087–1095.
   VC

460. Gamerman, D. (1991) Dynamic Bayesian models for survival data. *Journal of the Royal Statistical Society* **C40**, 63–79.
   DGLM SURV TS

461. Gamerman, D. (1992) A dynamic approach to the statistical analysis of point processes. *Biometrika* **79**, 39–50.
   DD DGLM PP TS

462. Ganio, L.M. and Schafer, D.W. (1992) Diagnostics for overdispersion. *Journal of the American Statistical Association* **87**, 795–804.
   DD OD

463. Gardner, M.J. and Osmond, C. (1984) Interpretation of time trends in disease rates in the presence of generation effects. *Statistics in Medicine* **3**, 113–130.
   DD

464. Gart, J.J. (1969) An exact test for comparing matched proportions in crossover designs. *Biometrika* **56**, 75–80.
   CO

465. Gaver, D.P. and Lewis, P.A.W. (1980) First-order autoregressive gamma sequences and point processes. *Advances in Applied Probability* **12**, 727–745.

AR DD PP RP TS

466. Gaynor, J.J. (1987) The use of time dependent covariates in modelling data from an occupational cohort study. *Journal of the Royal Statistical Society* **C36**, 340–351.
    SURV

467. Geary, D.N. (1988) Sequential testing in clinical trials with repeated measurements. *Biometrika* **75**, 311–318.
    TS

468. Geary, D.N. (1989) Modelling the covariance structure of repeated measurements. *Biometrics* **45**, 1183–1195.
    TS VC

469. Geisser, S. (1963) Multivariate analysis of variance for a special covariance case. *Journal of the American Statistical Association* **58**, 660–669.
    VC

470. Geisser, S. (1974) A predictive approach to the random effect model. *Biometrika* **61**, 101–107.
    VC

471. Gelfand, A.E. and Dalal, S.R. (1990) A note on overdispersed exponential families. *Biometrika* **77**, 55–64.
    GLM OD

472. Gelfand, A.E., Hills, S.E., Racine-Poon, A., and Smith, A.F.M. (1990) Illustration of Bayesian inference in normal data models using Gibbs sampling. *Journal of the American Statistical Association* **85**, 972–985.
    VC

473. Gelman, A. and Meng, X.L. (1991) A note on bivariate distributions that are conditionally normal. *American Statistician* **45**, 125–126.
    DIST MULT

474. Genest, C. (1987) Frank's family of bivariate distributions. *Biometrika* **74**, 549–555.
    DIST MULT

475. Genest, C. and MacKay, J. (1986a) Copules archimédiennes et familles de lois bidimensionnelles dont les marges sont données. *Canadian Journal of Statistics* **14**, 145–159.
    DIST MULT

476. Genest, C. and MacKay, J. (1986b) The joy of copulas: bivariate distributions with uniform marginals. *American Statistician* **40**, 280–283.
    DIST MULT

477. Gennings, C., Chinchilli, V.M., and Carter, W.H. (1989) Response surface analysis with correlated data: a nonlinear model approach. *Journal of the American Statistical Association* **84**, 805–809.
    GC TS VC

478. Ghosh, M., Grizzle, J.E., and Sen, P.K. (1973) Nonparametric methods in longitudinal studies. *Journal of the American Statistical Association* **69**, 29–36.
    GC NPAR TS

479. Gianola, D. and Fernando, R.L. (1986) Random effects models for binary responses. *Biometrics* **42**, 217–218.
     DD VC

480. Gilchrist, R. (1982, ed.) *GLIM82*. Berlin: Springer Verlag.
     GLM

481. Gilchrist, R., Francis, B., and Whittaker, J. (1985, ed.) *Generalized Linear Models*. Berlin: Springer Verlag.
     GLM

482. Gill, J.L. (1988) Repeated measurement: split-plot trend analysis versus analysis of first differences. *Biometrics* **44**, 289–297.
     VC

483. Gill, J.L. and Jensen, E.L. (1968) Probability of obtaining negative estimates of heritability. *Biometrics* **24**, 517–526.
     VC

484. Gill, P.S. (1992) A note on modelling the covariance structure of repeated measurements. *Biometrics* **48**, 965–968.
     AR TS VC

485. Gill, R.D. (1984) Understanding Cox's regression model: a martingale approach. *Journal of the American Statistical Association* **79**, 441–447.
     DD PP SURV TS

486. Gill, R.D. (1986) On estimating transition intensities of a Markov process with aggregate data of a certain type: 'occurrences but no exposures'. *Scandinavian Journal of Statistics* **13**, 113–134.
     MP TS

487. Gilmour, A.R., Anderson, R.D., and Rae, A.L. (1985) The analysis of binomial data by a generalized linear mixed model. *Biometrika* **72**, 593–599.
     DD VC

488. Gladen, B. (1979) The use of the jackknife to estimate proportions from toxicological data in the presence of litter effects. *Journal of the American Statistical Association* **74**, 278–283.
     DD OD

489. Glasbey, C.A. (1979) Correlated residuals in non-linear regression applied to growth data. *Journal of the Royal Statistical Society* **C28**, 251–259.
     AR GC TS

490. Glasbey, C.A. (1980) Nonlinear regression with autoregressive time series errors. *Biometrics* **36**, 135–140.
     AR GC TS

491. Glasbey, C.A. (1986) Conservative estimates of the variances of regression parameter estimates for classes of error model. *Biometrika* **73**, 746–750.
     TS

492. Glasbey, C.A. (1988) Standard errors resilient to error variance misspecification. *Biometrika* **75**, 201–206.
     GC TS

493. Godambe, V.P. and Heyde, C.C. (1987) Quasi-likelihood and optimal estimation. *International Statistical Review* **55**, 231–244.

GEE GLM

494. Goldberg, J.D. and Wittes, J.T. (1978) The estimation of false negatives in medical screening. *Biometrics* **34**, 77–86.
     DD TS

495. Goldstein, H. (1986) Multilevel mixed model analysis using iterative generalized least squares. *Biometrika* **73**, 43–56.
     GLM VC

496. Goldstein, H. (1991) Nonlinear multilevel models, with an application to discrete response data. *Biometrika* **78**, 45–51.
     DD VC

497. Goldstein, S. (1964) The extent of repeated migration: an analysis based on the Danish population register. *Journal of the American Statistical Association* **59**, 1121–1132.
     DD MC TS

498. Goodman, L.A. (1958) Simplified runs tests and likelihood ratio tests for Markoff chains. *Biometrika* **45**, 181–197.
     DD MC TS

499. Goodman, L.A. (1961) Statistical methods for the mover–stayer model. *Journal of the American Statistical Association* **56**, 841–868.
     DD MC TS

500. Goodman, L.A. (1962) Statistical methods for analyzing processes of change. *American Journal of Sociology* **68**, 57–78.
     DD MC TS

501. Goodman, L.A. (1964) The analysis of persistence in a chain of multiple events. *Biometrika* **51**, 405–411.
     DD MC TS

502. Goodman, L.A. (1965) On the statistical analysis of mobility tables. *American Journal of Sociology* **70**, 564–585.
     DD MC TS

503. Goodman, L.A. (1969) How to ransack social mobility tables and other kinds of cross-classification tables. *American Journal of Sociology* **75**, 1–40.
     DD MC TS

504. Goodman, L.A. (1979) Multiplicative models for the analysis of occupational mobility tables and other kinds of cross-classification tables. *American Journal of Sociology* **84**, 804–819.
     DD MC TS

505. Goodman, M.M. (1968) A measure of 'overall variability' in populations. *Biometrics* **23**, 189–192.
     VC

506. Gordon, K. and Smith, A.F.M. (1990) Modeling and monitoring biomedical time series. *Journal of the American Statistical Association* **85**, 328–337.
     DGLM TS

507. Gornbein, J.A., Lazaro, C.G., and Little, R.J.A. (1992) Incomplete data in repeated measures analysis. *Statistical Methods in Medical Research* **1**,

275–295.
AR VC VC

508. Göttlein, A. and Pruscha, H. (1992) Ordinal time series models with application to forest damage data. In Fahrmeir *et al.*, pp. 113–118.
DD TS

509. Gottschau, A. (1992) Exchangibility in multivariate Markov chain models. *Biometrics* **48**, 751–763.
DD MC TS

510. Gould, A.L. (1980) A new approach to the analysis of clinical drug trials with withdrawals. *Biometrics* **36**, 721–727.
TS

511. Gouriéroux, C. and Montfort, A. (1979) On the characterization of a joint probability distribution by conditional distributions. *Journal of Econometrics* **10**, 115–118.
MULT

512. Gouriéroux, C. and Montfort, A. (1991) Simulation based inference in models with heterogeneity. *Annales d'Economie et de Statistique* **20/21**, 69–107.
GLM VC

513. Gouriéroux, C., Montfort, A., and Trognon, A. (1984) Pseudo maximum likelihood methods: applications to Poisson models. *Econometrica* **52**, 701–720.
DD OD

514. Grender, J.M., Johnson, W.D., and Elston, R.C. (1992) Regression toward the mean in 2x2 crossover designs with baseline measurements. *Statistics in Medicine* **11**, 727–741.
CO

515. Grieve, A.P. (1985) A Bayesian analysis of the two-period crossover design for clinical trials. *Biometrics* **41**, 979–990, **42**, 459.
CO

516. Griffiths, D.A. (1973) Maximum likelihood estimation for the beta-binomial distribution and an application to the household distribution of the total number of cases of disease. *Biometrics* **29**, 637–648.
CD DD OD

517. Griffiths, R.C., Milne, R.K., and Wood, R. (1979) Aspects of correlation in bivariate Poisson distributions and processes. *Australian Journal of Statistics* **21**, 238–255.
DD MULT PP TS

518. Griffiths, W.E. (1972) Estimation of actual response coefficients in the Hildreth–Houck random coefficient model. *Journal of the American Statistical Association* **67**, 633–635.
VC

519. Grimson, R.C., Aldrich, T.E., and Drane, J.W. (1992) Clustering in sparse data and an analysis of rhabdomyosarcoma incidence. *Statistics in Medicine* **11**, 761–768.
DD PP TS

520. Grizzle, J.E. (1965) The two period change-over design and its uses in clinical trials. *Biometrics* **21**, 467–480, **30**, 72.
CO DES

521. Grizzle, J.E. and Allen, D.M. (1969) Analysis of growth and dose response curves. *Biometrics* **25**, 357–382.
GC MULT TS

522. Grogger, J.T. (1990) The deterrent effect of capital punishment: an analysis of daily homicide counts. *Journal of the American Statistical Association* **85**, 295–303.
DD TS

523. Grogger, J.T. and Carson, R.T. (1991) Models for truncated counts. *Journal of Applied Econometrics* **6**, 225–238.
DD OD

524. Grüger, J., Kay, R., and Schumacher, M. (1991) The validity of inferences based on incomplete observations in disease state models. *Biometrics* **47**, 595–605.
MP SURV TS

525. Gumbel, E.J. (1960) Bivariate exponential distributions. *Journal of the American Statistical Association* **55**, 698–707.
DIST MULT SURV

526. Gumbel, E.J. (1961) Bivariate logistic distributions. *Journal of the American Statistical Association* **56**, 335–349.
DIST MULT SURV

527. Gumbel, E.J. and Goldstein, N. (1964) Analysis of empirical bivariate distributions. *Journal of the American Statistical Association* **59**, 794–816.
DIST MULT

528. Gumbel, E.J. and Mustafi, C.K. (1967) Some analytic properties of bivariate extremal distributions. *Journal of the American Statistical Association* **62**, 569–588.
DIST MULT

529. Gumpertz, M.L. and Pantula, S.G. (1992) Nonlinear regression with variance components. *Journal of the American Statistical Association* **87**, 201–209.
VC

530. Guo, G. and Rodriguez, G. (1992) Estimating a multivariate proportional hazards model for clustered data using the EM algorithm, with an application to child survival in Guatemala. *Journal of the American Statistical Association* **87**, 969–976.
SURV VC

531. Gurmu, S. (1991) Tests for detecting overdispersion in the positive Poisson regression model. *Journal of Business and Economic Statistics* **9**, 215–222.
DD OD

532. Guttorp, P. and Thompson, M.L. (1990) Nonparametric estimation of intensities for sampled counting processes. *Journal of the Royal Statistical*

Society **B52**, 157–173.
DD PP TS

533. Haber, M., Chen, C.C.H., and Williamson, G.D. (1991) Analysis of repeated categorical responses from fully and partially cross-classified data. *Communications in Statistics* **A20**, 3293–3313.
DD

534. Haber, M., Longini, I.M., and Cotsonis, G.A. (1988) Models for the statistical analysis of infectious disease data. *Biometrics* **44**, 163–173.
DD PP TS

535. Hagenaars, J.A. (1990) *Categorical Longitudinal Data. Log-Linear Panel, Trend, and Cohort Analysis.* Newbury Park: Sage.
DD TS

536. Hamerle, A. (1989) Multiple-spell regression models for duration data. *Journal of the Royal Statistical Society* **C38**, 127–138.
SURV TS

537. Hamerle, A. (1990) On a simple test for neglected heterogeneity in panel studies. *Biometrics* **46**, 193–199.
DD OD VC

538. Hamilton, D.C. (1992) Analysis of fish behaviour data. *Canadian Journal of Statistics* **20**, 228–233.
DD GC MC OD TS

539. Hand, D.J. and Taylor, C.C. (1987) *Multivariate Analysis of Variance and Repeated Measures.* London: Chapman and Hall.
MULT VC

540. Handa, B.R. (1972) Choice of initial states in estimating transition probabilities of a finite ergodic Markov chain. *Biometrika* **59**, 407–414.
DD MC TS

541. Hannan, M.T. (1989) Macrosociological applications of event history analysis: state transitions and event recurrences. *Quality and Quantity* **23**, 351–383.
EH MP SURV TS

542. Hannan, M.T. and Carroll, G.R. (1981) Dynamics of formal political structure: an event-history analysis. *American Sociological Review* **46**, 19–35.
EH MP SURV TS

543. Hannan, M.T., Tuma, N.B., and Groeneveld, L.P. (1977) Income and marital events: evidence from an income–maintenance experiment. *American Journal of Sociology* **82**, 1186–1211.
EH MP SURV TS

544. Hannan, M.T., Tuma, N.B., and Groeneveld, L.P. (1978) Income and independence effects on marital dissolution: results from the Seattle and Denver income-maintenance experiments. *American Journal of Sociology* **84**, 611–633.
EH MP SURV TS

545. Harris, P. (1985) Testing for variance homogeneity of correlated variables. *Biometrika* **72**, 103–107.

VC

546. Harris, R. (1970) A multivariate definition for increasing hazard rate distribution functions. *Annals of Mathematical Statistics* **41**, 713–717.
DIST MULT SURV

547. Harrison, P.J. and Stevens, C.F. (1976) Bayesian forecasting. *Journal of the Royal Statistical Society* **B38**, 205–247.
DGLM TS

548. Hart, J.D. and Wehrly, T.E. (1986) Kernel regression estimation using repeated measurements data. *Journal of the American Statistical Association* **81**, 1080–1089.
GC TS

549. Hartley, H.O. and Rao, J.N.K. (1967) Maximum-likelihood estimation for the mixed analysis of variance model. *Biometrika* **54**, 93–108.
VC

550. *Harvey, A.C. (1989) *Forecasting, Structural Time Series Models and the Kalman Filter*. Cambridge: Cambridge University Press.
AR DGLM TS

551. Harvey, A.C. and Durbin, J. (1986) The effects of seat belt legislation on British road casualties: a case study in structural time series modelling. *Journal of the Royal Statistical Society* **A149**, 187–227.
AR DGLM TS

552. Harvey, A.C. and Fernandes, C. (1989) Time series models for count or qualitative observations. *Journal of Business and Economic Statistics* **7**, 407–423.
DD DGLM TS

553. Harvey, A.C. and Phillips, G.D.A. (1979) Maximum likelihood estimation of regression models with autoregressive-moving average disturbances. *Biometrika* **66**, 49–58.
AR DGLM TS

554. Harville, D.A. (1974) Bayesian inference for variance components using only error contrasts. *Biometrika* **61**, 383–385.
VC

555. Harville, D.A. (1977) Maximum likelihood approaches to variance component estimation and to related problems. *Journal of the American Statistical Association* **72**, 320–340.
VC

556. Harville, D.A. and Mee, R.W. (1984) A mixed-model procedure for analyzing ordered categorical data. *Biometrics* **40**, 393–408.
DD VC

557. Haseman, J.K. and Kupper, L.L. (1979) Analysis of dichotomous response data from certain toxicological experiments. *Biometrics* **35**, 281–203.
DD OD

558. Hatzinger, R. (1989) The Rasch model, some extensions and their relation to the class of generalized linear models. In Decarli *et al.*, pp. 172–179.
DD

559. Hausman, J., Hall, B.H., and Griliches, Z. (1984) Econometric models for count data and an application to the patents — R and D relationship. *Econometrica* **52**, 909–938.
DD OD

560. Hausman, J. and Taylor, W.E. (1981) Panel data and unobservable individual effects. *Econometrica* **49**, 1377–1398.
AR TS VC

561. Hawkes, A.G. (1972) A bivariate exponential distribution with applications to reliability. *Journal of the Royal Statistical Society* **B34**, 129–131.
DIST MULT SURV

562. Healy, M.J.R. (1988) *Glim: An Introduction.* Oxford: Oxford University Press.
GLM

563. Healy, M.J.R. and Tillett, H.E. (1988) Short-term extrapolation of the AIDS epidemic. *Journal of the Royal Statistical Society* **A151**, 50–61.
GC TS

564. Hearne, E.M., Clark, G.M., and Hatch, J.P. (1983) A test for serial correlation in univariate repeated-measure analysis. *Biometrics* **39**, 237–243.
TS

565. *Heckman, J.J. (1978) Simple statistical models for discrete panel data developed and applied to test the hypothesis of true state dependence against the hypothesis of spurious state dependence. *Annales de l'INSEE* **30-31**, 227–269.
DD VC

566. *Heckman, J.J. and Borjas, G. (1980) Does unemployment cause future unemployment? definitions, questions and answers from a continuous time model of heterogeneity and state dependence. *Economica* **47**, 247–284.
EH MP SURV TS VC

567. Heckman, J.J. and Honoré, B.E. (1989) The identifiability of the competing risks model. *Biometrika* **76**, 325–330.
SURV VC

568. Heckman, J.J. and MaCurdy, T.E. (1980) A life cycle model of female labor supply. *Review of Economic Studies* **47**, 47–74.
SURV TS VC

569. Heckman, J.J., Robb, R., and Walker, J.R. (1990) Testing the mixture of exponentials hypothesis and estimating the mixing distribution by the method of moments. *Journal of the American Statistical Association* **85**, 582–589.
CD SURV VC

570. Heckman, J.J. and Singer, B. (1984a) Econometric duration analysis. *Journal of Econometrics* **24**, 63–132.
SURV VC

571. Heckman, J.J. and Singer, B. (1984b) A method for minimizing the impact of distributional assumptions in econometric models for duration

data. *Econometrica* **52**, 271–319.
SURV VC

572. Heckman, J.J. and Singer, B. (1984c) The identifiability of the proportional hazards model. *Review of Economic Studies* **60**, 231–241.
SURV VC

573. Heckman, J.J. and Singer, B. (1985, ed.) *Longitudinal Analysis of Labor Market Data.* Cambridge: Cambridge University Press.
SURV VC

574. Heckman, J.J. and Walker, J.R. (1990a) Estimating fecundability from data on waiting times to first conception. *Journal of the American Statistical Association* **85**, 283–294.
SURV VC

575. Heckman, J.J. and Walker, J.R. (1990b) The relationship between wages and income and the timing and spacing of births: evidence from Swedish longitudinal data. *Econometrica* **58**, 1411–1441.
SURV VC

576. Heckman, J.J. and Willis, R.J. (1977) A beta-logistic model for the analysis of sequential labor force participation by married women. *Journal of Political Economy* **85**, 27–58.
CD DD OD

577. Heitjan, D.F. (1991a) Generalized Norton–Simon models of tumour growth. *Statistics in Medicine* **10**, 1075–1088.
AR GC TS VC

578. Heitjan, D.F. (1991b) Nonlinear modeling of serial immunologic data: a case study. *Journal of the American Statistical Association* **86**, 891–898.
AR GC TS VC

579. Henderson, C.R. (1971) Comment on the use of error components models in combining cross section with time series data. *Econometrica* **39**, 397–401.
VC

580. Henderson, C.R. (1982) Analysis of covariance in the mixed model: higher-level, nonhomogeneous, and random regressions. *Biometrics* **38**, 623–640.
VC

581. Henstridge, J.D. and Tweedie, R.L. (1984) A model for the growth pattern of muttonbirds. *Biometrics* **40**, 917–925.
GC GLM TS

582. Heyting, A., Tolboom, J.T.B.M., and Essers, J.G.A. (1992) Statistical handling of drop-outs in longitudinal clinical trials. *Statistics in Medicine* **11**, 2043–2061.
DD

583. Hildreth, C. and Houck, J.P. (1968) Some estimators for a linear model with random coefficients. *Journal of the American Statistical Association* **63**, 584–595.
VC

584. Hill, B.M. (1967) Correlated errors in the random model. *Journal of the American Statistical Association* **62**, 1387–1400.
VC

585. Hill, J.R. and Tsai, C.L. (1988) Calculating the efficiency of maximum quasi-likelihood estimation. *Journal of the Royal Statistical Society* **C37**, 219–230.
GEE

586. Hills, M. (1968) A note on the analysis of growth curves. *Biometrics* **24**, 189–196.
GC TS

587. Hinde, J. (1982) Compound Poisson regression models. In Gilchrist, pp. 109–121.
CD DD VC

588. Hinkley, D.V. (1979) Likelihood inference for a simple growth model. *Biometrika* **66**, 659–662.
GC TS

589. Hirotsu, C. (1991) An approach to comparing treatments based on repeated measures. *Biometrika* **78**, 583–594.
VC

590. Hjort, N.L. (1990) Goodness of fit tests in models for life history data based on cumulative hazard rates. *Annals of Statistics* **18**, 1221–1258.
EH SURV

591. Hoch, I. (1962) Estimation of production function parameters combining time-series and cross-section data. *Econometrica* **30**, 34–53.
TS

592. Hocking, R.R. (1990) A new approach to variance component estimation with diagnostic implications. *Communications in Statistics* **A19**, 4591–4617.
VC

593. Hoel, P.G. (1954) A test for Markoff chains. *Biometrika* **41**, 430–433.
DD MC TS

594. Hoel, P.G. (1964) Methods for comparing growth type curves. *Biometrics* **20**, 859–872.
GC TS

595. Hoem, J.M. (1971) Point estimation of forces of transition in demographic models. *Journal of the Royal Statistical Society* **B33**, 275–289.
DD MC TS

596. Hoem, J.M. (1972) On the statistical theory of analytic graduation. *Proceedings of the Sixth Berkeley Symposium* **1**, 569–600.
DD MC TS

597. Hoem, J.M. (1976) The statistical theory of demographic rates. A review of current developments. *Scandinavian Journal of Statistics* **3**, 169–185.
DD MC TS

598. Hoem, J.M. (1985) Weighting, misclassification and other issues in the analysis of survey samples of life histories. In Heckman and Singer (1985),

pp. 249–293.
EH DD SURV TS

599. Holford, T.R. (1976) Life tables with concomitant information. *Biometrics* **32**, 587–597.
SURV

600. Holford, T.R. (1983) The estimation of age, period and cohort effects for vital rates. *Biometrics* **39**, 311–324.
DD SURV

601. Holford, T.R. (1992) Analyzing the temporal effects of age, period and cohort. *Statistical Methods in Medical Research* **1**, 317–337.
DD

602. Holgate, P. (1964) Estimation for the bivariate Poisson distribution. *Biometrika* **51**, 241–245.
DD DIST MULT

603. Holland, P.W. and Wang, Y.J. (1987a) Regional dependence for continuous bivariate densities. *Communications in Statistics* **A16**, 193–206.
DIST MULT

604. Holland, P.W. and Wang, Y.J. (1987b) Dependence function for continuous bivariate densities. *Communications in Statistics* **A16**, 863–876.
DIST MULT

605. Holt, J.D. (1978) Competing risk analysis with special reference to matched pair experiments. *Biometrika* **65**, 159–166.
SURV VC

606. Holt, J.D. and Prentice, R.L. (1974) Survival analysis in twin studies and matched pair experiments. *Biometrika* **61**, 17–30.
SURV VC

607. Honoré, B.E. (1990) Simple estimation of a duration model with unobserved heterogeneity. *Econometrica* **58**, 453–473.
SURV VC

608. Hooper, P. and Larin, K.A. (1989) International comparisons of labor costs in manufacturing. *Review of Income and Wealth* **35**, 335-355.

609. Hopper, J.L. and Young, G.P. (1988) A random walk model for evaluating clinical trials involving serial observations. *Statistics in Medicine* **7**, 581–590.
DD MC TS

610. Hougaard, P. (1984) Life table methods for heterogeneous populations: distributions describing the heterogeneity. *Biometrika* **71**, 75–84.
DIST MULT SURV

611. Hougaard, P. (1986a) Survival models for heterogeneous populations derived from stable distributions. *Biometrika* **73**, 387–396.
DIST MULT SURV VC

612. Hougaard, P. (1986b) A class of multivariate failure time distributions. *Biometrika* **73**, 671–678.
DIST MULT SURV

613. Hougaard, P. (1987) Modelling multivariate survival. *Scandinavian Journal of Statistics* **14**, 291–304.

DIST MULT SURV

614. Hougaard, P., Harvald, B., and Holm, N.V. (1992) Measuring the similarities between the lifetimes of adult Danish twins born between 1881-1930. *Journal of the American Statistical Association* **87**, 17–24.
SURV VC

615. Hougaard, P. and Madsen, E.B. (1985) Dynamic evaluation of short-term prognosis after myocardial infarction. *Statistics in Medicine* **4**, 29–38.
SURV

616. Hsaio, C. (1985) *Analysis of Panel Data.* Cambridge: Cambridge University Press.
AR DD TS VC

617. Hudson, I.L. (1983) Asymptotic tests for growth curve models with autoregressive errors. *Australian Journal of Statistics* **25**, 413–424.
AR TS

618. Hui, S.L. (1984) Curve fitting for repeated measurements made at irregular time points. *Biometrics* **40**, 691–697.
GC TS VC

619. Hui, S.L. and Berger, J.O. (1983) Empirical Bayes estimation of rates in longitudinal studies. *Journal of the American Statistical Association* **78**, 753–761.
GC TS VC

620. Hui, S.L. and Rosenberg, S.H. (1985) Multivariate slope ratio assay with repeated measurements. *Biometrics* **41**, 11–18.
MULT

621. Hulting, F.L. and Harville, D.A. (1991) Some Bayesian and non-Bayesian procedures for the analysis of comparative experiments and for small-area estimation: computational aspects, frequentist properties, and relationships. *Journal of the American Statistical Association* **86**, 557–568.
VC

622. Huster, W.J., Brookmeyer, R., and Self, S.G. (1989) Modelling paired survival data with covariates. *Biometrics* **45**, 145–156.
SURV VC

623. Huynh, H. and Feldt, L.S. (1970) Conditions under which mean square ratios in repeated measurements designs have exact F-distributions. *Journal of the American Statistical Association* **65**, 1582–1589.
VC

624. Im, S. (1984) Plan mixte équilibré à deux facteurs hiérarchiques pour les données binaires. *Biometrics* **40**, 383–392.
DD VC

625. Im, S. (1992) Mixed linear model with uncertain paternity. *Journal of the Royal Statistical Society* **C41**, 109–116.
VC

626. Im, S. and Gianola, D. (1988) Mixed models for binomial data with application to lamb mortality. *Journal of the Royal Statistical Society* **C37**, 196–204.
DD VC

627. Iosifescu, M. (1980) *Finite Markov Processes and their Applications.* New York: John Wiley.
     DD MC MP

628. Izeman, A.J. and Williams, J.S. (1989) A class of linear spectral models and analyses for the study of longitudinal data. *Biometrics* **45**, 831–849.
     TS

629. Jacobs, P.A. and Lewis, P.A.W. (1977) A mixed autoregressive moving average exponential sequence and point process (EARMA(1,1)). *Advances in Applied Probability* **9**, 87–104.
     AR CD DD TS

630. Jacobs, P.A. and Lewis, P.A.W. (1978a) Discrete time series generated by mixtures. I. Correlational and runs properties. *Journal of the Royal Statistical Society* **B40**, 94–105.
     CD DD TS

631. Jacobs, P.A. and Lewis, P.A.W. (1978b) Discrete time series generated by mixtures. II. Asymptotic properties. *Journal of the Royal Statistical Society* **B40**, 222–228.
     CD DD TS

632. Jagers, P. (1991) The growth and stabilization of populations. *Statistical Science* **6**, 269–283.
     GC TS

633. James, I.R. (1975) Multivariate distributions which have beta conditional distributions. *Journal of the American Statistical Association* **70**, 681–684.
     DIST MULT

634. James, I.R. (1983) Analysis of nonagreements among multiple raters. *Biometrics* **39**, 651–657.
     DD VC

635. James, I.R. (1991) Estimation of von Bertalanffy growth curve parameters from recapture data. *Biometrics* **47**, 1519–1530.
     GC TS

636. Jansen, J. (1990) On the statistical analysis of ordinal data when extravariation is present. *Journal of the Royal Statistical Society* **C39**, 75–84.
     DD OD

637. Jarjoura, D. and Logue, E. (1990) Variation in heart disease mortality across census tracts as a function of overdispersion and social class mixture. *Statistics in Medicine* **9**, 1199–1209.
     DD OD

638. Jarrett, R.G. (1979) A note on the intervals between coal- mining disasters. *Biometrika* **66**, 191–193.
     DD EH PP RP TS

639. Jennrich, R.I. and Schluchter, J.R. (1986) Unbalanced repeated measures with structured covariance matrices. *Biometrics* **42**, 805–820.
     GC TS

640. Jensen, D.R. (1982) Efficiency and robustness in the use of repeated measurements. *Biometrics* **38**, 813–825.
VC

641. Jensen, J.L. (1987) On asymptotic expansions in non-ergodic models. *Scandinavian Journal of Statistics* **14**, 305–318.
DD TS

642. Jeuland, A.P., Bass, F.M., and Wright, G.P. (1980) A multibrand stochastic model compounding heterogenous Erlang timing and multinomial choice processes. *Operations Research* **28**, 255–277.
DD PP RP TS

643. Joe, H. (1990) Families of min-stable multivariate exponential and multivariate extreme value distributions. *Statistics and Probability Letters* **9**, 75–82.
DIST MULT SURV

644. Jogdeo, K. (1968) Characterizations of independence in certain families of bivariate and multivariate distributions. *Annals of Mathematical Statistics* **39**, 433–441.
MULT

645. Johansen, S. (1982) Asymptotic inference in random coefficients regression models. *Scandinavian Journal of Statistics* **9**, 201–207.
VC

646. Johansen, S. (1983a) An extension of Cox's regression model. *International Statistical Review* **51**, 165–174.
DD SURV

647. Johansen, S. (1983b) Some topics in regression. *Scandinavian Journal of Statistics* **10**, 161–194.
VC

648. Johnson, M.E. (1987) *Multivariate Statistical Simulation*. New York: John Wiley.
DIST MULT

649. Johnson, M.E. and Tenenbein, A. (1981) A bivariate distribution family with specified marginals. *Journal of the American Statistical Association* **76**, 198–201.
DIST MULT

650. Johnson, N.L. (1957) Uniqueness of a result in the theory of accident proneness. *Biometrika* **44**, 530–531.
CD DD

651. Johnson, N.L. and Kotz, S. (1969) *Distributions in Statistics. I. Discrete Distributions*. New York: John Wiley.
DIST

652. Johnson, N.L. and Kotz, S. (1970) *Distributions in Statistics. II. Continuous Distributions 1*. New York: John Wiley.
DIST

653. Johnson, N.L. and Kotz, S. (1971) *Distributions in Statistics. III. Continuous Distributions 2*. New York: John Wiley.
DIST

654. Johnson, N.L. and Kotz, S. (1972) *Distributions in Statistics. IV. Continuous Multivariate Distributions.* New York: John Wiley.
   DIST MULT

655. Jolicoeur, P. and Heusner, A.A. (1986) Log-normal variation belts for growth curves. *Biometrics* **42**, 785–794.
   GC TS

656. Jolicoeur, P., Pontier, J., Pernin, M.O., and Sempé, M. (1988) A lifetime asymptotic growth curve for human height. *Biometrics* **44**, 995–1003.
   GC TS

657. Jolly, G.M. (1965) Explicit estimates from capture-recapture data with both death and immigration — stochastic models. *Biometrika* **52**, 225–247.
   DD TS

658. Jones, B. and Kenward, M.G. (1987) Modelling binary data from a three-period cross-over trial. *Statistics in Medicine* **6**, 555–564.
   CO DD

659. *Jones, B. and Kenward, M.G. (1989) *Design and Analysis of Cross-over Trials.* London: Chapman and Hall.
   CO DES DD

660. Jones, R.H. (1966) Exponential smoothing for multivariate time series. *Journal of the Royal Statistical Society* **B28**, 241–251.
   DGLM TS

661. Jones, R.H. (1984) Fitting multivariate models to unequally spaced data. In Parzen, E. (ed.) *Time Series Analysis of Irregularly Observed Data.* Berlin: Springer Verlag, pp. 158–188.
   DGLM TS

662. Jones, R.H. (1985) Time series analysis with unequally spaced data. *Handbook of Statistics* **5**, 157–177.
   DGLM TS

663. Jones, R.H. (1987) Serial correlation in unbalanced mixed models. *Bulletin of the International Statistical Institute* **46**, 105–122.
   DGLM TS VC

664. *Jones, R.H. and Ackerson, L.M. (1990) Serial correlation in unequally spaced longitudinal data. *Biometrika* **77**, 721–731.
   DGLM TS VC

665. Jones, R.H. and Boadi-Boateng, F. (1991) Unequally spaced longitudinal data with AR(1) serial correlation. *Biometrics* **47**, 161–175.
   AR DGLM TS

666. Jones, R.H., Ford, P.M., and Hamman, R.F. (1988) Seasonality comparisons among groups using incidence data. *Biometrics* **44**, 1131–1144.
   DD TS

667. Jöreskog, K.L. (1981) Analysis of covariance structures. *Scandinavian Journal of Statistics* **8**, 65–92.
   MULT

668. Jørgensen, B. (1982) *Statistical Properties of the Generalized Inverse Gaussian Distribution.* Berlin: Springer Verlag.

DIST

669. Jørgensen, B. (1985) Estimation of interobserver variation for ordinal rating scales. In Gilchrist, pp. 93–104.
DD VC

670. Jørgensen, B. (1986) Some properties of exponential dispersion models. *Scandinavian Journal of Statistics* **13**, 187–198.
GLM

671. Jørgensen, B. (1987) Exponential dispersion models. *Journal of the Royal Statistical Society* **B49**, 127–162.
GLM

672. *Jørgensen, B. (1992) Exponential dispersion models and extensions: a review. *International Statistical Review* **60**, 5–20.
GLM TS

673. Jørgensen, B., Seshadri, V., and Whitmore, G.A. (1991) On the mixture of the inverse Gaussian distribution with its complementary reciprocal. *Scandinavian Journal of Statistics* **18**, 77–89.
CD

674. Jørgensen, M., Keiding, N., and Sakkebak, N.E. (1991) Estimation of spermarche from longitudinal spermaturia data. *Biometrics* **47**, 177–193.
DD PP TS VC

675. Jorgenson, D.W. (1961) Multiple regression analysis of a Poisson process. *Journal of the American Statistical Association* **56**, 235–245.
DD PP TS

676. Kackar, R.N. and Harville, D.A. (1984) Approximations for standard errors of estimators of fixed and random effects in mixed linear models. *Journal of the American Statistical Association* **79**, 853–862.
VC

677. Kalbfleisch, J.D., Krewski, D.R., and van Ryzin, J. (1983) Dose-response models for time-to-response toxicity data. *Canadian Journal of Statistics* **11**, 25–49.
MP SURV TS

678. Kalbfleisch, J.D. and Lawless, J.F. (1984) Least-squares estimation of transition probabilities from aggregate data. *Canadian Journal of Statistics* **12**, 169–182.
DD MC TS

679. Kalbfleisch, J.D. and Lawless, J.F. (1985) The analysis of panel data under Markov assumptions. *Journal of the American Statistical Association* **80**, 863–871.
DD MC TS

680. Kalbfleisch, J.D. and Lawless, J.F. (1988) Likelihood analysis of multi-state models for disease incidence and mortality. *Statistics in Medicine* **7**, 149–160.
DD PP SURV TS

681. Kalbfleisch, J.D., Lawless, J.F., and Robinson, J.A. (1991) Methods for the analysis and prediction of warranty claims. *Technometrics* **33**, 273–285.

DD OD PP TS

682. Kalbfleisch, J.D., Lawless, J.F., and Vollmer, W.M. (1983) Estimation in Markov models from aggregate data. *Biometrics* **39**, 907–919.

DD MC TS

683. Kalbfleisch, J.D. and Prentice, R.L. (1980) *The Statistical Analysis of Failure Time Data*. New York: John Wiley.

SURV

684. Kalbfleisch, J.D. and Sprott, D.A. (1970) Application of likelihood methods to models involving large numbers of parameters. *Journal of the Royal Statistical Society* **B32**, 175–208.

685. Kalbfleisch, J.G. (1985) *Probability and Statistical Inference*. Vol. 2. Berlin: Springer Verlag.

686. Kanter, M. (1975) Autoregression for discrete processes mod 2. *Journal of Applied Probability* **12**, 371–375.

AR DD TS

687. Kao, E.P.C. (1974) Modelling the movement of coronary patients within a hospital by semi-Markov processes. *Operations Research* **22**, 683–699.

DD EH MP SURV TS

688. Karim, M.R. and Zeger, S.L. (1992) Generalized linear models with random effects: salamander mating revisited. *Biometrics* **48**, 631–644.

DD GEE VC

689. *Karlin, S. and Taylor, H.M. (1975) *A First Course in Stochastic Processes*. New York: Academic Press.

DD MC MP PP RP TS

690. Karlin, S. and Taylor, H.M. (1981) *A Second Course in Stochastic Processes*. New York: Academic Press.

DD MC MP PP TS VC

691. Karr, A.F. (1987) Maximum likelihood estimation in the multiplicative intensity model via sieves. *Annals of Statistics* **15**, 473–490.

DD PP SURV TS

692. Karr, A.F. (1991, 2nd ed.) *Point Processes and their Statistical Inference*. Basel: Marcel Dekker.

DD PP SURV TS

693. Kashiwagi, N. and Yanagimoto, T. (1992) Smoothing serial count data through a state-space model. *Biometrics* **48**, 1187–1194.

DD DGLM TS

694. Kaufmann, H. (1987) Regression models for nonstationary categorical time series: asymptotic estimation theory. *Annals of Statistics* **15**, 79–98.

AR DD TS

695. Kay, R. (1986) A Markov model for analyzing cancer markers and disease states in survival studies. *Biometrics* **42**, 855–865.

EH MP SURV TS

696. Keane, M.P. and Runkle, D.E. (1992) On the estimation of panel-data models with serial correlation when instruments are not strictly exogenous. *Journal of Business and Economic Statistics* **10**, 1–29.

TS VC

697. Keen, A., Thissen, J.T.M.M., Hoekstra, J.A., and Jansen, J. (1986) Successive measurements experiments. *Statistica Neerlandica* **40**, 205–223.
GC TS

698. Keenan, D.M. (1982) A time series analysis of binary data. *Journal of the American Statistical Association* **77**, 816–821.
DD DGLM PP TS

699. Keiding, N. (1975) Maximum likelihood estimation in the birth-and-death process. *Annals of Statistics* **3**, 363–372.
DD TS

700. Keiding, N. (1991) Age-specific incidence and prevalence: a statistical perspective. *Journal of the Royal Statistical Society* **A154**, 371–412.
DD PP SURV TS

701. Keiding, N. and Andersen, P.K. (1989) Nonparametric estimation of transition intensities and transition probabilities: a case study of a two-state Markov process. *Journal of the Royal Statistical Society* **C38**, 319–329.
DD EH MP NPAR PP SURV TS

702. Keiding, N., Andersen, P.K., and Frederiksen, K. (1990) Modelling excess mortality of the unemployed: choice of scale and extra-Poisson variability. *Journal of the Royal Statistical Society* **C39**, 63–74.
DD OD

703. Keiding, N. and Gill, R.D. (1990) Random truncation models and Markov processes. *Annals of Statistics* **18**, 582–602.
DD MP PP SURV TS

704. Kelton, W.D. and Kelton, C.M.L. (1984) Hypothesis tests for Markov process models estimated from aggregate frequency data. *Journal of the American Statistical Association* **79**, 922–928.
DD MC TS

705. Kempthorne, O. (1977) Why randomize? *Journal of Statistical Planning and Inference* **1**, 1–25.
DES

706. Kennan, J. (1985) The duration of contract strikes in US manufacturing. *Journal of Econometrics* **28**, 5–28.
SURV VC

707. Kenward, M.G. (1985) The use of fitted higher-order polynomial coefficients as covariates in the analysis of growth curves. *Biometrics* **41**, 19–28.
GC TS

708. Kenward, M.G. (1987) A method for comparing profiles of repeated measurements. *Journal of the Royal Statistical Society* **C36**, 296–308, **40**, 379.
GC TS

709. Kenward, M.G. and Jones, B. (1987a) A log-linear model for binary crossover data. *Journal of the Royal Statistical Society* **C36**, 192–204.
CO DD

710. Kenward, M.G. and Jones, B. (1987b) The analysis of data from 2x2 cross-over trials with baseline measurements. *Statistics in Medicine* **6**, 911–926.
CO

711. Kenward, M.G. and Jones, B. (1991) The analysis of categorical data from cross-over trials using a latent variable model. *Statistics in Medicine* **10**, 1607–1619.
CO DD

712. Kepner, J.L. and Robinson, D.H. (1988) Nonparametric methods for detecting treatment effects in repeated-measures designs. *Journal of the American Statistical Association* **83**, 456–461.
NPAR

713. Keramidas, E.M. and Lee, J.C. (1990) Forecasting technological substitutions with concurrent short time series. *Journal of the American Statistical Association* **85**, 625–632.
GC TS

714. Kershner, D.P. and Federer, W.T. (1981) Two-treatment crossover designs for estimating a variety of effects. *Journal of the American Statistical Association* **76**, 612–619.
CO DES

715. Keselman, H.J. and Keselman, J.C. (1984) The analysis of repeated measures designs in medical research. *Statistics in Medicine* **3**, 185–195.
VC

716. Key, P.B. and Godolphin, E.J. (1980) On the Bayesian steady forecasting model. *Journal of the Royal Statistical Society* **B43**, 92–96.
DGLM TS

717. Khatri, C.G. (1988) Robustness study for a linear growth model. *Journal of Multivariate Analysis* **24**, 66–87.
GC MULT TS

718. Kiefer, N.M. (1985) Specification diagnostics based on LaGuerre alternatives for econometric models of duration. *Journal of Econometrics* **28**, 135–154.
SURV VC

719. Kiefer, N.M. (1988) Economic duration data and hazard functions. *Journal of Economic Literature* **26**, 646–679.
SURV VC

720. Kiefer, N.M. and Neumann, G.R. (1981) Individual effects in a nonlinear model: explicit treatment of heterogeneity in the empirical job-search model. *Econometrica* **49**, 965–979.
SURV VC

721. Kim, B.S. and Margolin, B.H. (1992) Testing goodness of fit of a multinomial model against overdispersed alternatives. *Biometrics* **48**, 711–719.
DD OD

722. Kingman, J.F.C. (1963) Poisson counts for random sequences of events. *Annals of Mathematical Statistics* **34**, 1217–1232.
DD PP RP TS

723. Kingman, J.F.C. (1966) An approach to the study of Markov processes. *Journal of the Royal Statistical Society* **B28**, 417–447.
DD MC TS

724. Kingman, J.F.C. (1969) Markov population processes. *Journal of Applied Probability* **6**, 1–18.
DD MC TS

725. Kitagawa, G. (1987) Non-Gaussian state-space modeling of nonstationary time series. *Journal of the American Statistical Association* **82**, 1032–1063.
DD DGLM TS

726. Kitagawa, G. (1988) Numerical approach to non-Gaussian smoothing and its applications. In Wegman, E.J., Gantz, D.T., and Miller, J.J. (ed.) *Computing Science and Statistics*, Alexandria, American Statistical Association, pp. 379–388.
DGLM TS

727. Kitagawa, G. (1989) Non-Gaussian seasonal adjustment. *Computers and Mathematical Applications* **18**, 503–514.
DGLM TS

728. Kitagawa, G. and Gersch, W. (1984) A smoothness priors — state space modeling of time series with trend and seasonality. *Journal of the American Statistical Association* **79**, 378–389.
DGLM TS

729. Kjaergaard-Andersen, P., Christensen, F., Schmidt, S.A., Pedersen, N.W., and Jørgensen, B. (1988) A new method of estimation of inter-observer variation and its application to the radiological assessment of osteoarthrosis in hip joints. *Statistics in Medicine* **7**, 639–647.
DD VC

730. Kleffe, J., Prasad, N.G.N., and Rao, J.N.K. (1991) 'Optimal' estimation of correlated response variance under additive models. *Journal of the American Statistical Association* **86**, 144–150.
VC

731. Klein, J.P. (1992) Semiparametric estimation of random effects using the Cox model based on the EM algorithm. *Biometrics* **48**, 795–806.
SURV VC

732. Klein, J.P., Keiding, N., and Kamby, C. (1989) Semiparametric Marshall-Olkin models applied to the occurrence of metastases at multiple sites after breast cancer. *Biometrics* **45**, 1073–1086.
MULT SURV

733. Klein, J.P., Klotz, J.H., and Grever, M.R. (1984) A biological marker model for predicting disease transition. *Biometrics* **40**, 927–936.
DD MC TS

734. Kleinman, J.C. (1973) Proportions with extraneous variance: single and independent samples. *Journal of the American Statistical Association* **68**, 46–54.
DD VC

735. Kleinman, J.C. (1975) Proportions with extraneous variance: two dependent samples. *Biometrics* **31**, 737–743.
DD VC

736. Klotz, J. (1972) Markov chain clustering of births by sex. *Proceedings of*

the Sixth Berkeley Symposium **4**, 173–185.
DD MC TS

737. Klotz, J. (1973) Statistical inference in Bernoulli trials with dependence. *Annals of Statistics* **1**, 373–379.
DD MC TS

738. Klotz, J. (1990) Maximum likelihood estimators for the two way mixed model. *Communications in Statistics* **A19**, 3741–3749.
VC

739. Knoebel, B.R. and Burkhart, H.E. (1991) A bivariate distribution approach to modeling forest diameter distributions at two points in time. *Biometrics* **47**, 241–253.
DIST MULT

740. Koch, G.G. (1969) Some aspects of the statistical analysis of 'split plot' experiments in completely randomized layouts. *Journal of the American Statistical Association* **64**, 485–505.
VC

741. Koch, G.G. (1970) The use of non-parametric methods in the statistical analysis of a complex split plot experiment. *Biometrics* **26**, 105–128.
NPAR VC

742. Koch, G.G., Amara, I.A., Stokes, M.E., and Gillings, D.B. (1980) Some views on parametric and nonparametric analysis for repeated measurements and selected bibliography. *International Statistical Review* **48**, 249–265.
P VC

743. Koch, G.G., Elashoff, J.D., and Amara, I.A. (1988) Repeated measurements — design and analysis. *Encyclopedia of Statistical Science* **8**,46–73.
CO DES TS VC

744. Koch, G.G., Landis, J.R., Freeman, J.L., Freeman, D.H., and Lehnen, R.G. (1977) A general methodology for the analysis of experiments with repeated measurement of categorical data. *Biometrics* **33**, 133–158.
DD VC

745. Koch, G.G., Tolley, H.D., and Freeman, J.L. (1976) An application of the clumped binomial model to the analysis of clustered attribute data. *Biometrics* **32**, 337–354.
DD

746. Kocherlakota, S. and Kocherlakota, K. (1990) The bivariate logarithmic series distribution. *Communications in Statistics* **A19**, 3387–3432.
DD DIST MULT

747. Kodell, R.L. and Matis, J.H. (1976) Estimating the rate constants in a two-compartmental stochastic model. *Biometrics* **32**, 377–400.
MP SURV TS

748. Kodell, R.L. and Nelson, C.J. (1980) An illness–death model for the study of the carcinogenesis process using survival/sacrifice data. *Biometrics* **36**, 267–277.
SURV

749. Korn, E.L. and Whittemore, A.S. (1979) Methods for analyzing panel studies of acute health effects of air pollution. *Biometrics* **35**, 795–802.
DD VC

750. Koziol, J.A., Maxwell, D.A., Fukushima, M., Colmerauer, M.E.M., and Pilch, Y.H. (1981) A distribution-free test for tumour-growth curve analyses with application to an animal tumor immunotherapy experiment. *Biometrics* **37**, 383–390.
GC TS

751. Küchler, U. and Sørensen, M. (1989) Exponential families of stochastic processes a unifying semimartingale approach. *International Statistical Review* **57**, 123–144.
MP SURV TS

752. Kuh, E. (1959) The validity of cross-sectionally estimated behavior equations in time series applications. *Econometrica* **27**, 197–214.
TS

753. Kunert, J. (1985) Optimal repeated measurements designs for correlated observations and analysis by weighted least squares. *Biometrika* **72**, 375–389.
AR DES TS

754. Kupper, L.L. and Haseman, J.K. (1978) The use of a correlated binomial model for the analysis of certain toxicological experiments. *Biometrics* **34**, 69–76.
DD OD

755. Kupper, L.L., Portier, C., Hogan, M.D., and Yamamoto, E. (1986) The impact of litter effects on dose-response modeling in teratology. *Biometrics* **42**, 85–98.
DD OD

756. Lachin, J.M. and Wei, L.J. (1988) Estimators and tests in the analysis of multiple nonindependent 2x2 tables with partially missing observations. *Biometrics* **44**, 513–528.
DD TS

757. Lagakos, S.W. (1976) A stochastic model for censored-survival data in the presence of an auxiliary variable. *Biometrics* **32**, 551–559.
MP SURV TS

758. Lagakos, S.W. (1977) Using auxiliary variables for improved estimates of survival time. *Biometrics* **33**, 399–404.
MP SURV TS

759. Lagakos, S.W., Sommer, C.J., and Zelen, M. (1978) Semi-Markov models for partially censored data. *Biometrika* **65**, 311-317.
DD EH MP MULT SURV TS

760. Laird, N.M. (1988) Missing data in longitudinal studies. *Statistics in Medicine* **7**, 305–315.
DD

761. Laird, N.M. (1991) Topics in likelihood-based methods for longitudinal data analysis. *Statistica Sinica* **1**, 33–50.
DD VC

762. Laird, N.M., Donnelly, C., and Ware, J.H. (1992) Longitudinal studies with continuous responses. *Statistical Methods in Medical Research* **1**, 225–247.
   AR GC GEE TS VC

763. Laird, N.M., Lange, N., and Stram, D. (1987) Maximum likelihood computations with repeated measures: application of the EM algorithm. *Journal of the American Statistical Association* **82**, 97–105.
   GC TS VC

764. Laird, N.M., Skinner, J., and Kenward, M. (1992) An analysis of two-period crossover designs with carry-over effects. *Statistics in Medicine* **11**, 1967–1979.
   CO VC

765. Laird, N.M. and Ware, J.H. (1982) Random-effects models for longitudinal data. *Biometrics* **38**, 963–974.
   GC TS VC

766. Lancaster, H.O. (1958) The structure of bivariate distributions. *Annals of Mathematical Statistics* **29**, 719–736.
   MULT

767. Lancaster, H.O. (1963) Correlations and canonical forms of bivariate distributions. *Annals of Mathematical Statistics* **34**, 532–538.
   MULT

768. Lancaster, T. (1972) A stochastic model for the duration of a strike. *Journal of the Royal Statistical Society* **A135**, 257–271.
   DIST SURV

769. Lancaster, T. (1979) Econometric methods for the duration of unemployment. *Econometrica* **47**, 939–956.
   SURV VC

770. Lancaster, T. (1985) Generalized residuals and heterogeneous duration models with applications to the Weibull model. *Journal of Econometrics* **28**, 155–169.
   SURV VC

771. *Lancaster, T. (1990) *The Econometric Analysis of Transition Data.* Cambridge: Cambridge University Press.
   SURV VC

772. Lancaster, T. and Nickell, S. (1980) The analysis of re-employment probabilities for the unemployed. *Journal of the Royal Statistical Society* **A143**, 141–165.
   SURV VC

773. Landis, J.R. and Koch, G.G. (1975) A review of statistical methods in the analysis of data arising from observer reliability studies. I and II. *Statistica Neerlandica* **29**, 101–123, 151–161.
   DD VC

774. Landis, J.R. and Koch, G.G. (1977a) The measurement of observer agreement for categorical data. *Biometrics* **33**, 159–174.
   DD VC

775. Landis, J.R. and Koch, G.G. (1977b) An application of hierarchical kappa-type statistics in the assessment of majority agreement among multiple observers. *Biometrics* **33**, 363–374.
DD VC

776. Landis, J.R. and Koch, G.G. (1977c) A one-way components of variance model for categorical data. *Biometrics* **33**, 671–679.
DD VC

777. Landis, J.R., Miller, M.E., Davis, C.S., and Koch, G.G. (1988) Some general methods for the analysis of categorical data in longitudinal studies. *Statistics in Medicine* **7**, 109–137.
DD TS

778. Lange, N. (1992) Graphs and stochastic relaxation for hierarchical Bayes modelling. *Statistics in Medicine* **11**, 2001–2016.
VC

779. Lange, N., Carlin, B.P., and Gelfand, A.E. (1992) Hierarchical Bayes models for the progression of HIV infection using longitudinal CD4 T-cell numbers. *Journal of the American Statistical Association* **87**, 615–632.
GC TS VC

780. Lange, N. and Laird, N.M. (1989) The effect of covariance structure on variance estimation in balanced growth curve models with random parameters. *Journal of the American Statistical Association* **84**, 241–247.
GC TS VC

781. Larsen, W.A. (1969) The analysis of variance for the two-way classification fixed-effects model with observations within a row serially correlated. *Biometrika* **56**, 509–515.
AR TS

782. Laska, E.M. and Meisner, M. (1985) A variational approach to optimal two-treatment crossover designs: application to carryover-effect models. *Journal of the American Statistical Association* **80**, 704–710.
CO DES

783. Lauer, R.M. and Clarke, W.R. (1988) A longitudinal view of blood pressure during childhood: the Muscatine study. *Statistics in Medicine* **7**, 47–57.
TS

784. Lauritzen, S.L. (1975) General exponential models for discrete observations. *Scandinavian Journal of Statistics* **2**, 23–33.
DD

785. Lawless, J.F. (1982) *Statistical Models and Methods for Lifetime Data*. New York: John Wiley.
DIST SURV

786. Lawless, J.F. (1987a) Negative binomial and mixed Poisson regression. *Canadian Journal of Statistics* **15**, 209–225.
CD DD OD

787. Lawless, J.F. (1987b) Regression methods for Poisson process data. *Journal of the American Statistical Association* **82**, 808–815.
DD TS

788. Lawless, J.F. and McLeish, D.L. (1984) The information in aggregate data from Markov chains. *Biometrika* **71**, 419–430.
DD MC TS

789. Lawrance, A.J. (1970a) Selective interaction of a Poisson and renewal process: first-order stationary results. *Journal of Applied Probability* **7**, 359–372.
DD RP TS

790. Lawrance, A.J. (1970b) Selective interaction of a stationary point process and a renewal process. *Journal of Applied Probability* **7**, 483–489.
DD RP TS

791. Lawrance, A.J. (1971) Selective interaction of a Poisson and renewal process: the dependency structure of the intervals between responses. *Journal of Applied Probability* **8**, 170–183.
DD RP TS

792. Lawrance, A.J. (1973) Dependency of intervals between events in superposition processes. *Journal of the Royal Statistical Society* **B35**, 306–315.
AR SURV TS

793. Lawrance, A.J. (1982) The innovation distribution of a gamma distributed autoregressive process. *Scandinavian Journal of Statistics* **9**, 234–236.
AR SURV TS

794. Lawrance, A.J. (1991) Directionality and reversibility in time series. *International Statistical Review* **59**, 67–79.
TS

795. Lawrance, A.J. and Kottegoda, N.T. (1977) Stochastic modelling of river-flow time series. *Journal of the Royal Statistical Society* **A140**, 1–47.
AR SURV TS

796. Lawrance, A.J. and Lewis, P.A.W. (1977) An exponential moving-average sequence and point process (EMA1). *Journal of Applied Probability* **14**, 98–113.
AR DD PP SURV TS

797. Lawrance, A.J. and Lewis, P.A.W. (1981a) A new autoregressive time series model in exponential variables (NEAR(1)). *Advances in Applied Probability* **13**, 826–845.
AR SURV TS

798. Lawrance, A.J. and Lewis, P.A.W. (1981b) The exponential autoregressive moving EARMA(p,q) process. *Journal of the Royal Statistical Society* **B42**, 150–161.
AR SURV TS

799. Lawrance, A.J. and Lewis, P.A.W. (1985) Modelling and residual analysis of nonlinear autoregressive time series in exponential variables. *Journal of the Royal Statistical Society* **B47**, 165–202.
AR SURV TS

800. Lawrance, A.J. and Lewis, P.A.W. (1987) Higher-order residual analysis for nonlinear time series with autoregressive correlation structures. *International Statistical Review* **55**, 21–35.
AR SURV TS VC

801. Lawrence, R.J. (1984) The lognormal distribution of the duration of strikes. *Journal of the Royal Statistical Society* **A147**, 464–483.
DIST SURV

802. Layard, M.W.J. and Arvesen, J.N. (1978) Analysis of Poisson data in crossover experimental designs. *Biometrics* **34**, 421–428.
CO DD

803. Laycock, P.J. and Gott, G.F. (1989) Markov modelling for extra-binomial variation in HF (radio) spectral occupancy. In Decarli *et al.*, pp. 198–205.
DD MC OD TS

804. Le, N.D. (1988) Testing for linear trends in proportions using correlated otolaryngology or ophthalmology data. *Biometrics* **44**, 299–303.
DD TS

805. Le, N.D., Leroux, B.G., and Puterman, M.L. (1992) Exact likelihood evaluation in a Markov mixture model for time series of seizure counts. *Biometrics* **48**, 317–323.
CD DD MC TS

806. Leach, D. (1981) Re-evaluation of the logistic curve for human populations. *Journal of the Royal Statistical Society* **A144**, 94–103.
GC TS

807. Lee, J.C. (1977) Bayesian classification of data from growth curves. *South African Statistical Journal* **11**, 155–166.
GC MULT TS

808. Lee, J.C. (1988) Prediction and estimation of growth curves with special covariance structures. *Journal of the American Statistical Association* **83**, 432–440.
GC TS

809. Lee, J.C. (1991) Tests and model selection for the general growth curve model. *Biometrics* **47**, 147–159.
GC MULT TS

810. Lee, J.C.S. and Geisser, S. (1975) Applications of growth curve prediction. *Sankhya* **A37**, 239–256.
GC TS

811. Lee, J.W. and DeMets, D.L. (1991) Sequential comparison of changes with repeated measurements data. *Journal of the American Statistical Association* **86**, 757–762.
NPAR

812. Lee, J.W. and DeMets, D.L. (1992) Sequential rank tests with repeated measurements in clinical trials. *Journal of the American Statistical Association* **87**, 136–142.
NPAR

813. Lee, K.R. and Kapadia, C.H. (1984) Variance components estimators for the balanced two-way mixed model. *Biometrics* **40**, 507–512.
VC

814. Lee, L.F. (1986) Specification test for Poisson regression models. *International Economic Review* **27**, 689–706.
DD OD

815. Lee, R.D. (1974) Forecasting births in post-transition populations: stochastic renewal with serially correlated fertility. *Journal of the American Statistical Association* **69**, 607–617.
GC TS

816. Lee, Y.H.K. (1974) A note on Rao's reduction of Potthoff and Roy's generalized linear model. *Biometrika* **61**, 349–351.
MULT

817. Leech, F.B. and Healy, M.J.R. (1959) The analysis of experiments on growth rate. *Biometrics* **15**, 98–106.
GC TS

818. Lefkopoulou, M., Moore, D., and Ryan, L. (1989) The analysis of multiple correlated binary outcomes: application to rodent teratology experiments. *Journal of the American Statistical Association* **84**, 810–815.
DD OD VC

819. Lehmacher, W. (1991) Analysis of the crossover design in the presence of residual effects. *Statistics in Medicine* **10**, 891–899.
CO

820. Lehmann, E.L. (1966) Some concepts of dependence. *Annals of Mathematical Statistics* **37**, 1137–1153.
MULT

821. Leiter, R.E. and Hamdan, M.A. (1973) Some bivariate probability models applicable to traffic accidents and fatalities. *International Statistical Review* **41**, 87–100.
DIST MULT

822. Leroux, B.G. (1992) Maximum-likelihood estimation for hidden Markov models. *Stochastic Processes and their Applications* **40**, 127–143.
CD MP TS

823. Leroux, B.G. and Puterman, M.L. (1992) Maximum-penalized-likelihood estimation for independent and Markov-dependent mixture models. *Biometrics* **48**, 545–558.
CD MP OD TS

824. Leslie, R.T. (1967) Recurrent composite events. *Journal of Applied Probability* **4**, 34–61.
DD PP TS

825. Leslie, R.T. (1969) Recurrence times of clusters of Poisson points. *Journal of Applied Probability* **6**, 372–388.
DD PP TS

826. Lewis, P.A.W. (1964) A branching Poisson process model for the analysis of computer failure patterns. *Journal of the Royal Statistical Society* **B26**, 398–456.
DD PP SURV TS

827. Lewis, P.A.W. (1965) Some results on tests for Poisson processes. *Biometrika* **52**, 67–77.
DD PP RP TS

828. Lewis, P.A.W. (1967) Non-homogeneous branching Poisson processes. *Journal of the Royal Statistical Society* **B29**, 343–354.

DD PP SURV TS

829. Lewis, P.A.W. and McKenzie, E. (1991) Minification processes and their transformations. *Journal of Applied Probability* **28**, 45–57.

AR SURV TS

830. Lewis, T. (1961) The intervals between regular events displaced in time by independent random deviations of large dispersion. *Journal of the Royal Statistical Society* **B23**, 476–483.

RP TS

831. Lewis, T. and Govier, L.J. (1964) Some properties of counts of events for certain types of point processes. *Journal of the Royal Statistical Society* **B26**, 325–337.

DD PP TS

832. Li, S.H. and Klotz, J.H. (1978) Components of variance estimation for the split-plot design. *Journal of the American Statistical Association* **73**, 147–152.

VC

833. Li, W.K. (1991) Testing model adequacy for some Markov regression models for time series. *Biometrika* **78**, 83–89.

GLM MC TS

834. Liang, K.Y. (1985) Odds ratio inference with dependent data. *Biometrika* **72**, 678–682.

DD VC

835. Liang, K.Y., Self, S.G., and Liu, X. (1990) The Cox proportional hazards model with change point: an epidemiologic application. *Biometrics* **46**, 783–793.

SURV

836. Liang, K.Y. and Zeger, S.L. (1986) Longitudinal data analysis using generalized linear models. *Biometrika* **73**, 13–22.

GEE GLM TS

837. Liang, K.Y. and Zeger, S.L. (1989) A class of logistic regression models for multivariate binary time series. *Journal of the American Statistical Association* **84**, 447–451.

DD MC TS

838. Liang, K.Y., Zeger, S.L., and Qaqish, B. (1992) Multivariate regression analyses for categorical data. *Journal of the Royal Statistical Society* **B54**, 3–40.

DD GEE TS VC

839. Lianto, S. and McGilchrist, C.A. (1988) Cholesky decomposition of a variance matrix in repeated measures analysis. *Australian Journal of Statistics* **30**, 228–234.

AR TS

840. Lin, J.S. and Wei, L.J. (1992) Linear regression analysis for multivariate failure time observations. *Journal of the American Statistical Association* **87**, 1091–1097.

MULT SURV TS

841. Lin, T.H. and Harville, D.A. (1991) Some alternatives to Wald's confidence interval and test. *Journal of the American Statistical Association* **86**, 179–187.
VC

842. Lindley, D.V. and Singpurwalla, N.D. (1986) Multivariate distributions for the life lengths of components of a system sharing a common environment. *Journal of Applied Probability* **23**, 418–431.
MULT SURV VC

843. Lindley, R.M. (1976) Inter-industry mobility of male employees in Great Britain, 1959-68. *Journal of the Royal Statistical Society* **A139**, 56–79.
DD MC TS

844. Lindsay, B.G., Clogg, C.C., and Grego, J. (1991) Semiparametric estimation in the Rasch model and related exponential response models, including a simple latent class model for item analysis. *Journal of the American Statistical Association* **86**, 96–107.
DD VC

845. Lindsey, J.K. (1970) Exact statistical inferences about the parameter for an exponential growth curve following a Poisson distribution. *Journal of the Fisheries Research Board of Canada* **27**, 172–174.
GC TS

846. Lindsey, J.K. (1974a) Comparison of probability distributions. *Journal of the Royal Statistical Society* **B36**, 38–47.
GLM

847. Lindsey, J.K. (1974b) Construction and comparison of statistical models. *Journal of the Royal Statistical Society* **B36**, 418–425.
GLM

848. Lindsey, J.K. (1975) Likelihood analysis and tests for binary data. *Journal of the Royal Statistical Society* **C24**, 1–16.
DD

849. Lindsey, J.K. (1989) *The Analysis of Categorical Data Using GLIM.* Berlin: Springer Verlag.
DD

850. Lindsey, J.K. (1992) *The Analysis of Stochastic Processes Using GLIM.* Berlin: Springer Verlag.
AR CO DD GC GLM MC PP RP SURV TS VC

851. Lindsey, J.K., Alderdice, D.F., and Pienaar, L.V. (1970) Analysis of nonlinear models — the nonlinear response surface. *Journal of the Fisheries Research Board of Canada* **27**, 765–791.
GLM

852. Lindsey, J.K. and Mersch, G. (1992) Fitting and comparing probability distributions with log linear models. *Computational Statistics and Data Analysis* **13**, 373–384.
DD GLM

853. Lindstrom, M.J. and Bates, D.M. (1988) Newton–Raphson and EM algorithms for linear mixed-effects models for repeated-measures data. *Journal of the American Statistical Association* **83**, 1014–1022.

GC TS VC

854. Lindstrom, M.J. and Bates, D.M. (1990) Nonlinear mixed effects models for repeated measures data. *Biometrics* **46**, 673–687.

GC TS VC

855. Linhart, H. (1970) The gamma point process. *South African Statistical Journal* **4**, 1–17.

DD DIST MULT PP RP TS

856. Lipsitz, S.R., Laird, N.M., and Harrington, D.P. (1990) Using the jackknife to estimate the variance of regression estimators from repeated measures studies. *Communications in Statistics* **19**, 821–845.

DD VC

857. Lipsitz, S.R., Laird, N.M., and Harrington, D.P. (1991) Generalized estimating equations for correlated binary data: using the odds ratio as a measure of association. *Biometrika* **78**, 153–160.

DD VC

858. Lipsitz, S.R., Laird, N.M., and Harrington, D.P. (1992) A three-stage estimator for studies with repeated and possibly missing binary outcomes. *Journal of the Royal Statistical Society* **C41**, 203–213.

DD VC

859. Liski, E.P. (1991) Detecting influential measurements in a growth curves model. *Biometrics* **47**, 659–668.

GC MULT TS

860. Little, R.J.A. (1988) Commentary. *Statistics in Medicine* **7**, 347–355.

VC

861. Liu, X.L. and Liang, K.Y. (1992) Efficacy of repeated measures in regression models with measurement error. *Biometrics* **48**, 645–654.

GEE

862. Ljung, G.M. and Box, G.E.P. (1980) Analysis of variance with autocorrelated observations. *Scandinavian Journal of Statistics* **7**, 172–180.

AR TS

863. Lloyd, F.T., Muse, H.D., and Hafley, W.L. (1982) A regression application for comparing growth potential of environments at different points in the growth cycle. *Biometrics* **38**, 479–484.

GC TS

864. Logan, J.A. (1983) A multivariate model for mobility tables. *American Journal of Sociology* **89**, 324–349.

DD MC TS

865. Lomnicki, Z.A. and Zaremba, S.K. (1955) Some applications of zero-one processes. *Journal of the Royal Statistical Society* **B17**, 243–255.

AR DD PP TS

866. Longford, N.T. (1985a) Statistical modelling of data from hierarchical structures using variance component analysis. In Gilchrist, pp. 112–119.

GLM VC

867. Longford, N.T. (1985b) Intrasubject covariance estimation in longitudinal studies. *Biometrics* **41**, 1075–1076.

VC

868. Longford, N.T. (1987) A fast scoring algorithm for maximum likeli-
     hood estimation in unbalanced mixed models with nested random effects.
     Biometrika **74**, 817–827.
     VC

869. Longini, I.M., Byers, R.H., Hessol, N.A., and Tan, W.Y. (1992) Estimat-
     ing the stage-specific numbers of HIV infection using a Markov model
     and back-calculation. *Statistics in Medicine* **11**, 831–843.
     DD MP PP TS

870. Longini, I.M., Clark, W.S., Byers, R.H., Ward, J.W., Darrow, W.W., and
     Lemp, G.F. (1989) Statistical analysis of the stages of HIV infection using
     a Markov model. *Statistics in Medicine* **8**, 831–843.
     DD MP PP TS

871. Lorenzen, G. (1990) A unified approach to the calculation of growth rates.
     AmericanGuo Statistician **44**, 148–150.

872. Louis, T.A. (1988) General methods for analyzing repeated measures.
     Statistics in Medicine **7**, 29–45.
     TS VC

873. Lui, K.J. (1989) A discussion on the conventional estimator of sensitivity
     and specificity in multiple tests. *Statistics in Medicine* **8**, 1231–1240.
     DD OD

874. Lui, K.J. (1991) Sample sizes for repeated measurements in dichotomous
     data. *Statistics in Medicine* **10**, 463–472.
     DD DES OD TS

875. Lui, K.J. and Cumberland, W.G. (1992) Sample size requirements for
     repeated measurements in continuous data. *Statistics in Medicine* **11**,
     633–641.
     DES

876. Lundbye-Christensen, S. (1991) A multivariate growth curve model for
     pregnancy. *Biometrics* **47**, 637–657.
     GC MULT TS

877. Mackenzie, G. (1988) A proportional hazard model for accident data.
     Journal of the Royal Statistical Society **A149**, 366–375.
     MP SURV TS

878. MacRae, E.C. (1977) Estimation of time-varying Markov processes with
     aggregate data. *Econometrica* **45**, 183–198.
     DD MC TS

879. Maddala, G.S. (1971a) Generalized least squares with an estimated vari-
     ance covariance matrix. *Econometrica* **39**, 23–33.
     AR TS VC

880. Maddala, G.S. (1971b) The use of variance components models in pooling
     cross-section and time series data. *Econometrica* **39**, 341–358.
     AR TS VC

881. Maddala, G.S. and Mount, T.D. (1973) A comparative study of alter-
     native estimators for variance components models used in econometrics
     applications. *Journal of the American Statistical Association* **68**, 324–328.
     VC

882. Maguire, B.A., Pearson, E.S., and Wynn, A.H.A. (1952) The time intervals between industrial accidents. *Biometrika* **39**, 168–180.
    EH RP SURV TS

883. Mak, T.K. (1988) Analyzing intraclass correlation for dichotomous variables. *Journal of the Royal Statistical Society* **C37**, 344–352.
    DD OD

884. Malik, H.J. and Abraham, B. (1973) Multivariate logistic distributions. *Annals of Statistics* **1**, 588–590.
    DIST MULT

885. Manski, C. (1987) Semiparametric analysis of random effects linear models from binary panel data. *Econometrica* **55**, 357–362.
    DD TS

886. Mansour, H., Nordheim, E.V., and Rutledge, J.J. (1985) Maximum likelihood estimation of variance components in repeated measures designs assuming autoregressive errors. *Biometrics* **41**, 287–294.
    AR TS VC

887. Mantel, N. and Byar, D.P. (1974) Evaluation of response-time data involving transient states: an illustration using heart-transplant data. *Journal of the American Statistical Association* **69**, 81–86.
    SURV

888. Manton, K.G., Stallard, E., and Riggan, W. (1982) Strategies for analyzing ecological health data: models for the biological risk of individuals. *Statistics in Medicine* **1**, 163–181.
    DD

889. Manton, K.G., Stallard, E., and Vaupel, J.W. (1986) Alternative models for the heterogeneity of mortality risks among the aged. *Journal of the American Statistical Association* **81**, 635–644.
    SURV VC

890. Manton, K.G., Woodbury, M.A., and Stallard, E. (1981) A variance components approach to categorical data models with heterogeneous cell populations: analysis of spatial gradients in lung cancer mortality rates in North Carolina counties. *Biometrics* **37**, 259–269.
    CD DD OD

891. Manton, K.G., Woodbury, M.A., and Stallard, E. (1988) Models of the interaction of mortality and the evolution of risk factor distribution: a general stochastic process formulation. *Statistics in Medicine* **7**, 239–256.
    MP SURV TS

892. Marshall, A.W. and Goldhamer, H. (1955) An application of Markov processes to the study of the epidemiology of mental disease. *Journal of the American Statistical Association* **50**, 99–129.
    MP SURV TS

893. Marshall, A.W. and Olkin, I. (1967) A multivariate exponential distribution. *Journal of the American Statistical Association* **62**, 30–44.
    DIST MULT SURV VC

894. Marshall, A.W. and Olkin, I. (1988) Families of multivariate distributions. *Journal of the American Statistical Association* **83**, 834–841.

DIST MULT SURV VC

895. Marshall, A.W. and Olkin, I. (1991) Functional equations for multivariate exponential distributions. *Journal of Multivariate Analysis* **39**, 209–215.
DIST MULT SURV

896. Martin, R.J. (1990) The use of time-series models and methods in the analysis of agricultural field trials. *Communication in Statistics* **A19**, 55–81.
AR TS

897. Martin, R.J. (1992) Leverage, influence and residuals in regression models when observations are correlated. *Communications in Statistics* **A21**, 1183–1212.
AR TS

898. Mathew, T. and Sinha, B.K. (1992) Exact and optimum tests in unbalanced split-plot designs under mixed and random models. *Journal of the American Statistical Association* **87**, 192–200.
VC

899. Matis, J.H. (1972) Gamma time-dependency in Blaxter's compartmental model. *Biometrics* **28**, 597–602.
MP SURV TS

900. Matis, J.H. and Hartley, H.O. (1971) Stochastic compartment analysis: model and least squares estimation from time series data. *Biometrics* **27**, 77–102.
MP SURV TS

901. Matis, J.H. and Wehrly, T.E. (1979) Stochastic models of compartmental systems. *Biometrics* **35**, 199–220.
MP SURV TS

902. Matthews, D.E. (1984) Some observations on semi-Markov models for partially censored data. *Canadian Journal of Statistics* **12**, 201–205.
MP SURV TS

903. Matthews, J.N.S. (1990) The analysis of data from crossover designs: the efficiency of ordinary least squares. *Biometrics* **46**, 689–696.
CO

904. Mau, J. and Steinke, B. (1986) A counting process approach to the analysis of the course of non-Hodgkin's lymphoma. *Statistics in Medicine* **5**, 491–495.
DD SURV

905. McCarthy, C. and Ryan, T.M. (1977) Estimates of voter transition probabilities from the British General Elections of 1974. *Journal of the Royal Statistical Society* **A140**, 78–85.
DD MC TS

906. McCullagh, P. (1977) A logistic model for paired comparisons with ordered categorical data. *Biometrika* **64**, 449–453.
DD VC

907. McCullagh, P. (1982) Some applications of quasi-symmetry. *Biometrika* **69**, 303–308.
DD MC

908. McCullagh, P. (1983) Quasi-likelihood functions. *Annals of Statistics* **11**, 59–67.
GEE

909. McCullagh, P. and Nelder, J.A. (1989, 2nd ed.) *Generalized Linear Models*. London: Chapman and Hall.
GEE GLM VC

910. McDonald, J.B. and Butler, R.J. (1990) Regression models for positive random variables. *Journal of Econometrics* **43**, 227–251.
SURV VC

911. McFadden, J.A. (1962) On the lengths of intervals in a stationary point process. *Journal of the Royal Statistical Society* **B24**, 364–382.
DD PP RP TS

912. McFarland, D.D. (1970) Intergenerational social mobility as a Markov process: including a time-stationary Markovian model that explains observed declines in mobility rates over time. *American Sociological Review* **35**, 463–476.
DD MC TS

913. McGilchrist, C.A. and Aisbett, C.W. (1991) Regression with frailty in survival analysis. *Biometrics* **47**, 461–466.
SURV VC

914. McGilchrist, C.A. and Hills, L.J. (1991) A semi-Markov model for ear infection. *Australian Journal of Statistics* **33**, 5–16.
MP TS

915. McGilchrist, C.A. and Sandland, R.L. (1979) Recursive estimation of the general linear model with dependent errors. *Journal of the Royal Statistical Society* **B41**, 65–68.
AR TS VC

916. McGinnis, R. (1968) A stochastic model of social mobility. *American Sociological Review* **33**, 712–722.
DD MC TS

917. McKeague, I.W. and Utikal, K.J. (1990) Inference for a nonlinear counting process regression model. *Annals of Statistics* **18**, 1172–1187.
EH MP SURV TS

918. McKenzie, E. (1981) Extending the correlation structure of exponential autoregressive-moving-average processes. *Journal of Applied Probability* **17**, 181–189.
AR SURV TS

919. McKenzie, E. (1982) Product autoregression: a time series characterization of the gamma distribution. *Journal of Applied Probability* **19**, 463–468.
AR SURV TS

920. McKenzie, E. (1986) Autoregressive moving-average processes with negative binomial and geometric marginal distributions. *Advances in Applied Probability* **18**, 679–705.
AR CD DD TS

921. McKenzie, E. (1988a) The distribution structure of finite moving-average processes. *Journal of Applied Probability* **25**, 313–321.
     AR DD TS

922. McKenzie, E. (1988b) Some ARMA models for dependent sequences of Poisson counts. *Advances in Applied Probability* **20**, 822–835.
     AR DD TS

923. McLean, R.A., Sanders, W.L., and Stroup, W.W. (1991) A unified approach to mixed linear models. *American Statistician* **45**, 54–64.
     VC

924. McLeish, D.L. (1984) Estimation for aggregate models. The aggregate Markov chain. *Canadian Journal of Statistics* **12**, 265–282.
     DD MC TS

925. McMahon, C.A. (1981) An index of tracking. *Biometrics* **37**, 447–455.
     GC TS

926. Meade, N. (1988) A modified logistic model applied to human populations. *Journal of the Royal Statistical Society* **A151**, 491–498.
     GC TS

927. Meinhold, R.J. and Singpurwalla, N.D. (1983) Understanding the Kalman filter. *American Statistician* **37**, 123–127.
     DGLM TS

928. Mendoza, J.L., Toothaker, L.E., and Crain, B.R. (1976) Necessary and sufficient conditions for F ratios in the $L \times J \times K$ factorial design with two repeated factors. *Journal of the American Statistical Association* **71**, 992–993.
     MULT

929. Mersch, G., Nassogne, C., and Havelange, A. (1990) Flower bud and root neoformation in thin cell layers of *Nicotiana tabacum* var. Samsum: a statistical analysis of spatial and temporal changes in mitotic activity. *Canadian Journal of Botany* **68**, 2501–2508.
     DD OD

930. Metzler, C.M. (1971) Usefulness of the two-compartment open model in pharmacokinetics. *Journal of the American Statistical Association* **66**, 49–53.
     MP SURV TS

931. Meyer, B.D. (1990) Unemployment insurance and unemployment spells. *Econometrica* **58**, 775–782.
     SURV VC

932. Micciolo, R. (1989) Regression models for repeated event data: an application to accident frequencies of railway workers. In Decarli *et al.*, pp. 214–221.
     DD OD PP TS

933. Michelini, C. (1972) Estimating the exponential growth function by direct least square: a comment. *Journal of the Royal Statistical Society* **C21**, 333–335.
     GC TS

934. Mihram, G.A. and Hultquist, R.A. (1967) A bivariate warning-time/failure-time distribution. *Journal of the American Statistical Association* **62**, 589–599.
DIST MULT SURV

935. Miller, M.E. and Landis, J.R. (1991a) Generalized variance components models for clustered categorical response variables. *Biometrics* **47**, 33–44.
DD VC

936. Miller, M.E. and Landis, J.R. (1991b) Evaluation of an analysis approach used to account for extra-variation in clustered categorical responses. *Communications in Statistics* **A20**, 2645–2661.
DD OD

937. Milne, R.K. and Westcott, M. (1972) Further results for the Gauss-Poisson process. *Advances in Applied Probability* **4**, 151–176.
DD OD PP VC

938. Minder, C.H.E. and McMillan, I. (1977) Estimation of linear compartment model parameters using marginal likelihood. *Biometrics* **33**, 333–342.
MP SURV TS

939. Mislevy, R. (1985) Estimation of latent group effects. *Journal of the American Statistical Association* **80**, 993–997.
VC

940. Mode, C.J., Fife, D., and Troy, S.M. (1991) Stochastic methods for short term projections of symptomatic HIV disease. *Statistics in Medicine* **10**, 1427–1440.
MP SURV TS

941. Moffitt, R. (1985) Unemployment insurance and the distribution of unemployment spells. *Journal of Econometrics* **28**, 85–101.
EH MP SURV TS

942. Monlezun, C.J., Blouin, D.C., and Malone, L.C. (1984) Contrasting split plot and repeated measures experiments and analyses. *American Statistician* **38**, 21–31.
VC

943. Montgomery, M.R., Richards, T., and Braun, H.I. (1986) Child health, breast-feeding, and survival in Malaysia: a random-effects logit approach. *Journal of the American Statistical Association* **81**, 297–309.
DD VC

944. Moore, D.F. (1986) Asymptotic properties of moment estimators for overdispersed counts and proportions. *Biometrika* **73**, 583–588.
DD OD

945. Moore, D.F. (1987) Modelling the extraneous variance in the presence of extra-binomial variation. *Journal of the Royal Statistical Society* **C36**, 8–14.
DD OD

946. Moore, D.F. and Tsiatis, A. (1991) Robust estimation of the variance in moment methods for extra-binomial and extra-Poisson variation. *Biometrics* **47**, 383–401.

DD OD

947. Morgan, B.J.T. (1976) Markov properties of sequences of behaviours. *Journal of the Royal Statistical Society* **C25**, 31–36.
MC TS

948. Morgan, B.J.T. and Smith, D.M. (1992) A note on Wadley's problem with overdispersion. *Journal of the Royal Statistical Society* **C41**, 349–354.
DD OD

949. Morris, C.N. (1983) Natural exponential families with quadratic variance functions: statistical theory. *Annals of Statistics* **11**, 515–529.
CD GLM

950. Morrison, D.F. (1970) The optimal spacing of repeated measurements. *Biometrics* **26**, 281–290.
DES TS

951. Morrison, D.F. (1972) The analysis of a single sample of repeated measurements. *Biometrics* **28**, 55–71.
TS VC

952. Morton, R. (1981) Efficiency of estimating equations and the use of pivots. *Biometrika* **68**, 227–233.
GEE

953. Morton, R. (1987) A generalized linear model with nested strata of extra-Poisson variation. *Biometrika* **74**, 247–257.
DD GEE VC

954. Morton, R. (1989) On the efficiency of the quasi-likelihood estimators for exponential families with extra variation. *Australian Journal of Statistics* **31**, 194–199.
DD GEE OD

955. Morton, R. (1991) Analysis of extra-multinomial data derived from extra-Poisson variables conditional on their total. *Biometrika* **74**, 1–6.
DD OD

956. Mosimann, J.E. (1962) On the compound multinomial distribution, the multivariate beta-distribution and correlations among proportions. *Biometrika* **49**, 65–82.
CD DD OD

957. Mosimann, J.E. (1963) On the compound negative multinomial distribution and correlations among inversely sampled pollen counts. *Biometrika* **50**, 47–54.
CD DD OD

958. Moulton, L.H. and Zeger, S.L. (1989) Analyzing repeated measures on generalized linear models via the bootstrap. *Biometrics* **45**, 381–394.
DD GLM TS

959. Muenz, L.R. and Rubinstein, L.V. (1985) Markov models for covariate dependence of binary sequences. *Biometrics* **41**, 91–101.
DD MC TS

960. Mukerjee, H. (1988) Order restricted inference in a repeated measures model. *Biometrika* **75**, 616–617.
VC

961. Muller, K.E. and Barton, C.N. (1989) Approximate power for repeated-measures ANOVA lacking sphericity. *Journal of the American Statistical Association* **84**, 540–555, **86**, 255–256.
VC

962. Muller, K.E., LaVange, L.M., Ramey, S.L., and Ramey, C.T. (1992) Power calculations for general linear multivariate models including repeated measures applcations. *Journal of the American Statistical Association* **87**, 1209–1226.
MULT VC

963. Mundlak, Y. (1978) On the pooling of time series and cross-section data. *Econometrica* **46**, 69–86.
VC

964. Mundlak, Y. and Yahav, J.A. (1981) Random effects, fixed effects, convolution, and separation. *Econometrica* **49**, 1399–1416.
VC

965. Munholland, P.L. and Kalbfleisch, J.D. (1991) A semi-Markov model for insect life history data. *Biometrics* **47**, 1117–1126.
EH MP SURV TS

966. Muñoz, A., Carey, V., Schouten, J.P., Segal, M., and Rosner, B. (1992) A parametric family of correlation structures for the analysis of longitudinal data. *Biometrics* **48**, 733–742.
AR TS VC

967. Murtaugh, P.A. and Fisher, L.D. (1990) Bivariate binary models for efficacy and toxicity in dose-ranging trials. *Communications in Statistics* **A19**, 2003–2020.
DIST DD MULT

968. Nagel, P.J.A. and de Waal, D.J. (1979) Bayesian classification, estimation and prediction of growth curves. *South African Statistical Journal* **13**, 127–137.
GC MULT TS

969. Nash, J.C. (1977) A discrete alternative to the logistic growth function. *Journal of the Royal Statistical Society* **C26**, 9–14.
GC TS

970. Nelder, J.A. (1954) The interpretation of negative components of variance. *Biometrika* **41**, 544–548.
VC

971. Nelder, J.A. (1961) The fitting of a generalization of the logistic curve. *Biometrika* **17**, 89–100.
GC TS

972. Nelder, J.A. (1962) An alternative form of a generalized logistic equation. *Biometrics* **18**, 614–616.
GC TS

973. Nelder, J.A. (1985) Quasi-likelihood and GLIM. In Gilchrist *et al.*, pp. 120–127.
GEE GLM

974. Nelder, J.A. and Lee, Y. (1992) Likelihood, quasi-likelihood and pseudolikelihood: some comparisons. *Journal of the Royal Statistical Society* **B54**, 273–284.
GEE GLM

975. Nelder, J.A. and Pregibon, D. (1987) An extended quasi-likelihood function. *Biometrika* **74**, 221–232.
GEE GLM

976. *Nelder, J.A. and Wedderburn, R.W.M. (1972) Generalized linear models. *Journal of the Royal Statistical Society* **A135**, 370–384.
GLM

977. Nelson, J.F. (1980) Multiple victimization in American cities: a statistical analysis of rare events. *American Journal of Sociology* **85**, 870–891.
CD DD

978. Nelson, J.F. (1985) Multivariate gamma-Poisson models. *Journal of the American Statistical Association* **80**, 828–834.
CD DD MULT OD

979. Nerlove, M. (1971a) Further evidence on the estimation of dynamic relations from a time series of cross sections. *Econometrica* **39**, 359–382.
AR TS VC

980. Nerlove, M. (1971b) A note on error components models. *Econometrica* **39**, 383–396.
VC

981. Neuhaus, J.M. (1992) Statistical methods for longitudinal and clustered designs with binary responses. *Statistical Methods in Medical Research* **1**, 249–273.
DD GEE MC TS VC

982. Neuhaus, J.M. and Jewell, N.P. (1990a) Some comments on Rosner's multiple logistic model for clustered data. *Biometrics* **46**, 523–534.
DD VC

983. Neuhaus, J.M. and Jewell, N.P. (1990b) The effect of retrospective sampling on binary regression models for clustered data. *Biometrics* **46**, 977–990.
DD OD

984. Neuhaus, J.M., Kalbfleisch, J.D., and Hauck, W.W. (1991) A comparison of cluster-specific and population-averaged approaches for analyzing correlated binary data. *International Statistical Review* **59**, 25–35.
DD VC

985. Newby, M. and Winterton, J. (1983) The duration of industrial stoppages. *Journal of the Royal Statistical Society* **A146**, 62–70.
DIST SURV

986. Newman, D.S. (1970) A new family of point processes which are characterized by their second moments. *Journal of Applied Probability* **7**, 338–358.
DD PP TS

987. Newman, J.L. and McCulloch, C.E. (1984) A hazard rate approach to the timing of births. *Econometrica* **52**, 939–961.
SURV VC

988. Nielsen, G.G., Gill, R.D., Andersen, P.K., and Sørensen, T.I.A. (1992) A counting process approach to maximum likelihood estimation in frailty models. *Scandinavian Journal of Statistics* **19**, 25–43.
SURV VC

989. Norros, I. (1985) Systems weakened by failures. *Stochastic Processes and their Applications* **20**, 181–196.
DD PP SURV TS

990. Norros, I. (1986) A compensator representation of multivariate life length distributions, with applications. *Scandinavian Journal of Statistics* **13**, 99–112.
DD DIST MULT PP SURV TS

991. Oakes, D. (1982) A model for association in bivariate survival data. *Journal of the Royal Statistical Society* **B44**, 414–422.
DIST MULT SURV VC

992. Oakes, D. (1986) Semiparametric inference in a model for association in bivariate survival data. *Biometrika* **73**, 353–361.
DIST MULT SURV VC

993. Oakes, D. (1989) Bivariate survival models induced by frailties. *Journal of the American Statistical Association* **84**, 487–493.
SURV VC

994. Ochi, Y. and Prentice, R.L. (1984) Likelihood inference in a correlated probit regression model. *Biometrika* **71**, 531–543.
DD OD

995. O'Connell, D.L. and Dobson, A.J. (1984) General observer-agreement measures on individual subjects and groups of subjects. *Biometrics* **40**, 973–983.
DD VC

996. Ogata, S. (1988) Statistical models for earthquake occurrences and residual analysis for point processes. *Journal of the American Statistical Association* **83**, 9–27.
DD PP TS

997. Ogata, S. and Akaike, H. (1982) On linear intensity models for mixed doubly stochastic Poisson and self-exciting point processes. *Journal of the Royal Statistical Society* **B44**, 102–107.
CD DD PP TS VC

998. Oliver, F.R. (1964) Methods of estimating the logistic growth function. *Journal of the Royal Statistical Society* **C13**, 57–66.
GC TS

999. Oliver, F.R. (1966) Aspects of maximum likelihood estimation of the logistic growth function. *Journal of the American Statistical Association* **61**, 697–705.
GC TS

1000. Oliver, F.R. (1969) Another generalization of the logistic growth function. *Econometrica* **37**, 144–147.
GC TS

1001. Oliver, F.R. (1970) Estimating the exponential growth function by direct least squares. *Journal of the Royal Statistical Society* **C19**, 92–100.
      GC TS

1002. Oliver, F.R. (1982) Notes on the logistic curve for human populations. *Journal of the Royal Statistical Society* **A145**, 359–363.
      GC TS

1003. Olkin, I. and Vaeth, M. (1981) Maximum likelihood estimation in a two-way analysis of variance with correlated errors in one classification. *Biometrika* **68**, 653–660.
      AR TS

1004. Olschewski, M. and Schumacher, M. (1990) Statistical analysis of quality of life data. *Statistics in Medicine* **9**, 749–763.
      EH MP SURV TS

1005. Oman, S.D. (1991) Multiplicative effects in mixed model analysis of variance. *Biometrika* **78**, 729–739.
      VC

1006. Ong, S.H. (1990) Mixture formulations of a bivariate negative binomial distribution with applications. *Communications in Statistics* **A19**, 1303–1322.
      CD DD DIST MULT

1007. Orav, E.J., Louis, T.A., Palmer, R.H., and Wright, E.A. (1991) Variance components and their implications for statistical information in medical data. *Statistics in Medicine* **10**, 599–616.
      VC

1008. Osmond, C. and Gardner, M.J. (1982) Age, period and cohort models applied to cancer mortality rates. *Statistics in Medicine* **1**, 245–259.
      DD

1009. Pack, S.E. (1986) Hypothesis testing for proportions with overdispersion. *Biometrics* **42**, 967–972.
      DD OD

1010. Paik, M.C. (1992) Parametric variance function estimation for nonnormal repeated measurement data. *Biometrics* **48**, 19–30.
      GEE GLM

1011. Palmer, M.J., Phillips, B.F., and Smith, G.T. (1991) Application of nonlinear models with random coefficients to growth data. *Biometrics* **47**, 623–635.
      AR GC TS VC

1012. Palta, M. and Cook, T. (1987) Some considerations in the analysis of rates of change in longitudinal studies. *Statistics in Medicine* **6**, 599–611.
      MULT

1013. Palta, M. and Yao, T.J. (1991) Analysis of longitudinal data with unmeasured confounders. *Biometrics* **47**, 1355–1369.
      VC

1014. Pantula, S.G. and Pollock, K.H. (1985) Nested analysis of variance with autocorrelated errors. *Biometrics* **41**, 909–920.
      AR TS VC

1015. Patel, H.I. (1986) Analysis of repeated measures designs with changing covariates in clinical trials. *Biometrika* **73**, 707–715.
      CO

1016. Patel, H.I. (1991) Analysis of incomplete data from a clinical trial with repeated measurements. *Biometrika* **78**, 609–619.
      AR TS

1017. Patterson, H.D. (1951) Change-over trials. *Journal of the Royal Statistical Society* **B13**, 256–271.
      CO

1018. Patterson, H.D. (1964) Theory of cyclic rotation experiments. *Journal of the Royal Statistical Society* **B26**, 1–45.
      VC

1019. Patterson, H.D. and Thompson, R. (1971) Recovery of inter-block information when block sizes are unequal. *Biometrika* **58**, 545–554.
      VC

1020. Paul, S.R. (1979) A clumped beta-binomial model for the analysis of clustered attribute data. *Biometrics* **35**, 821–824.
      CD DD OD

1021. Paul, S.R. (1982) Analysis of proportions of affected foetuses in teratological experiments. *Biometrics* **38**, 361–370.
      DD OD

1022. Paul, S.R. (1990) Maximum likelihood estimation of intraclass correlation in the analysis of familial data. *Biometrika* **77**, 549–555.
      VC

1023. Paul, S.R., Liang, K.Y., and Self, S.G. (1989) On testing departure from the binomial and multinomial assumptions. *Biometrics* **45**, 231–236.
      DD OD

1024. Paul, S.R. and Plackett, R.L. (1978) Inference sensitivity for Poisson mixtures. *Biometrika* **61**, 509–515.
      CD DD OD

1025. Payne, C.D. (1985) *The GLIM System. Release 3.77.* Oxford: NAG.
      GLM

1026. Pegram, G.G.S. (1980) An autoregressive model for multilag Markov chains. *Journal of Applied Probability* **17**, 350–362.
      AR DD MC TS

1027. Perrin, E.B. and Sheps, M.C. (1964) Human reproduction: a stochastic process. *Biometrics* **20**, 28–45.
      EH MP SURV TS

1028. Petersen, T. (1986) Fitting parametric survival models with time-dependent covariates. *Journal of the Royal Statistical Society* **C35**, 281–288.
      SURV

1029. Petkau, A.J. and Sitter, R.R. (1989) Models for quantal response experiments over time. *Biometrics* **45**, 1299–1307.
      DD OD TS VC

1030. Peto, R. (1987) Why do we need systematic overviews of randomized trials? *Statistics in Medicine* **6**, 233–240.
DES VC

1031. Pettitt, A.N. (1984) Fitting a sinusoid to biological rhythm data using ranks. *Biometrics* **40**, 295–300.
GC TS

1032. Pierce, D.A. and Sands, B.R. (1975) Extra-Bernoulli variation in binary data. *Technical Report 46*. Department of Statistics, Oregon State University.
DD VC

1033. Pierce, D.A., Stram, D.O., Vaeth, M., and Schafer, D.W. (1992) The errors-in-variables problem: considerations provided by radiation dose-response analyses of the A-bomb survivor data. *Journal of the American Statistical Association* **87**, 351–359.
GLM VC

1034. Plackett, R.L. (1965) A class of bivariate distributions. *Journal of the American Statistical Association* **60**, 516–522.
DIST MULT

1035. Pocock, S.J., Cook, D.G., and Beresford, S.A.A. (1981) Regression of area mortality rates on explanatory variables: what weighting is appropriate? *Journal of the Royal Statistical Society* **C30**, 286–295.
DD VC

1036. Pocock, S.J., Cook, D.G., and Shaper, A.G. (1982) Analyzing geographic variation in cardiovascular mortality: methods and results. *Journal of the Royal Statistical Society* **A145**, 313–341.
DD

1037. Poirier, D.J. and Ruud, P.A. (1988) Probit with dependent observations. *Review of Economic Studies* **55**, 593–614.
DD TS

1038. Pons, O. and de Turckheim, E. (1988) Cox's periodic regression model. *Annals of Statistics* **16**, 678–693.
MP SURV TS

1039. Pons, O. and de Turckheim, E. (1991) Tests of independence for bivariate censored data based on the empirical joint hazard function. *Scandinavian Journal of Statistics* **18**, 21–37.
MULT SURV

1040. Porteous, B.T. (1987) The mutual independence hypothesis for categorical data in complex sampling schemes. *Biometrika* **74**, 857–862.
DD VC

1041. Potthoff, R.F. and Roy, S.N. (1964) A generalized multivariate analysis of variance model useful especially for growth curve problems. *Biometrika* **51**, 313–326.
GC MULT TS

1042. Potthoff, R.F. and Whittinghill, M. (1966) Testing for homogeneity. I. The binomial and multinomial distributions. II. The Poisson distribution.

Biometrika **53**, 167–190.
DD OD

1043. Pregibon, D. (1982) Score tests with applications. In Gilchrist, pp. 87–97.
GLM

1044. Preisler, H.K. (1988a) Maximum likelihood estimates for binary data with random effects. *Biometrical Journal* **30**, 339–350.
DD VC

1045. Preisler, H.K. (1988b) Assessing insecticide bioassay data with extra-binomial variation. *Journal of Economic Entomology* **81**, 759–765.
DD VC

1046. Preisler, H.K. (1989a) Analysis of a toxicological experiment using a generalized linear model with nested random random effects. *International Statistical Review* **57**, 145–159.
DD VC

1047. Preisler, H.K. (1989b) Fitting dose-response data with non-zero background within generalized linear and generalized additive models. *Computational Statistics and Data Analysis* **7**, 279–290.
DD VC

1048. Prentice, R.L. (1986) Binary regression using an extended beta-binomial distribution, with discussion of correlation induced by covariate measurement error. *Journal of the American Statistical Association* **81**, 321–327.
CD DD OD

1049. Prentice, R.L. (1988) Correlated binary regression with covariates specific to each binary observation. *Biometrics* **44**, 1033–1048.
DD VC

1050. Prentice, R.L. and Cai, J. (1992) Covariance and survivor function estimation using censored multivariate failure time data. *Biometrika* **79**, 495–512.
MULT SURV

1051. Prentice, R.L. and Gloeckler, L.A. (1978) Regression analysis of grouped survival data with application to breast cancer data. *Biometrics* **34**, 57–68.
SURV

1052. Prentice, R.L. and Kalbfleisch, J.D. (1979) Hazard rate models with covariates. *Biometrics* **35**, 25–39.
SURV

1053. Prentice, R.L. and Self, S.G. (1988) Aspects of the use of relative risk models in the design and analysis of cohort studies and prevention trials. *Statistics in Medicine* **7**, 275–287.
SURV

1054. Prentice, R.L., Williams, B.J., and Peterson, B.J. (1981) On the regression analysis of multivariate failure time data. *Biometrika* **68**, 373–379.
MP SURV TS

1055. Prentice, R.L. and Zhao, L.P. (1991) Estimating equations for parameters in means and covariances of multivariate discrete and continuous

responses. *Biometrics* **47**, 825–839.
GEE

1056. Prescott, R.J. (1981) The comparison of success rates in cross-over trials in the presence of an order effect. *Journal of the Royal Statistical Society* **C30**, 9–15.
CO DD

1057. Priestley, M.B. (1981) *Spectral Analysis and Time Series.* San Diego: Academic Press.
TS

1058. Proschan, F. and Sullo, P. (1976) Estimating the parameters of a multivariate exponential distribution. *Journal of the American Statistical Association* **71**, 465–472.
DIST MULT SURV

1059. Province, M.A. and Rao, D.C. (1988) Familial aggregation in the presence of temporal trends. *Statistics in Medicine* **7**, 185–198.
MULT

1060. Pyke, R. (1958) On renewal processes related to Type I and Type II counter models. *Annals of Mathematical Statistics* **29**, 737–754.
MP RP SURV TS

1061. Pyke, R. (1961a) Markov renewal processes: definitions and preliminary properties. *Annals of Mathematical Statistics* **32**, 1231–1242.
EH MP RP SURV TS

1062. Pyke, R. (1961b) Markov renewal processes with finitely many states. *Annals of Mathematical Statistics* **32**, 1243–1259.
EH MP RP SURV TS

1063. Pyke, R. and Schaufele, R. (1966) The existence and uniqueness of stationary measures for Markov renewal processes. *Annals of Mathematical Statistics* **37**, 1439–1462.
MP RP SURV TS

1064. Qaqish, B.F. and Liang, K.Y. (1992) Marginal models for correlated binary responses with multiple classes and multiple levels of nesting. *Biometrics* **48**, 939–950.
DD GEE VC

1065. Qu, Y., Williams, G.W., and Beck, G.J. (1992) Latent variable models for clustered dichotomous data with multiple subclusters. *Biometrics* **48**, 1095–1102.
DD VC

1066. Quenouille, M.H. (1949) A relation between the logarithmic, Poisson and negative binomial series. *Biometrics* **5**, 162–164.
CD DD

1067. Quenouille, M.H. (1958) The comparison of correlations in time series. *Journal of the Royal Statistical Society* **B20**, 158–164.
TS

1068. Raab, G.M. (1981) Estimation of a variance function, with application to immunoassay. *Journal of the Royal Statistical Society* **C30**, 32–40.
GEE

1069. Raboud, J. and Pintilie, M. (1992) The effect of environment on the growth of fish. *Canadian Journal of Statistics* **20**, 233–239.
GC GEE TS

1070. Raeside, R. (1988) The use of sigmoids in modelling and forecasting human populations. *Journal of the Royal Statistical Society* **A151**, 499–513.
GC TS

1071. Rafail, S.Z. (1971) A new growth model for fishes and the estimation of optimum age of fish populations. *Marine Biology* **10**, 13–21.
GC TS

1072. Raftery, A.E. (1985) A model for high-order Markov chains. *Journal of the Royal Statistical Society* **B47**, 528–539.
DD MC TS

1073. Ramakrishnan, A. (1951) Some simple stochastic processes. *Journal of the Royal Statistical Society* **B13**, 131–140.
MP TS

1074. Rampey, A.H., Longini, I.M., Haber, M., and Monto, A.S. (1992) A discrete-time model for the statistical analysis of infectious disease incidence data. *Biometrics* **48**, 117–128.
DD PP TS

1075. Rao, C.R. (1958) Some statistical methods for comparison of growth curves. *Biometrics* **14**, 1–17.
GC MULT TS

1076. Rao, C.R. (1959) Some problems involving linear hypotheses in multivariate analysis. *Biometrika* **46**, 49–58.
GC MULT

1077. Rao, C.R. (1965) The theory of least squares when the parameters are stochastic and its application to the analysis of growth curves. *Biometrika* **52**, 447–458.
GC TS VC

1078. Rao, C.R. (1970) Estimation of heteroscedastic variances in linear models. *Journal of the American Statistical Association* **65**, 161–172.
VC

1079. Rao, C.R. (1972) Estimation of variance and covariance components in linear models. *Journal of the American Statistical Association* **67**, 112–115.
VC

1080. Rao, C.R. (1975) Simultaneous estimation of parameters in different linear models and applications to biometric problems. *Biometrics* **31**, 545–555.
VC

1081. Rao, C.R. (1987) Prediction of future observations in growth curve models. *Statistical Science* **2**, 434–471.
GC TS VC

1082. Rao, C.R. and Kshirsagar, A.M. (1975) A semi-Markovian model for predator-prey interaction. *Biometrics* **34**, 611–619.
MP SURV TS

1083. Rao, J.N.K. and Scott, A.J. (1992) A simple method for the analysis of clustered binary data. *Biometrics* **48**, 577–585.
DD OD VC

1084. Rao, P.S.R.S., Kaplan, J., and Cochran, W.G. (1981) Estimators for one-way random effects model with unequal error variances. *Journal of the American Statistical Association* **76**, 89–97.
VC

1085. Rasch, G. (1960) *Probabilistic Models for some Intelligence and Attainment Tests.* Copenhagen: Danish Institute for Educational Research.
DD

1086. Ratkowsky, D.A. and Dolby, G.R. (1975) Taylor series linearization and scoring for parameters in nonlinear regression. *Journal of the Royal Statistical Society* **C24**, 109–111.
GC

1087. Read, K.L.Q. and Ashford, J.R. (1968) A system of models for the life cycle of a biological organism. *Biometrika* **55**, 211–221.
EH MP SURV TS

1088. Regier, M.H. (1968) A two-state Markov model for behavior change. *Journal of the American Statistical Association* **63**, 993–999.
DD MC TS

1089. Reinsel, G.C. (1982) Multivariate repeated measurement or growth curve models with multivariate random-effects covariance structure. *Journal of the American Statistical Association* **77**, 190–195.
GC TS VC

1090. Reinsel, G.C. (1984) Estimation and prediction in a multivariate random effects generalized linear model. *Journal of the American Statistical Association* **79**, 406–414.
GC TS VC

1091. Reinsel, G.C. (1985) Mean squared error properties of empirical Bayes estimators in a multivariate random effects general linear model. *Journal of the American Statistical Association* **80**, 642–650.
GC TS VC

1092. Revankar, N.S. (1980) Analysis of regression containing serially correlated and serially uncorrelated error components. *International Economic Review* **21**, 185–199.
AR TS VC

1093. Ridout, M.S. (1991) Testing for random dropouts in repeated measurement data. *Biometrics* **47**, 1617–1621.
DES

1094. Ripley, B.D. (1976) The second-order analysis of stationary point processes. *Journal of Applied Probability* **13**, 255–266.
DD PP TS

1095. Roberts, E.A. and Raison, J.M. (1983) An analysis of moisture content of soil cores in a designed experiment. *Biometrics* **39**, 1097–1105.
GC TS VC

1096. Robertson, C. (1990) A matrix regression model for the transition probabilities in a finite state stochastic process. *Journal of the Royal Statistical Society* **C39**, 1–19.
      DD MC TS

1097. Robertson, C. and Boyle, P. (1986) Age, period and cohort models: the use of individual records. *Statistics in Medicine* **5**, 527–538.
      DD OD

1098. Robinson, G.K. (1991) The BLUP is a good thing: the estimation of random effects. *Statistical Science* **6**, 15–51.
      VC

1099. Rochon, J. (1991) Sample size calculations for two-group repeated-measures experiments. *Biometrics* **47**, 1383–1398.
      DES

1100. Rochon, J. (1992) ARMA covariance structures with time heteroscedasticity for repeated measures experiments. *Journal of the American Statistical Association* **87**, 777–784.
      AR TS

1101. Rochon, J. and Helms, R.W. (1989) Maximum likelihood estimation for incomplete repeated measures experiments under an ARMA covariance structure. *Biometrics* **45**, 207–218.
      AR TS

1102. Ronning, G. and Jung, R.C. (1992) Estimation of first-order autoregressive process with Poisson marginals for count data. In Fahrmeir *et al.*, pp. 188–194.
      AR DD TS

1103. Rosenberg, P.S. and Gail, M.H. (1991) Backcalculation of flexible linear models of the human immunodeficiency virus infection curve. *Journal of the Royal Statistical Society* **C40**, 269–282.
      GC TS

1104. Rosenwaike, I. (1966) Seasonal variation of deaths in the United States, 1951–1960. *Journal of the American Statistical Association* **61**, 706–719.
      GC TS

1105. Rosner, B. (1982) Statistical methods in ophthalmology: an adjustment for the intraclass correlation between eyes. *Biometrics* **38**, 105–114.
      VC

1106. Rosner, B. (1984) Multivariate methods in ophthalmology with application to other paired data situations. *Biometrics* **40**, 1025–1035.
      DD VC

1107. Rosner, B. (1989) Multivariate methods for clustered binary data with more than one level of nesting. *Journal of the American Statistical Association* **84**, 373–380.
      DD VC

1108. Rosner, B. (1992a) Multivariate methods for clustered binary data with multiple subclasses, with application to binary longitudinal data. *Biometrics* **48**, 721–731.
      DD TS VC

1109. Rosner, B. (1992b) Multivariate methods for binary longitudinal data with heterogeneous correlation over time. *Statistics in Medicine* **11**, 1915–1928.
   DD TS VC

1110. Rosner, B. and Milton, R.C. (1988) Significance testing for correlated binary outcome data. *Biometrics* **44**, 505–512.
   DD TS VC

1111. Rosner, B. and Muñoz, A. (1988) Autoregressive modelling for the analysis of longitudinal data with unequally spaced examinations. *Statistics in Medicine* **7**, 59–71.
   AR TS

1112. Rosner, B., Muñoz, A., Tage, I., Speizer, F., and Weiss, S. (1985) The use of an autoregressive model for the analysis of longitudinal data in epidemiological studies. *Statistics in Medicine* **4**, 457–467.
   AR TS

1113. *Ross, S.M. (1989) *Introduction to Probability Models*. New York: Academic Press.
   DIST DD MC MP PP RP TS

1114. Rotnitzky, A. and Jewell, N.P. (1990) Hypothesis testing of regression parameters in semiparametric generalized linear models for cluster correlated data. *Biometrika* **77**, 485–497.
   GLM VC

1115. Rubin, D.B. (1992) Computational aspects of analyzing random effects/longitudinal models. *Statistics in Medicine* **11**, 1809–1821.
   VC

1116. Rubin, G., Umbach, D., Shyu, S.F., and Castillo-Chavez, C. (1992) Using mark-recapture methodology to estimate the size of a population at risk for sexually transmitted diseases. *Statistics in Medicine* **11**, 1533–1549.
   DD TS

1117. Rudemo, M., Ruppert, D., and Steibig, J.C. (1989) Random-effects models in nonlinear regression with applications in bioassay. *Biometrics* **45**, 349–362.
   VC

1118. Rudolfer, S.M. (1990) A Markov chain model of extrabinomial variation. *Biometrika* **77**, 255–264.
   DD MC OD TS

1119. Rugg, D.J. and Buech, R.R. (1990) Analyzing time budgets with Markov chains. *Biometrics* **46**, 1123–1131.
   DD MC TS

1120. Sahai, H. (1975) Bayes equivariant estimators in high order hierarchical random effects models. *Journal of the Royal Statistical Society* **B37**, 193–197.
   VC

1121. Sahai, H. (1976) A comparison of estimators of variance components in the balanced three-stage nested random effects model using mean squared error criterion. *Journal of the American Statistical Association* **71**, 435–

444.
VC

1122. Sahai, H. and Anderson, R.L. (1973) Confidence regions for variance ratios of random models for balanced data. *Journal of the American Statistical Association* **68**, 951-952.
VC

1123. Sallas, W.M. and Harville, D.A. (1981) Best linear recursive estimation
for mixed linear models. *Journal of the American Statistical Association*
76, 860–869.
AR TS VC

1124. Sampson, M. (1990) A Markov chain model for unskilled workers and the
highly mobile. *Journal of the American Statistical Association* **85**, 177–
180.
DD MC TS

1125. Samuels, M.L., Casella, G., and McCabe, G.P. (1991) Interpreting blocks
and random factors. *Journal of the American Statistical Association* **86**,
798–821.
VC

1126. Sandland, R.L. and Cormack, R.M. (1984) Statistical inference for
Poisson and multinomial models for capture–recapture experiments.
Biometrika **71**, 27–33.
DD TS

1127. Sandland, R.L. and McGilchrist, C.A. (1979) Stochastic growth curve
analysis. *Biometrics* **35**, 255–271.
GC TS

1128. Scallan, A.J. (1987) A GLIM model for repeated measurements. *GLIM
Newsletter* **15**, 10–22.
DIST MULT SURV VC

1129. Scallon, C.V. (1985) Fitting autoregressive processes in GLIM. *GLIM
Newsletter* **9**, 17–22.
AR TS

1130. Schaalje, B., Zhang, J., Pantula, S.G., and Pollock, K.H. (1991) Analy-
sis of repeated measurements data from randomized block experiments.
Biometrics **47**, 813–824.
AR TS VC

1131. Schall, R. (1991) Estimation in generalized linear models with random
effects. *Biometrika* **78**, 719–727.
GLM VC

1132. Schifflers, E., Smans, M., and Muir, C.S. (1985) Birth cohort analysis using irregular cross-sectional data: a technical note. *Statistics in Medicine*
4, 63–75.
DD

1133. Schlain, B.R., Lavin, P.T., and Hayden, C.L. (1992) Using an autoregressive model to detect departures from steady states in unequally spaced
tumour biomarker data. *Statistics in Medicine* **11**, 515–532.
AR DGLM TS

1134. Schluchter, M.D. (1988) Analysis of incomplete multivariate data using linear models with structured covariance matrices. *Statistics in Medicine* **7**, 317–324.
AR GC TS VC

1135. Schluchter, M.D. (1992) Methods for the analysis of informatively censored longitudinal data. *Statistics in Medicine* **11**, 1861–1870.
VC

1136. Schouten, H.J.A. (1982) Measuring pairwise interobserver agreement when all subjects are judged by the same observers. *Statistica Neerlandica* **36**, 45–61.
DD VC

1137. Schumacher, M., Olschewski, M., and Schmoor, C. (1987) The impact of heterogeneity on the comparison of survival times. *Statistics in Medicine* **6**, 773–784.
SURV VC

1138. Schweder, T. (1970) Composable Markov processes. *Journal of Applied Probability* **7**, 400–410.
MP SURV TS

1139. Schweder, T. (1979) A statistical analysis of the mating behaviour of *Euchaeta norvegica* (Copepoda: Calanoida). *Scandinavian Journal of Statistics* **6**, 71–76.
DD OD

1140. Schweder, T. (1982) On the dispersion of mixtures. *Scandinavian Journal of Statistics* **9**, 165–169.
CD DD OD

1141. Schwertman, N.C. (1978) A note on the Greenhouse-Geisser correction for incomplete data split-plot analysis. *Journal of the American Statistical Association* **73**, 393–396.
VC

1142. Schwertman, N.C. and Heilbrun, L.K. (1986) A successive differences method for growth curves with missing data and random observation times. *Journal of the American Statistical Association* **81**, 912–916.
TS

1143. Searle, S.R. (1971a) *Linear Models*. New York: John Wiley.
VC

1144. Searle, S.R. (1971b) Topics in variance components estimation. *Biometrics* **27**, 1–76.
VC

1145. Seber, G.A.F. (1965) A note on the multiple-recapture census. *Biometrika* **49**, 339–349.
DD TS

1146. Segreti, A.C. and Munso, A.E. (1981) Estimation of the median lethal dose when responses within a litter are correlated. *Biometrics* **37**, 153–156.
DD OD

1147. Selby, B. (1965) The index of dispersion as a test statistic. *Biometrika* **52**, 627–629.
DD OD

1148. Senn, S.J. (1988) Cross-over trials, carry-over effects and the art of self-delusion. *Statistics in Medicine* **7**, 1099–1101.
CO

1149. Senn, S.J. and Hildebrand, H. (1991) Crossover trials, degrees of freedom, the carryover problem and its dual. *Statistics in Medicine* **10**, 1361–1374.
CO

1150. Seshadri, V. (1988) Exponential models, Brownian motion, and independence. *Canadian Journal of Statistics* **16**, 209–221.
MP SURV TS

1151. Seshadri, V. (1991) Finite mixtures of natural exponential families. *Canadian Journal of Statistics* **19**, 437–445.
CD DD GLM SURV

1152. Shaffer, J.P. (1981) The analysis of variance mixed model with allocated observations: application to repeated measurements designs. *Journal of the American Statistical Association* **76**, 607–611.
VC

1153. Shaked, M. (1977) A family of concepts of dependence for bivariate distributions. *Journal of the American Statistical Association* **72**, 642–650.
DIST MULT

1154. Shaked, M. (1980) On mixtures from exponential families. *Journal of the Royal Statistical Society* **B42**, 192–198.
CD

1155. Shaked, M. and Shanthikumar, J.G. (1987) The multivariate hazard construction. *Stochastic Processes and their Applications* **24**, 241–258.
MULT SURV

1156. Sheps, M.C., Menken, J.A., Ridley, J.C., and Lingner, J.W. (1970) Truncation effect in closed and open birth interval data. *Journal of the American Statistical Association* **65**, 678–693.
RP TS

1157. Shiboski, S.C. and Jewell, N.P. (1992) Statistical analysis of the time dependence of HIV infectivity based on partner study data. *Journal of the American Statistical Association* **87**, 360–372.
MP SURV TS

1158. Shoukri, M.M., Mian, I.U.H., and Tracy, D.S. (1991) Correlated linear models for the analysis of familial correlations. *Canadian Journal of Statistics* **19**, 79–91.
VC

1159. Shumway, R.H. (1970) Applied regression and analysis of variance for stationary time series. *Journal of the American Statistical Association* **65**, 1527–1546.
TS

1160. Sichel, H.S. (1982) Repeat-buying and the generalized inverse Gaussian-Poisson distribution. *Journal of the Royal Statistical Society* **C31**, 193–

204.
CD

1161. Sikkel, D. and Jelierse, G. (1988) Renewal theory and retrospective questions. *Journal of the Royal Statistical Society* **C37**, 412–420.
RP TS

1162. Silvey, S.D. (1961) A note on maximum likelihood in the case of dependent random variables. *Journal of the Royal Statistical Society* **B23**, 444–452.
MP TS

1163. Simar, L. (1976) Maximum likelihood estimation of a compound Poisson process. *Annals of Statistics* **4**, 1200–1209.
CD PP TS

1164. *Singer, B. (1982) Aspects of non-stationarity. *Journal of Econometrics* **18**, 169–190.
EH MP SURV TS

1165. Singer, B. (1985) Longitudinal data analysis. *Encyclopedia of Statistical Science* **5**, 142–155.
EH MP SURV TS

1166. Singer, B. and Spilerman, S. (1976) The representation of social processes by Markov models. *American Journal of Sociology* **82**, 1–54.
EH MP SURV TS

1167. Singh, A.C. and Roberts, G.R. (1992) State space modelling of cross-classified time series of counts. *International Statistical Review* **60**, 321–335.
DD DGLM TS

1168. Skellam, J.G. (1948) A probability distribution derived from the binomial distribution by the probability of success as variable between the sets of trials. *Journal of the Royal Statistical Society* **B10**, 257–261.
CD DD

1169. Skellam, J.G. and Shenton, L.R. (1957) Distributions associated with random walks and recurrent events. *Journal of the Royal Statistical Society* **B19**, 64–118.
EH RP TS

1170. Skene, A.M. and White, S.A. (1992) A latent class model for repeated measurements experiments. *Statistics in Medicine* **11**, 2111–2122.
VC

1171. Slud, E.V. (1992) Partial likelihood for continuous-time stochastic processes. *Scandinavian Journal of Statistics* **19**, 97–109.
MP SURV TS

1172. Smith, A.F.M. and West, M. (1983) Monitoring renal transplants: an application of the multiprocess Kalman filter. *Biometrics* **39**, 867–878.
DGLM TS

1173. Smith, D.W. and Murray, L.W. (1984) An alternative to Eisenhart's Model II and mixed model in the case of negative variance estimates. *Journal of the American Statistical Association* **79**, 145–151.
VC

1174. Smith, J.Q. (1979) A generalization of the Bayesian steady forecasting model. *Journal of the Royal Statistical Society* **B41**, 375–387.
DGLM TS

1175. Smith, J.Q. (1981) The multiparameter steady model. *Journal of the Royal Statistical Society* **B43**, 256–260.
DGLM TS

1176. Smith, J.Q. (1992) A comparison of the characteristics of some Bayesian forecasting models. *International Statistical Review* **60**, 75–87.
DGLM TS

1177. Smith, R.L. (1986) Maximum likelihood estimation for the NEAR(2) model. *Journal of the Royal Statistical Society* **B48**, 251–257.
AR SURV TS

1178. Smith, R.L. (1988) Forecasting records by maximum likelihood. *Journal of the American Statistical Association* **83**, 331–338.
AR SURV TS

1179. Smith, R.L. and Miller, J.E. (1986) A non-Gaussian state space model and application to prediction of records. *Journal of the Royal Statistical Society* **B48**, 79–88.
DGLM TS

1180. Smith, W.L. (1958) Renewal theory and its ramifications. *Journal of the Royal Statistical Society* **B20**, 243–302.
RP TS

1181. Snee, R.D., Acuff, S.K., and Gibson, J.R. (1979) A useful method for the analysis of growth studies. *Biometrics* **35**, 835–848.
GC TS

1182. Snee, R.D. and Andrews, H.P. (1971) Statistical design and analysis of shape studies. *Journal of the Royal Statistical Society* **C20**, 250–258.
DES VC

1183. Solomon, P.J. (1985) Transformations of components of variance and covariance. *Biometrika* **72**, 233–239.
VC

1184. Solomon, P.J. and Cox, D.R. (1992) Nonlinear component of variance models. *Biometrika* **79**, 1–11.
VC

1185. Sørensen, M. (1984) Maximum likelihood estimation in the multiplicative intensity model: a survey. *International Statistical Review* **52**, 193–207.
MP SURV TS

1186. Sørensen, M. (1986) On sequential maximum likelihood estimation for exponential families of stochastic processes. *International Statistical Review* **54**, 191–210.
MP SURV TS

1187. Spilerman, S. (1972) The analysis of mobility processes by the introduction of independent variables into a Markov chain. *American Sociological Review* **37**, 277–294.
DD MC TS

1188. Spilerman, S. (1973) Extensions of the mover–stayer model. *American Journal of Sociology* **78**, 599–626.
DD MC TS

1189. Sprent, P. (1965) Fitting a polynomial to correlated equally spaced observations. *Biometrika* **52**, 275–276.
GC MULT TS

1190. Sprent, P. (1968) Linear relationships in growth and size studies. *Biometrics* **24**, 639–656.
GC MULT TS

1191. Srinivasan, S.K. and Rajamannar, G. (1970) Selective interaction between two independent stationary recurrent point processes. *Journal of Applied Probability* **7**, 476–482.
DD DD TS

1192. Stablein, D.M., Carter, W.H., and Wampler, G.L. (1980) Survival analysis of drug combinations using a hazards model with time-dependent covariates. *Biometrics* **36**, 537–546.
SURV

1193. Stanek, E.J. (1991) A two-step method for understanding and fitting growth curve models. *Statistics in Medicine* **9**, 841–851.
GC MULT TS

1194. Stanek, E.J. and Diehl, S.R. (1988) Growth curve models for repeated binary response. *Biometrics* **44**, 973–983.
DD GC TS

1195. Stanek, E.J. and Kline, G. (1991) Estimating prediction equations in repeated measures designs. *Statistics in Medicine* **10**, 119–130.
GC MULT TS

1196. Stanek, E.J. and Koch, G.G. (1985) The equivalence of parameter estimates from growth curve models and seemingly unrelated regression models. *American Statistician* **39**, 149–152.
GC MULT TS

1197. Stanish, W.M., Gillings, D.B., and Koch, G.G. (1978) An application of multivariate ratio methods for the analysis of a longitudinal clinical trials with missing data. *Biometrics* **34**, 305–317.
DD TS

1198. Stasny, E.A. (1986) Estimating gross flows using panel data: an example from the Canadian labour force survey. *Journal of the American Statistical Association* **81**, 42–47.
DD MC TS

1199. Stedinger, J.R., Shoemaker, C.A., and Tenga, R.F. (1985) A stochastic model of insect phenology for a population with spatially variable development rate. *Biometrics* **41**, 691–701.
DD OD

1200. Steele, J.M. (1973) When successes and failures are independent, a compound process is Poisson. *American Statistician* **27**, 232.
CD DD PP TS

1201. Stefanov, V.T. (1986) Efficient sequential estimation in exponential-type processes. *Annals of Statistics* **14**, 1606–1611.
MP TS

1202. Stefanov, V.T. (1988) On some stopping times for dependent Bernoulli trials. *Scandinavian Journal of Statistics* **15**, 39–50.
MC TS

1203. Stefanov, V.T. (1991) Noncurved exponential families associated with observations over finite state Markov chains. *Scandinavian Journal of Statistics* **18**, 353–356.
MC TS

1204. Steinijans, V.W. (1970) On the definition and specification of stationary point processes. *South African Statistical Journal* **4**, 33–40.
DD PP TS

1205. Stern, R.D. and Coe, R. (1984) A model fitting analysis of daily rainfall data. *Journal of the Royal Statistical Society* **A147**, 1–34.
DD MC SURV TS

1206. Sterne, J.A.C., Kingman, A., and Loe, H. (1992) Assessing the nature of periodontal disease progression — an application of covariance structure estimation. *Journal of the Royal Statistical Society* **C41**, 539–552.
GEE SURV

1207. Stewart, P.W. (1987) Line-segment confidence bands for repeated measures. *Biometrics* **43**, 629–640.
GC MULT TS

1208. Stewman, S. (1975) Two Markov models of open system occupational mobility: underlying conceptualizations and empirical tests. *American Sociological Review* **40**, 298–321.
DD MC TS

1209. Steyn, H.S. (1976) On the multivariate Poisson normal distribution. *Journal of the American Statistical Association* **71**, 233–236.
CD DD DIST MULT

1210. Stiratelli, R., Laird, N., and Ware, J. (1984) Random effects models for serial observations with binary response. *Biometrics* **40**, 961–971.
DD VC

1211. Stirling, W.D. (1985) Heteroscedastic models and an application to block designs. *Journal of the Royal Statistical Society* **C34**, 33–41.
VC

1212. Stock, J.H. (1988) Estimating continuous-time processes subject to time deformation. An application to postwar U.S. GNP. *Journal of the American Statistical Association* **83**, 77–85.
DGLM TS

1213. Stoffer, D.S. (1991) Walsh–Fourier analysis and its statistical applications. *Journal of the American Statistical Association* **86**, 461–485.
DD TS

1214. Stoffer, D.S., Scher, M.S., Richardson, G.A., Day, N.L., and Coble, P.A. (1991) A Walsh–Fourier analysis of the effect of moderate maternal alcohol consumption on neonatal sleep-state cycling. *Journal of the American*

Statistical Association **83**, 954–963.
DD TS

1215. Stokes, L. (1988) Estimation of interviewer effects for categorical items in a random digit dial telephone survey. *Journal of the American Statistical Association* **83**, 623–630.
DD VC

1216. Stram, D.O., Wei, L.J., and Ware, J.H. (1988) Analysis of repeated ordered categorical outcomes with possibly missing observations and time-dependent observations. *Journal of the American Statistical Association* **83**, 631–637.
DD TS

1217. Strenio, J.F., Weisberg, H.I., and Bryk, A.S. (1983) Empirical Bayes estimation of individual growth-curve parameters and their relationship to covariates. *Biometrics* **39**, 71–86.
VC

1218. Struthers, C.A. and Farewell, V.T. (1989) A mixture model for time to AIDS data with left truncation and uncertain origin. *Biometrika* **76**, 814–817.
CD SURV

1219. Stuart, A. (1955) A test for homogeneity of the marginal distribution in a two-way classification. *Biometrika* **42**, 412–416.
DD

1220. Sundberg, R. (1986) Tests for underlying Markovian structure from panel data with partially aggregated states. *Biometrika* **73**, 717–721.
DD MC TS

1221. Swamy, P.A.V.B. (1970) Efficient inference in a random coefficient regression model. *Econometrica* **38**, 311–323.
VC

1222. Swamy, P.A.V.B. (1975) Bayesian and non-Bayesian analysis of switching regressions and of random coefficient regression models. *Journal of the American Statistical Association* **70**, 593–602.
VC

1223. Swamy, P.A.V.B. and Arora, S.S. (1972) The exact finite sample properties of the estimators of coefficients in the error components regression models. *Econometrica* **40**, 261–276.
VC

1224. Sweeting, T.J. (1982) A Bayesian analysis of some pharmacological data using a random coefficient regression model. *Journal of the Royal Statistical Society* **C31**, 205–213.
GC TS VC

1225. Sweeting, T.J. (1983) On estimator efficiency in stochastic processes. *Stochastic Processes and their Applications* **15**, 93–98.
MP TS

1226. Sweeting, T.J. (1992) Asymptotic ancillarity and conditional inference for stochastic processes. *Annals of Statistics* **20**, 580–589.
DD PP TS

1227. Tam, S.M. (1987) Analysis of repeated surveys using a dynamic linear model. *International Statistical Review* **55**, 63–73.
    DGLM TS

1228. Tan, W.Y. (1969) Note on the multivariate and the generalized multivariate Beta distributions. *Journal of the American Statistical Association* **64**, 230–241.
    DIST MULT

1229. Tan, W.Y. (1976) On testing correlation between growth curves. *Canadian Journal of Statistics* **4**, 13–32.
    GC MULT TS

1230. Tanner, M.A. and Young, M.A. (1985) Modelling agreement among raters. *Journal of the American Statistical Association* **80**, 175–180.
    DD VC

1231. Tarone, R.E. (1979) Testing the goodness-of-fit of the binomial distribution. *Biometrika* **66**, 585–590.
    DD OD

1232. Tarone, R.E. (1982) The use of historical control information in testing for a trend in proportions. *Biometrics* **38**, 215–220.
    DD OD

1233. Taub, A.J. (1979) Prediction in the context of the variance-components model. *Journal of Econometrics* **10**, 103–107.
    VC

1234. Tavaré, S. (1983) Serial dependence in contingency tables. *Journal of the Royal Statistical Society* **B45**, 100–106.
    DD MC TS

1235. Tavaré, S. and Altham, P.M.E. (1983) Serial dependence of observations leading to contingency tables and corrections to chi-square statistics. *Biometrika* **70**, 139–144.
    DD MC TS

1236. Tavares, L.V. (1980) An exponential Markovian stationary process. *Journal of Applied Probability* **17**, 1117–1120.
    DIST MP

1237. Tawn, J.A. (1990) Modelling multivariate extreme value distributions. *Biometrika* **77**, 245–253.
    DIST MULT

1238. Taylor, W.E. (1980) Small sample considerations in estimation from panel data. *Journal of Econometrics* **13**, 203–223.
    VC

1239. Temkin, N.R. (1978) An analysis for transient states with application to tumor shrinkage. *Biometrics* **34**, 571–580.
    EH MP SURV TS

1240. Teugels, J.L. (1990) Some representations of the multivariate Bernoulli and binomial distributions. *Journal of Multivariate Analysis* **32**, 256–268.
    DD MULT

1241. Thall, P.F. (1988) Mixed Poisson likelihood regression models for longitudinal interval count data. *Biometrics* **44**, 197–209.

DD TS

1242. Thall, P.F. and Lachin, J.M. (1988) Analysis of recurrent events: non-parametric methods for random-interval count data. *Journal of the American Statistical Association* **83**, 339–347.
       DD NPAR TS

1243. Thall, P.F. and Vail, S.C. (1990) Some covariance models for longitudinal count data with overdispersion. *Biometrics* **46**, 657–671.
       DD OD VC

1244. Thompson, E.A. and Shaw, R.G. (1990) Pedigree analysis for quantitative traits: variance components without matrix inversion. *Biometrics* **46**, 399–413.
       VC

1245. Thompson, G.L. (1991) A unified approach to rank tests for multivariate and repeated measures designs. *Journal of the American Statistical Association* **86**, 410–419.
       MULT

1246. Thompson, M.E. (1981) Estimation of the parameters of a semi-Markov process for censored records. *Advances in Applied Probability* **13**, 804–825.
       MP SURV TS

1247. Thompson, W.A. (1962) The problem of negative estimates of variance components. *Annals of Mathematical Statistics* **3**, 273–289.
       VC

1248. Tiao, G.C. and Ali, M.M. (1971) Analysis of correlated random effect: linear model with two random components. *Biometrika* **58**, 37–51.
       VC

1249. Tiao, G.C. and Box, G.E.P. (1967) Bayesian analysis of a three-component hierarchical design model. *Biometrika* **54**, 109–125.
       VC

1250. Tiao, G.C. and Tan, W.Y. (1965) Bayesian analysis of random-effect models in the analysis of variance. I. Posterior distribution of variance-components. *Biometrika* **52**, 37–53.
       VC

1251. Tiao, G.C. and Tan, W.Y. (1966) Bayesian analysis of random-effect models in the analysis of variance. II. Effect of autocorrelated errors. *Biometrika* **53**, 477–495.
       AR TS VC

1252. Timm, N.H. (1980) Multivariate analysis of variance of repeated measurements. *Handbook of Statistics* **1**, 41–87.
       CO MULT

1253. Tjur, T. (1982) A connection between Rasch's item analysis model and a multiplicative Poisson model. *Scandinavian Journal of Statistics* **9**, 23–30.
       DD VC

1254. Tolley, H.D., Burdick, D., Manton, K.G., and Stallard, E. (1978) A compartment model approach to the estimation of tumor incidence and growth investigation of a model of cancer latency. *Biometrics* **34**, 377–

389.
MP SURV TS

1255. Tomberlin, T.J. (1988) Predicting accident frequencies for drivers classified by two factors. *Journal of the American Statistical Association* **83**,
309–321.
DD VC

1256. Tong, H. (1975) Determination of the order of a Markov chain by Akaike's
information criterion. *Journal of Applied Probability* **12**, 488–497.
DD MC TS

1257. Tong, Y.L. (1976) Parameter estimation in studying circadian rhythms.
Biometrics **32**, 85–94.
TS

1258. Tosteson, T.D., Rosner, B., and Redline, S. (1991) Logistic regression
for clustered binary data in proband studies with application to familial
aggregation of sleep disorders. *Biometrics* **47**, 1257–1265.
DD OD

1259. Truitt, J.T. and Smith, H.F. (1956) Adjustment by covariance and conse-
quent tests of significance in split-plot experiments. *Biometrics* **12**, 23–39.
VC

1260. Tsutakawa, R.K. (1988) Mixed model for analyzing geographic variability
in mortality rates. *Journal of the American Statistical Association* **83**, 37–
42.
DD OD VC

1261. Tuma, N.B. (1976) Rewards, resources and the rate of mobility: a nonstationary stochastic model. *American Sociological Review* **41**, 338–360.
EH MP SURV TS

1262. Tuma, N.B., Hannan, M.T., and Groeneveld, L.P. (1979) Dynamic analysis of event histories. *American Journal of Sociology* **84**, 820–854.
EH MP SURV TS

1263. Tuma, N.B. and Robins, P.K. (1980) A dynamic model of employment
behavior: an application to the Seattle and Denver income maintenance
experiments. *Econometrica* **48**, 1031–1052.
EH MP SURV TS

1264. Turnbull, B.W., Brown, B.W., and Hu, M. (1974) Survivorship analysis
of heart transplant data. *Journal of the American Statistical Association*
69, 74–80.
SURV

1265. Turney, E.A., Amara, I.A., Koch, G.G., and Stewart, W.H. (1992) Evaluation of alternative statistical methods for linear model analysis to compare two treatments for 24-hour blood pressure response. *Statistics in
Medicine* **11**, 1843–1860.
VC

1266. Tweedie, R.L. (1976) Criteria for classifying general Markov chains. *Advances in Applied Probability* **8**, 737–771.
DD MC TS

1267. Ullman, N.S. and Jacquez, J.A. (1973) Analysis of tests of chemothera-
peutic agents involving repeated drug treatment. *Biometrics* **29**, 677–693.
GC TS

1268. Upton, G.J.G. (1977) A memory model for voting transitions in British
elections. *Journal of the Royal Statistical Society* **A140**, 86–94.
DD MC TS

1269. Upton, G.J.G. (1978) A note on the estimation of voter transition probabilities. *Journal of the Royal Statistical Society* **A141**, 507–512.
DD MC TS

1270. Upton, G.J.G. and Sarlvik, B. (1981) A loyalty-distance model for voting
change. *Journal of the Royal Statistical Society* **A144**, 247–259.
DD MC TS

1271. Vaillant, J. (1991) Negative binomial distributions of individuals and
spatio-temporal Cox processes. *Scandinavian Journal of Statistics* **18**,
235–248.
CD DD OD PP TS

1272. Vaupel, J.W. and Yashin, A.I. (1985) Heterogeneity's ruses: some surprising effects of selection on population dynamics. *American Statistician*
39, 176–185.
SURV VC

1273. Verbyla, A.P. (1986) Conditioning in the growth curve model. *Biometrika*
73, 475–484.
GC TS

1274. Verbyla, A.P. (1988) Analysis of repeated measures designs with changing
covariates. *Biometrika* **75**, 172–174.
VC

1275. Verbyla, A.P. and Cullis, B.R. (1990) Modelling in repeated measures
experiments. *Journal of the Royal Statistical Society* **C39**, 341–356.
TS VC

1276. Verbyla, A.P. and Cullis, B.R. (1992) The analysis of multistratum and
spatially correlated repeated measures data. *Biometrics* **48**, 1015–1032.
TS VC

1277. Verbyla, A.P. and Venables, W.N. (1988) An extension of the growth
curve model. *Biometrika* **75**, 129–138.
GC MULT TS

1278. Verhelst, N. and Molenaar, I.W. (1988) Logit based parameter estimation
in the Rasch model. *Statistica Neerlandica* **42**, 273–295.
DD

1279. Vieira, S. and Hoffmann, R. (1977) Comparison of the logistic and the
Gompertz growth functions considering additive and multiplicative error
terms. *Journal of the Royal Statistical Society* **C26**, 143–148.
GC TS

1280. Visser, H. and Molenaar, J. (1988) Kalman filter analysis in dendrocli-
matology. *Biometrics* **44**, 929–940.
DGLM TS

1281. Vit, I. (1974) Testing for homogeneity: the geometric distribution. *Biometrika* **61**, 565–568.
DD OD

1282. Voelkel, J.G. and Crowley, J. (1984) Nonparametric inference for a class of semi-Markov processes with censored observations. *Annals of Statistics* **12**, 142–160.
MP NPAR SURV TS

1283. Vollmer, W.M., Johnson, L.R., McCamant, L.E., and Buist, A.S. (1988) Longitudinal versus cross-sectional estimation of lung function decline – further insights. *Statistics in Medicine* **7**, 685–696.
MULT

1284. von Rosen, D. (1991) The growth curve model: a review. *Communications in Statistics* **A20**, 2791–2822.
GC MULT TS

1285. Vonesh, E.F. (1992) Non-linear models for the analysis of longitudinal data. *Statistics in Medicine* **11**, 1929–1954.
AR TS VC

1286. Vonesh, E.F. and Carter, R.L. (1992) Mixed-effects nonlinear regression for unbalanced repeated measures. *Biometrics* **48**, 1–17.
GC TS VC

1287. Vonesh, E.F. and Carter, R.L. (1987) Efficient inference for random-coefficient growth curve models with unbalanced data. *Biometrics* **43**, 617–628.
GC TS VC

1288. Vonesh, E.F. and Schork, M.A. (1986) Sample sizes in the multivariate analysis of repeated measurements. *Biometrics* **42**, 601–610.
DES GC MULT TS

1289. Waldman, D.M. (1985) Computation in duration models with heterogeneity. *Journal of Econometrics* **28**, 127–134.
SURV VC

1290. Walker, S.H. and Duncan, D.B. (1967) Estimation of the probability of an event as a function of several independent variables. *Biometrika* **54**, 167–179.
DD DGLM TS

1291. Wallace, T.D. and Hussain, A. (1969) The use of error component models in combining cross-section with time series data. *Econometrica* **37**, 55–72.
VC

1292. Wallenstein, S. (1982) Regression models for repeated measurements. *Biometrics* **38**, 849–850.
VC

1293. Wallenstein, S. and Fisher, A.C. (1977) The analysis of the two-period repeated measurements crossover design with applications to clinical trials. *Biometrics* **33**, 261–269.
CO

1294. Wang, M.C. and See, L.C. (1992) N-estimation from retrospectively ascertained events with applications to AIDS. *Biometrics* **48**, 129–141.

DD GC TS

1295. Ware, J.H. (1982) Growth curves. *Encyclopedia of Statistical Science* **3**, 539–542.

GC TS

1296. Ware, J.H. (1985) Linear models for the analysis of longitudinal studies. *American Statistician* **39**, 95–101.

TS VC

1297. Ware, J.H. and Bowden, R.E. (1985) Circadian rhythm analysis when output is collected at intervals. *Biometrics* **33**, 566–571.

TS

1298. Ware, J.H., Lipsitz, S., and Speizer, F.E. (1988) Issues in the analysis of repeated categorical outcomes. *Statistics in Medicine* **7**, 95–107.

DD MC TS VC

1299. Ware, J.H. and Wu, M.C. (1981) Tracking: prediction of future values from serial measurements. *Biometrics* **37**, 427–437.

GC TS VC

1300. Waternaux, C., Laird, N.M., and Ware, J.H. (1989) Methods for analysis of longitudinal data: blood-lead concentrations and cognitive development. *Journal of the American Statistical Association* **84**, 33–41.

VC

1301. Watson, G.S. (1955) Serial correlation in regression analysis. I. *Biometrika* **42**, 327–341.

TS

1302. Wedderburn, R.W.M. (1974) Quasi-likelihood functions, generalized linear models and the Gauss-Newton method. *Biometrika* **61**, 439–447.

GEE

1303. Weerahandi, S. (1991) Testing variance components in mixed models with generalized p values. *Journal of the American Statistical Association* **86**, 151–153.

VC

1304. Wei, L.J. and Johnson, W.E. (1985) Combining dependent tests with incomplete repeated measurements. *Biometrika* **72**, 359–364.

DD

1305. Wei, L.J. and Lachin, J.M. (1984) Two-sample asymptotically distribution-free tests for incomplete multivariate observations. *Journal of the American Statistical Association* **79**, 653–661.

MULT SURV

1306. Wei, L.J., Lin, D.Y., and Weissfeld, L. (1989) Regression analysis of multivariate incomplete failure time data by modeling marginal distributions. *Journal of the American Statistical Association* **84**, 1065–1073.

MULT SURV

1307. Wei, L.J. and Stram, D.O. (1988) Analyzing repeated measurements with possibly missing observations by modelling marginal distributions. *Statistics in Medicine* **7**, 139–148.

GEE GLM TS

1308. Wei, L.J., Su, J.Q., and Lachin, J.M. (1990) Interim analyses with repeated measurements in a sequential clinical trial. *Biometrika* **77**, 359–364.
GLM TS

1309. Weiss, G.H. and Zelen, M. (1965) A semi-Markov model for clinical trials. *Journal of Applied Probability* **2**, 269–285.
EH MP TS

1310. Weiss, R.E. and Lazaro, C.G. (1992) Residual plots for repeated measures. *Statistics in Medicine* **11**, 115–124.
TS VC

1311. Weissfeld, L.A. and Kshirsagar, A.M. (1992) A modified growth curve model and its application to clinical studies. *Australian Journal of Statistics* **34**, 161–168.
CO GC

1312. West, M. (1981) Robust sequential approximate Bayesian estimation. *Journal of the Royal Statistical Society* **B43**, 157–166.
DGLM TS

1313. West, M. (1986) Bayesian model monitoring. *Journal of the Royal Statistical Society* **B48**, 70–78.
DGLM TS

1314. West, M. and Harrison, P.J. (1986) Monitoring and adaption in Bayesian forecasting models. *Journal of the American Statistical Association* **81**, 741–750.
DGLM TS

1315. West, M. and Harrison, P.J. (1989) *Bayesian Forecasting and Dynamic Models*. Berlin: Springer Verlag.
DGLM TS

1316. *West, M., Harrison, P.J., and Migon, H.S. (1985) Dynamic generalized linear models and Bayesian forecasting. *Journal of the American Statistical Association* **80**, 73–97.
DGLM TS

1317. White, A.A., Landis, J.R., and Cooper, M.M. (1982) A note on the equivalence of several marginal homogeneity test criteria for categorical data. *International Statistical Review* **50**, 27–34.
DD VC

1318. Whitehead, A. and Whitehead, J. (1991) A general parametric approach to the meta-analysis of randomized clinical trials. *Statistics in Medicine* **10**, 1665–1677.
GLM VC

1319. Whitmore, G.A. (1979) An inverse Gaussian model for labour turnover. *Journal of the Royal Statistical Society* **A142**, 468–478.
DIST SURV

1320. Whitmore, G.A. and Lee, M.L.T. (1991) A multivariate survival distribution generated by an inverse Gaussian mixture of exponentials. *Technometrics* **33**, 39–50.
CD DIST MULT SURV

1321. Whitmore, G.A. and Seshadri, V. (1987) A heuristic derivation of the inverse Gaussian distribution. *American Statistician* **41**, 280–281.
DIST MULT SURV

1322. Whitt, W. (1976) Bivariate distributions with given marginals. *Annals of Statistics* **4**, 1280–1289.
DIST MULT

1323. Whittemore, A.S. (1988) Effect of cigarette smoking in epidemiological studies of lung cancer. *Statistics in Medicine* **7**, 223–238.
DD SURV

1324. Whittle, P. (1955) Some distributions and moment formulae for the Markov chains. *Journal of the Royal Statistical Society* **B17**, 235–242.
DD MC TS

1325. Wild, C.J. (1983) Failure time models with matched data. *Biometrika* **70**, 633–641.
MULT SURV

1326. Wilkinson, G.N., Eckert, S.R., Hancock, T.W., and Mayo, O. (1983) Nearest neighbour (NN) analysis of field experiments. *Journal of the Royal Statistical Society* **B45**, 151–211.
VC

1327. Wilkinson, G.N. and Rogers, C.E. (1973) Symbolic description of factorial models for analysis of variance. *Journal of the Royal Statistical Society* **C22**, 392–399.
GLM

1328. Willan, A.R. and Pater, J.L. (1986) Carryover and the two- period crossover design. *Biometrics* **42**, 593–599.
CO

1329. Williams, D.A. (1975) The analysis of binary responses from toxicological experiments involving reproduction and teratogenicity. *Biometrics* **31**, 949–952.
DD OD

1330. Williams, D.A. (1982a) Extra-binomial variation in logistic linear models. *Journal of the Royal Statistical Society* **C31**, 144–148.
DD OD

1331. Williams, D.A. (1982b) The use of the deviance to test the goodness of fit of a logistic linear model to binary data. *GLIM Newsletter* **6**, 60–62.
DD

1332. Williams, D.A. (1988) Estimation bias using the beta-binomial distribution in teratology. *Biometrics* **44**, 305–309.
CD DD OD

1333. Williams, D.A. (1989) Hypothesis tests for overdispersed generalized linear models. *GLIM Newsletter* **18**, 29–30.
DD OD

1334. Williams, E.R. (1986) A neighbour model for field experiments. *Biometrika* **73**, 279–287.
VC

1335. Williams, G.W. (1976) Comparing the joint agreement of several raters with another rater. *Biometrics* **32**, 619–627.
DD VC

1336. Williams, J.S. (1962) A confidence interval for variance components. *Biometrika* **49**, 278–281.
VC

1337. Williams, J.S. (1978) Efficient analysis of Weibull survival data from experiments on heterogeneous patient populations. *Biometrics* **34**, 209–222.
SURV VC

1338. Williamson, M. (1985) Apparent systematic effects on species-area curves under isolation and evolution. In Morgan, B.J.T. and North, P.M. (ed.) *Statistics in Ornithology.* Berlin: Springer Verlag, pp. 171–178.
DD TS

1339. Willis, D.M. (1964) The statistics of a particular non-homogeneous Poisson process. *Biometrika* **51**, 399–404.
DD PP RP TS

1340. Wilson, J.R. (1989) Chi-square tests for overdispersion with multiparameter estimates. *Journal of the Royal Statistical Society* **C38**, 441–453.
DD OD

1341. Wilson, J.R. and Koehler, K.J. (1991) Hierarchical models for cross-classified overdispersed multinomial data. *Journal of Business and Economic Statistics* **9**, 103–110.
DD OD

1342. Wilson, P.D. (1988) Autoregressive growth curves and Kalman filtering. *Statistics in Medicine* **7**, 73–86.
AR DGLM GC TS

1343. Wilson, P.D., Hebel, J.R., and Sherwin, R. (1981) Screening and diagnosis when within-individual observations are Markov-dependent. *Biometrics* **37**, 553–565.
TS VC

1344. Winkler, W. and Franz, J. (1979) Sequential estimation problems for the exponential class of processes with independent increments. *Scandinavian Journal of Statistics* **6**, 129–139.
MP SURV TS

1345. Wisniewski, T.K.M. (1968) Testing for homogeneity of a binomial series. *Biometrika* **55**, 426–428.
DD OD

1346. Wittes, J.T. (1974) Applications of a multinomial capture-recapture model to epidemiological data. *Journal of the American Statistical Association* **69**, 93–97.
DD TS

1347. Wolfe, R.A., Petroni, G.R., McLaughlin, C.G., and McMahon, L.F. (1991) Empirical evaluation of statistical models for counts or rates. *Statistics in Medicine* **10**, 1405–1416.
DD OD

1348. Wong, G.Y. and Mason, W.M. (1985) The hierarchical logistic regression model for multilevel analysis. *Journal of the American Statistical Association* **80**, 513–524.
      DD VC

1349. Wong, G.Y. and Mason, W.M. (1991) Contextually specific effects and other generalizations of the hierarchical linear model for comparative analysis. *Journal of the American Statistical Association* **86**, 487–503.
      VC

1350. Wong, W.H. (1986) Theory of partial likelihood. *Annals of Statistics* **14**, 88–123.
      AR SURV TS

1351. Wong, W.K. and Miller, R.B. (1990) Repeated time series analysis of ARIMA-noise models. *Journal of Business and Economic Statistics* **8**, 243–250.
      AR TS

1352. Woodbury, M.A., Manton, K.G., and Stallard, E. (1979) Longitudinal analysis of the dynamics and risk of coronary heart disease in the Framingham study. *Biometrics* **353**, 575–585.
      DD SURV TS

1353. Woodbury, M.A., Manton, K.G., and Yashin, A.I. (1988) Estimating hidden morbidity via its effect on mortality and disability. *Statistics in Medicine* **7**, 325–336.
      DD SURV TS

1354. Woolson, R.F. and Clarke, W.R. (1984) Analysis of categorical incomplete longitudinal data. *Journal of the Royal Statistical Society* **A147**, 87–99.
      DD TS

1355. Woolson, R.F., Leeper, J.D., and Clarke, W.R. (1978) Analysis of incomplete data from longitudinal and mixed longitudinal studies. *Journal of the Royal Statistical Society* **A141**, 242–252.
      MULT

1356. Wu, C.F.J. (1985) Efficient sequential designs with binary data. *Journal of the American Statistical Association* **80**, 974–984.
      DES DD TS

1357. Wu, M.C. and Bailey, K.R. (1988) Analyzing changes in the presence of informative right censoring caused by death and withdrawal. *Statistics in Medicine* **7**, 337–346.
      VC

1358. Wu, M.C. and Bailey, K.R. (1989) Estimation and comparison of changes in the presence of informative right censoring: conditional linear model. *Biometrics* **45**, 939–955.
      VC

1359. Wu, M.C. and Carroll, R.J. (1988) Estimation and comparison of changes in the presence of informative right censoring by modelling the censoring process. *Biometrics* **44**, 175–188.
      VC

1360. Wu, M.C. and Ware, J.H. (1979) On the use of repeated measurements in

regression analysis with dichotomous responses. *Biometrics* **35**, 513–521.
DD TS

1361. Wun, L.M. (1991) Regression analysis of autocorrelated Poisson-distributed data. *Communications in Statistics* **A20**, 3083–3091.
AR DD TS

1362. Yakowitz, S.J. (1976) Small sample hypothesis tests of Markov order, with application to simulated and hydrologic chains. *Journal of the American Statistical Association* **71**, 132–136.
DD MC TS

1363. Yang, M.C.K. and Carter, R.L. (1983) One-way analysis of variance with time series data. *Biometrics* **39**, 747–751.
TS VC

1364. Yang, M.C.K. and Hursch, C.J. (1973) The use of a semi-Markov model for describing sleep patterns. *Biometrics* **29**, 667–676.
DD MC TS

1365. Yates, F. (1982) Regression models for repeated measurements. *Biometrics* **38**, 850–853.
VC

1366. Yip, P. (1991) Conditional inference on a mixture model for the analysis of count data. *Communications in Statistics* **A20**, 2045–2057.
CD DD OD

1367. Zeger, S.L. (1988a) A regression model for time series of counts. *Biometrika* **75**, 621–629.
DD TS

1368. Zeger, S.L. (1988b) Commentary. *Statistics in Medicine* **7**, 161–168.
DD GEE TS VC

1369. Zeger, S.L. and Edelstein, S. L. (1989) Poisson regression with a surrogate X; an analysis of vitamin A and Indonesian children's mortality. *Journal of the Royal Statistical Society* **C38**, 309–318.
DD OD

1370. Zeger, S.L. and Karim, M.R. (1991) Generalized linear models with random effects: a Gibbs sampling approach. *Journal of the American Statistical Association* **86**, 79–86.
GLM VC

1371. Zeger, S.L. and Liang, K.Y. (1986) Longitudinal data analysis for discrete and continuous outcomes. *Biometrics* **42**, 121–130.
GEE GLM TS

1372. Zeger, S.L. and Liang, K.Y. (1991) Feedback models for discrete and continuous time series. *Statistica Sinica* **1**, 51–64.
GEE GLM TS

1373. Zeger, S.L. and Liang, K.Y. (1992) An overview of methods for the analysis of longitudinal data. *Statistics in Medicine* **11**, 1825–1839.
DD GEE GLM MC TS

1374. Zeger, S.L., Liang, K.Y., and Albert, P.S. (1988) Models for longitudinal data: a generalized estimating equation approach. *Biometrics* **44**, 1049–

1060.
GEE GLM TS

1375. Zeger, S.L., Liang, K.Y., and Self, S.G. (1985) The analysis of binary longitudinal data with time-dependent covariates. *Biometrika* **72**, 31–38.
DD MC TS

1376. Zeger, S.L. and Qagish, B. (1988) Markov regression models for time series: a quasi-likelihood approach. *Biometrics* **44**, 1019–1031.
GEE GLM TS

1377. Zerbe, G.O. (1979) Randomization analysis of the completely randomized design extended to growth and response curves. *Journal of the American Statistical Association* **74**, 215–221.
GC NPAR TS

1378. Zerbe, G.O. and Jones, R.H. (1980) On application of growth curve techniques to time series data. *Journal of the American Statistical Association* **75**, 507–509.
GC TS

1379. Zerbe, G.O. and Murphy, J.R. (1986) On multiple comparisons in the randomization analysis of growth and response curves. *Biometrics* **42**, 795–804.
GC NPAR TS

1380. Zerbe, G.O. and Walker, S.H. (1977) A randomization test for comparison of groups of growth curves with different polynomial design matrices. *Biometrics* **33**, 653–657.
GC NPAR TS

1381. Zhao, L.P. and Prentice, R.L. (1990) Correlated binary regression using a quadratic exponential model. *Biometrika* **77**, 642–648.
DD VC

1382. Zucker, D. and Wittes, J. (1992) Testing the effect of treatment in experiments with correlated binary outcomes. *Biometrics* **48**, 695–710.
DD OD VC

Index